Strategic Fuzzy Extensions and Decision-making Techniques

The application of a novel correlation coefficient of linguistic intuitionistic fuzzy sets to medical diagnosis problems provides the topic for **Strategic Fuzzy Extensions and Decision-making Techniques.** It further explains neutrosophic cubic set-based aggregation operators for library ranking systems, and techniques for order performance by similarity to ideal solution. The text also introduces the new aggregation operators, similarity measures, and distance measures for the fuzzy sets and their extensions.

This book:

- Introduces the new aggregation operators, similarity measures, and distance measures for the fuzzy sets and their extensions.
- Covers recent studies in the field of fuzzy optimization and decision-making such as advanced decision-making algorithms.
- Highlights the application in the field of image processing and pattern recognition.
- Presents a technique for order performance by similarity to an ideal solution and VIKOR method for decision-making.
- Explores the limitations of existing fuzzy decision-making approaches such as the malfunctioning of existing formulations.

It is primarily written for senior undergraduate, graduate students, and academic researchers in fields including industrial engineering, manufacturing engineering, production engineering, mechanical engineering, and engineering mathematics.

Mathematical Engineering, Manufacturing, and Management Sciences

Series Editor: Mangey Ram, *Professor, Assistant Dean (International Affairs), Department of Mathematics, Graphic Era University, Dehradun, India*

The aim of this new book series is to publish the research studies and articles that bring up the latest development and research applied to mathematics and its applications in the manufacturing and management sciences areas. Mathematical tool and techniques are the strength of engineering sciences. They form the common foundation of all novel disciplines as engineering evolves and develops. The series will include a comprehensive range of applied mathematics and its application in engineering areas such as optimization techniques, mathematical modelling and simulation, stochastic processes and systems engineering, safety-critical system performance, system safety, system security, high assurance software architecture and design, mathematical modelling in environmental safety sciences, finite element methods, differential equations, reliability engineering, etc.

Advances in Sustainable Machining and Manufacturing Processes
Edited by Kishor Kumar Gajrani, Arbind Prasad and Ashwani Kumar

Advanced Materials for Biomechanical Applications
Edited by Ashwani Kumar, Mangey Ram and Yogesh Kumar Singla

Biodegradable Composites for Packaging Applications
Edited by Arbind Prasad, Ashwani Kumar and Kishor Kumar Gajrani

Computing and Stimulation for Engineers
Edited by Ziya Uddin, Mukesh Kumar Awasthi, Rishi Asthana and Mangey Ram

Advanced Manufacturing Processes
Edited by Yashvir Singh, Nishant K. Singh and Mangey Ram

Additive Manufacturing
Advanced Materials and Design Techniques
Pulak M. Pandey, Nishant K. Singh and Yashvir Singh

Advances in Mathematical and Computational Modeling of Engineering Systems
Mukesh Kumar Awasthi, Maitri Verma and Mangey Ram

Biowaste and Biomass in Biofuel Applications
Edited by Yashvir Singh, Vladimir Strezov, and Prateek Negi

Lean Manufacturing and Service: Fundamentals, Applications, and Case Studies
Kanchan Das and Miranda Dixon

For more information about this series, please visit: www.routledge.com/Mathematical-Engineering-Manufacturing-and-Management-Sciences/book-series/CRCMEMMS

Strategic Fuzzy Extensions and Decision-making Techniques

Edited by
Kamal Kumar, Rishu Arora, and
Gagandeep Kaur

CRC Press is an imprint of the
Taylor & Francis Group, an **informa** business

Front cover image: EtiAmmos/Shutterstock

First edition published 2025
by CRC Press
2385 NW Executive Center Drive, Suite 320, Boca Raton FL 33431

and by CRC Press
4 Park Square, Milton Park, Abingdon, Oxon, OX14 4RN

CRC Press is an imprint of Taylor & Francis Group, LLC

ISBN: 978-1-032-54798-5 (hbk)
ISBN: 978-1-032-80425-5 (pbk)
ISBN: 978-1-003-49721-9 (ebk)

DOI: 10.1201/9781003497219

Typeset in Sabon
by Deanta Global Publishing Services, Chennai, India

Contents

Preface

The book *Strategic Fuzzy Optimization and Decision-making Techniques* discusses creative and recent developments of optimization and multi-criteria decision-making by using the fuzzy set and its generalizations, i.e., intuitionistic fuzzy set, interval-valued intuitionistic fuzzy set, Type-2 intuitionistic fuzzy set, linguistic intuitionistic fuzzy set, q-rung orthopair fuzzy set, linguistic q-rung orthopair fuzzy set, and neutrosophic cubic set. This book provides various techniques of decision-making such as entropy measures, distance measures, correlation coefficient, possibility measures, and aggregating operators. These approaches are used in many different disciplines, including decision-making, medical diagnosis, cluster analysis, service quality management, e-learning management, and environmental management. This book also explores the applications of fuzzy sets and logic applied to science, technology, and everyday life to further provide research on the subject.

The contributed chapters will be based on the writers' extensive research experiences in real-world multi-criteria decision-making problems. We hope that this edited volume will serve as a comprehensive source of several decision-making problems as a reference for students, researchers, and practitioners interested in the current advancements and applications of soft computing, machine learning, and data science. This book also explores the applications of fuzzy sets and its generalization to science, technology, and everyday life to further provide research on the subject. This book is ideal for mathematicians, physicists, computer specialists, engineers, practitioners, researchers, academicians, and students who are looking to learn more about fuzzy optimization and decision-making with their applications.

Kamal Kumar
Gagandeep Kaur
Rishu Arora

Editors

Kamal Kumar is Assistant Professor at Amity University Haryana, Gurugram, India. Prior to joining this university, Dr Kumar received his doctoral degree at the Thapar Institute of Engineering & Technology Patiala, Punjab, India, in 2020. He obtained his master's degree in mathematics during 2012–2014. He has more than seven years of teaching experience in different subjects of Engineering Mathematics and Statistics. He teaches Fuzzy Set Theory, Optimization techniques, and Statistics at under-graduate and post-graduate levels. His current research interest is in fuzzy decision-making, aggregation operators, soft computing, and uncertainty theory. He has published more than 50 international articles in different reputed SCI journals, including *Information Sciences*, *Transactions on Fuzzy Systems*, *Soft Computing*, *Artificial Intelligence Review*, *Applied Intelligence*, *Arabian Journal for Science and Engineering*, etc. He has also published five book chapters and two patents. His Google citation is over 2000 with h-index 21 and i10 index 26. He is Associate Editor in the journal *Mathematical Problems in Engineering* and the *Journal of Applied Mathematics.* He has been a reviewer for some prestigious journals of Elsevier, Springer, and Taylor & Francis. He is listed in the top 2% of Scientists in the World in all disciplines in 2021, 2022, and 2023 (three consecutive years) based on Stanford University and Elsevier B.V.

Rishu Arora is working as Assistant Professor II at Amity University, Noida, Uttar Pradesh, India. She completed her PhD in Mathematics from the Thapar Institute of Engineering & Technology (Deemed University) Patiala, Punjab, India, in 2020. She received her master's degree in mathematics from the Department of Mathematics, Guru Nanak Dev University, Amritsar, India, in 2010. She has qualified GATE in 2012–2015. Her current research interests are in the areas of multi-criteria decision-making, aggregation operator, and intuitionistic fuzzy soft set. She has published 20 papers in internationally reputed SCI journals and completed one DST-sponsored project as PI. Her Google citations are over 1200 with an h-index of 15.

Gagandeep Kaur is working as Assistant Professor at the Department of Mathematics, Amity School of Applied Sciences, Amity University Haryana, India. She has completed her PhD in Mathematics with specialization in Applied Mathematics and Computing from the Thapar Institute of Engineering and Technology. Prior to that, she has completed her MSc in Mathematics from Punjabi University, Patiala. Her area of research is Uncertainty quantification using Fuzzy Sets and Logic. She is a dedicated researcher in her field and has 17 published research articles in SCI journals of international repute with 750+ citations till now. Her recent research based on Traveller's screening at airport terminals during the COVID-19 pandemic has been acknowledged by the World Health Organization. She is an Editorial Board member of the scientific Web of Sciences journal *Frontiers of Sustainability* and has reviewed several articles related to journals *IEEE Transactions of Fuzzy Sets*, *Kybernetes*, *Complex & Intelligent Systems*, and *Computational and Applied Mathematics* under publishing houses such as Springer, Elsevier, and Wiley. She is proficient in Mathematical Modelling and software packages like MATLAB, MAPLE, Mathematics, and GeoGebra.

Contributors

Rishu Arora
Department of Mathematics
Amity Institute of Applied Sciences
Amity University Noida, India

Abhilash Kangsha Banik
Department of Mathematics
Dibrugarh University
Dibrugarh, Assam, India

Reeta Bhardwaj
Amity School of Applied Sciences
Amity University Haryana
Gurugram, Haryana, India

Kaushik Debnath
Vidyasagar University, Midnapore
West Bengal, India

Abolape Deborah Akwu
Joseph Sarwuan Tarka University
Makurdi, Nigeria

Chirag Dhankhar
Amity School of Applied Sciences
Amity University Haryana
Gurugram, Haryana, India

Palash Dutta
Department of Mathematics
Dibrugarh University
Dibrugarh, Assam, India

Paul Augustine Ejegwa
Department of Mathematics
Joseph Sarwuan Tarka University
Makurdi, Nigeria

Gagandeep Kaur
Amity School of Applied Sciences
Amity University Haryana
Gurugram, Haryana, India

Binoy Krishna Giri
Vidyasagar University, Midnapore
West Bengal, India

Bhuvaneshvar Kumar
Thapar Institute of Engineering & Technology
Patiala, India

Kamal Kumar
Amity School of Applied Sciences
Amity University Haryana
Gurugram, Haryana, India

Pravesh Kumar
Rajkiya Engineering College
Bijnor, India

Vijay Kumar
Amity School of Applied Sciences
Amity University Haryana
Gurugram, Haryana, India

Neeraj Lather
Amity School of Applied Sciences
Amity University Haryana
Gurugram, Haryana, India

Ritu Malik
Amity School of Applied Sciences
Amity University Haryana
Gurugram, Haryana, India

Naveen Mani
Chandigarh University
Chandigarh, India

Arijit Mondal
Vidyasagar University, Midnapore
West Bengal, India

Neelam
Amity School of Applied Sciences
Amity University Haryana

Vikash Patel
Gurukula Kangri (Deemed to be University)
Haridwar, Uttarakhand, India

Rohit Khatri
Amity School of Applied Sciences
Amity University Haryana
Gurugram, Haryana, India

Sankar Kumar Roy
Vidyasagar University, Midnapore
West Bengal, India

Amit Sharma
Amity School of Applied Sciences
Amity University Haryana
Gurugram, India

Devansh Sharma
Victorious kidss educares, Kharadi
Pune, India

Lokesh Singh
Villa College, Maldives

Priya Yadav
Amity University Haryana
Gurugram, India

Pooja Yadav
Amity School of Applied Sciences
Amity University Haryana
Gurugram, Haryana, India

Doonen Zuakwagh
Joseph Sarwuan Tarka University
Makurdi, Nigeria

Chapter 1

Fuzzy decision-making

Introduction and features

Rishu Arora, Gagandeep Kaur, and Kamal Kumar

1.1 FUZZY DECISION-MAKING: INTRODUCTION AND FEATURES

The cognitive process known as "decision-making" is typically applied in academia and industry to choose a course of action from a range of possible outcomes. Stated differently, decision-making is the process of determining and selecting options in accordance with the decision-maker's preferences and values. The analysis of individual decisions focuses on the rationale or logic behind the decisions, which might be irrational or rational depending on the explicit assumptions made. All science-based professions depend heavily on logical decision-making, where experts use their subject–matter expertise to make well-informed choices. It has been demonstrated, nevertheless, that group decisions are typically more successful than individual decisions. Consequently, group decision-making is a method of collaborative decision-making whereby decisions made by individual members of the group are combined to address a specific problem. However, there may occasionally be a propensity to show prejudice when discussing shared knowledge as compared to unknown info when people make judgements as a group. It is true that highly experienced, dynamic, and clever specialists or practitioners are needed to engage in order to overcome this type of inaccuracy in the decision-making process, and they should possess extensive expertise in the relevant field of judgement.

Furthermore, because most judgements are formed by alternating between the selection of criteria and the identification of alternatives, decision-making is a complex and cyclical process. Each choice takes place in the context of the decision-making environment, that is the set of facts, options, figures, and personal preferences that are at hand while a judgement is being made. Given that alternatives and information are restricted since there isn't enough opportunity or time to find alternatives or gather knowledge. In actuality, choices must be taken in this restricted setting. Currently, the main obstacle to reducing uncertainty is a primary objective of decision analysis, as decision-making involves unpredictability. Variability and criterion sensitivity have been formally incorporated into recent robust decision

DOI: 10.1201/9781003497219-1

efforts for the process of making decisions. Because there is so much subjectivity and ambiguity in fuzziness, an evaluation criterion has entered the picture. As an analytical instrument for intricate structures, Zadeh (1965) recommended using fuzzy set theory to address the types of qualitative, uncertain, and ambiguous data decision problems. A form of ambiguity known as fuzziness is connected to the application of fuzzy sets or classes where there isn't a clear break between membership and non-membership (Zimmermann, 1991).

The fuzzy set theory has been widely employed in numerous aspects of modern civilization since its introduction by Zadeh. The expansion of the characteristic function, which may take a value of 0 or 1, to the membership function, which can take any value from the closed interval [0,1], is central to the fuzzy set. However, because the membership function is a single-valued function, it cannot be utilized to describe pieces of evidence of both approval and disapproval at the same time in many real scenarios. Due to the rising complexity of socioeconomic contexts, people may lack an essential level of understanding of the problem's domain in the processes of cognition of things. In such instances, people frequently have some doubt in stating their preferences for the things under consideration, which causes the outcomes of cognitive performance to display affirmation, denial, and hesitation features. In a voting event, for example, in addition to support and opposition, there is generally abstention, which expresses the voter's reluctance and indecision about the object. Because the fuzzy set cannot be utilized to describe all the information in such an issue, it has a few limitations in practical implementations.

Atanassov (1983) extends the fuzzy set with a membership function to the intuitionistic fuzzy set (IFS), which has a membership function, a non-membership function, and a hesitancy function. As a result, the IFS can more precisely and fully represent the fuzzy characteristics of objects, which has been proven to be more helpful in dealing with ambiguity and uncertainty. The IFS theory has received increasing attention from both academics and practitioners over the past decade or so, and has been employed in a variety of fields, including the decision-making process, programming using logic, diagnosis in medicine, recognizing patterns, artificial intelligence, fuzzy topological structure, machine learning, and market prediction. The IFS theory is being researched thoroughly on a regular basis, as is the extent of its applications. As a result, effective collection and processing of intuitionistic fuzzy information has become increasingly vital. Information processing technologies, such as aggregation approaches for intuitionistic fuzzy information, correlation measures, distance measures, and similarity measures for IFSs, offer a wide range of potential applications and also provide a number of intriguing yet resilient research challenges.

Further development of fuzzy set theory termed interval-valued fuzzy set theory (Atanassov, 2006; Gorzalczany, 1987) assigns the closed subinterval of the unit interval, which roughly corresponds to the degree of membership that is undetermined, to each component in the entire universe. Since fuzzy sets are limited to describing imprecise information (variations in the concept of membership), their expressiveness is constrained. Conversely, interval-valued fuzzy sets may also handle uncertainty since they use an interval to approximate the actual membership degree that is only partially known. Moreover, the computational complexity is not significantly greater than that of using regular fuzzy sets. In order to overcome DM difficulties, Wang et al. (2012) developed a multi-criteria decision-making (MCDM) approach and suggested the weighted average AOs for IVIFS. Prioritized AOs were used by Arora and Garg (2019) to extend the linguistic IFS. Using their provided AOs, Garg and Rani (2022) resolved the MULTIMOORA method under IFS info. The weighted geometric and hybrid weighted geometric AOs for IVIFS were created by Xu and Chen (2007). They also developed the multi-attribute decision-making (MADM) method to address DM problems by utilizing their pre-existing AOs. Jia and Zhang (2019) introduced the multi-attribute group decision-making (MAGDM) model and extended the weighted arithmetic AOs for IVIFS. Within the IVIFS framework, Xu and Gou (2017) created a number of DM techniques and applied them to a range of real-world issues. The weighted arithmetic and geometric AOs for IVIFS were proposed by Xu (2007). These average and geometric AOs for IVIFS were extended by Mu et al. (2018). Additionally, they developed a few DM strategies to use AOs to overcome MADM obstacles. Zhang (2018) introduced the MAGDM method and created the Bonferroni mean geometric AOs in the IVIFS setup. The hybrid geometric aggregation operator for IVIFS was suggested by Park et al. (2009) and used for MAGDM issues. A correction model was created by Gupta et al. (2018) to ascertain the weight of experts. Specialists express their weight values using interval-valued intuitionistic fuzzy numbers.

The further development of the regular fuzzy set, or type-1 fuzzy set (T1FS), is a type-2 fuzzy set (T2FS). The capacity of the type-2 fuzzy set is to capture the membership of important membership values where the uncertainty is handled more accurately and has the basic advantage over the type-1 fuzzy set. A type-1 fuzzy set's membership value is a real integer in the interval [0, 1]. On the contrary, a type-1 fuzzy set is a T2FS's membership value. It was Zadeh (1999) who first presented the T2FS idea. The Mendel (2002) reference provided an overview of type-2 fuzzy sets. Considering type-2 fuzzy sets are specific examples of ordinary fuzzy sets and interval-valued fuzzy sets, Takac (2014) suggested that type-2 fuzzy sets are particularly useful in situations with higher levels of uncertainty. Using type-2 fuzzy parameters and considering both type reduction and the centroid, Kundu et al. (2014) presented a fixed charge transportation issue. Two studies looked at the logical processes of T2FS: Dubois and Prade (1980) and Mizumoto and Tanaka (1976; 1981). Many studies were later conducted on the theoretical (Coupland and John, 2008; Greenfield et al., 2005; Karnik and Mendel, 2001; Kar et al., 2019) and on different application areas (García, 2009; Hasuike and Ishii, 2009; Hidalgo, 2012; Kundu et al., 2019; Pramanik et al., 2015) of T2FS by other scholars.

In addition to this, Yager (2016) introduced a new idea, *q*-rung orthopair fuzzy sets (q-ROFs), in response to the ongoing theoretical advancements and sociological complexity. In these sets, the total of the qth power of the degrees of non-membership and the qth power of the membership degree is limited to one. Since FSs and IFSs are all their special instances, we may conclude that the q-ROFs are general. It is important to note that more orthopairs fulfil the bounding restriction as the *q*-rung grows, expanding the universe of admissible orthopairs. Thus, by employing q-ROFs, we can communicate a greater range of fuzzy information. In recent years, a lot of researchers have concentrated on solving DM problems in the q-ROFS setting. In this regard, the modified BM AOs and knowledge measure-based entropy measure for the q-ROFV were established by Liu (2018). Liu (2018a) provided the geometric and averaging AOs for q-ROFV. Riaz et al. (2020) propose Einstein AOs for the q-ROF environment. The Einstein-prioritized weighted average AOs for q-ROFV were first presented by Riaz et al. (2020a). For q-ROFVs, the possibility degree measure was defined by Garg (2021). The proposed knowledge measure for q-ROFVs was made by Khan (2021).

Despite the previously mentioned fuzzy tools that may be effectively used in solving difficulties that are classified as quantitative circumstances (Rodriguez et al., 2011), researchers frequently encounter ambiguity in challenges that have more qualitative character due to the imprecise nature of the meanings they employ. For instance, in the procedure of generating decisions under an intuitionistic fuzzy environment, a decision-maker might discover it challenging to communicate the degree of membership and non-membership with precise information due to the growing complexities of the decision-making environment, time constraints, and the absence of data or knowledge regarding the problem domain. Instead, he or she may believe that using linguistic values is a more direct and appropriate way to express the degree of membership and non-membership. Linguistic intuitionistic fuzzy set (LIFS) (Zhang, 2014) has characteristics as IFS in that it has two degrees of linguistic membership and non-membership, respectively. Since its appearance, many researchers have paid attention towards its application in decision-making problems (Chen et al., 2015; Zhang et al.,2017; Garg & Kumar, 2019; Ou et al., 2018). In addition to this, the fundamental ideas of linguistic q-rung orthopair fuzzy set (Lq-ROFS) were presented by Wang et al. (2019), in which the membership and non-membership degrees of an element to a linguistic variable are represented by a q-rung orthopair fuzzy number (q-ROFN). The author also created novel operational rules, comparison techniques, and distance measurement methods using Lq-ROFS. Based on the q-ROFSs and uncertain linguistic variables, Liu et al. (2019) created uncertain linguistic aggregation operations and presented a new class of fuzzy sets (Zadeh, 1965) called *q*-rung orthopair uncertain linguistic sets. A number of novel operational rules pertaining to probabilistic linguistic word sets were proposed by Yue et al. (2020) and their application in MADM. The MADM technique was created by Amin et al. (2020) using triangular cubic linguistic uncertain fuzzy aggregation operators.

A more comprehensive platform that expands on the ideas of the classic set and fuzzy set, intuitionistic fuzzy set, and interval-valued intuitionistic fuzzy set is the neutrosophic set (NS) notion, which was created by Sarandache (1999; 2005). Different parts are used with the theory of neutrosophic sets (see http://fs.gallup.unm.edu/neutrosophy.htm). Further, Jun et al. (2017) extended it to neutrosophic cubic sets. Ye (2018) went on to discuss the fundamental functions and NCS aggregation technique. Next, for decision-making applications in the NCS context, the Dombi (Shi & Ye, 2018), hybrid weighted arithmetic and geometric aggregation operators of NCSs (Shi & Yuan, 2019), and cosine (Lu & Ye, 2017), Jaccard, and Dice similarity measures of NCSs (Tu et al., 2018) were proposed. Fu et al. (2018) introduced the concept of cubic hesitant fuzzy sets (CHFSs) to represent hybrid information consisting of uncertain and hesitation fuzzy values. They also developed generalized measurements of CHFSs' distance and similarity that may be used in medical diagnostics.

REFERENCES

Amin, F., Fahmi, A., & Aslam, M. (2020). Approaches to multiple attribute group decision making based on triangular cubic linguistic uncertain fuzzy aggregation operators. *Soft Computing*, *24*, 11511–11533.

Arora, R., & Garg, H. (2019). Group decision-making method based on prioritized linguistic intuitionistic fuzzy aggregation operators and its fundamental properties. *Computational and Applied Mathematics*, *38*, 1–32.

Atanassov, K. T. (1983). Intuitionistic fuzzy sets, 1983. *VII ITKR's Session, Sofia (deposed in Central Sci.-Technical Library of Bulgarian Academy of Sciences, 1697/84) (in Bulgarian).*

Atanassov, K. (2006). Intuitionistic fuzzy sets and interval valued fuzzy sets. In *First International Workshop on IFSs, GNs, KE, London* (pp. 1–7).

Chen, Z., Liu, P., & Pei, Z. (2015). An approach to multiple attribute group decision making based on linguistic intuitionistic fuzzy numbers. *International Journal of Computational Intelligence Systems*, *8*(4), 747–760.

Coupland, S., & John, R. (2008). A fast geometric method for defuzzification of type-2 fuzzy sets. *IEEE Transactions on Fuzzy Systems*, *16*(4), 929–941.

Dubois, D. J. (1980). *Fuzzy sets and systems: Theory and applications* (Vol. 144). Academic Press.

Fu, J., Ye, J., & Cui, W. (2018). An evaluation method of risk grades for prostate cancer using similarity measure of cubic hesitant fuzzy sets. *Journal of Biomedical Informatics*, *87*, 131–137.

García, J. C. F. (2009, October). Solving fuzzy linear programming problems with interval type-2 RHS. In *2009 IEEE International Conference on Systems, Man and Cybernetics* (pp. 262–267). IEEE.

Garg, H., & Kumar, K. (2019). Linguistic interval-valued Atanassov intuitionistic fuzzy sets and their applications to group decision making problems. *IEEE Transactions on Fuzzy Systems*, *27*(12), 2302–2311.

Garg, H. (2021). A new possibility degree measure for interval-valued q-rung ortho-pair fuzzy sets in decision-making. *International Journal of Intelligent Systems*, *36*(1), 526–557.

Garg, H., & Rani, D. (2022). An efficient intuitionistic fuzzy MULTIMOORA approach based on novel aggregation operators for the assessment of solid waste management techniques. *Applied Intelligence*, *52*, 4330–4363.

Gorzalczany, M. B. (1987). A method of inference in approximate reasoning based on interval-valued fuzzy sets. *Fuzzy Sets and Systems*, *21*(1), 1–17.

Greenfield, S., John, R., & Coupland, S. (2005). A novel sampling method for type-2 defuzzification. *Proc. UKCI*, *6*, 120–127.

Gupta, P., Mehlawat, M. K., Grover, N., & Pedrycz, W. (2018). Multi-attribute group decision making based on extended TOPSIS method under interval-valued intuitionistic fuzzy environment. *Applied Soft Computing*, *69*, 554–567.

Hasuike, T., & Ishii, H. (2009, July). A type-2 fuzzy portfolio selection problem considering possibility measure and crisp possibilistic mean value. In *IFSA/EUSFLAT Conference* (pp. 1120–1125).

Hidalgo, D., Melin, P., & Castillo, O. (2012). An optimization method for designing type-2 fuzzy inference systems based on the footprint of uncertainty using genetic algorithms. *Expert Systems with Applications*, *39*(4), 4590–4598.

Jia, Z., & Zhang, Y. (2019). Interval-valued intuitionistic fuzzy multiple attribute group decision making with uncertain weights. *Mathematical Problems in Engineering*, *2019*.

Jun, Y. B., Smarandache, F., & Kim, C. S. (2017). Neutrosophic cubic sets. *New Mathematics and Natural Computation*, *13*(01), 41–54.

Kar, M. B., Roy, B., Kar, S., Majumder, S., & Pamucar, D. (2019). Type-2 multi-fuzzy sets and their applications in decision making. *Symmetry*, *11*(2), 170.

Karnik, N. N., & Mendel, J. M. (2001). Centroid of a type-2 fuzzy set. *Information Sciences*, *132*(1–4), 195–220.

Khan, M. J., Kumam, P., & Shutaywi, M. (2021). Knowledge measure for the q-rung orthopair fuzzy sets. *International Journal of Intelligent Systems*, *36*(2), 628–655.

Kundu, P., Kar, S., & Maiti, M. (2014). Fixed charge transportation problem with type-2 fuzzy variables. *Information Sciences*, *255*, 170–186.

Kundu, P., Majumder, S., Kar, S., & Maiti, M. (2019). A method to solve linear programming problem with interval type-2 fuzzy parameters. *Fuzzy Optimization and Decision Making*, *18*, 103–130.

Liu, P., & Liu, J. (2018). Some q-rung orthopair fuzzy Bonferroni mean operators and their application to multi-attribute group decision making. *International Journal of Intelligent Systems*, *33*(2), 315–347.

Liu, P., & Wang, P. (2018a). Some q-rung orthopair fuzzy aggregation operators and their applications to multiple-attribute decision making. *International Journal of Intelligent Systems*, *33*(2), 259–280.

Liu, Z., Xu, H., Yu, Y., & Li, J. (2019). Some q-rung orthopair uncertain linguistic aggregation operators and their application to multiple attribute group decision making. *International Journal of Intelligent Systems*, *34*(10), 2521–2555.

Lu, Z., & Ye, J. (2017). Cosine measures of neutrosophic cubic sets for multiple attribute decision-making. *Symmetry*, *9*(7), 121.

Mendel, J. M., & John, R. B. (2002). Type-2 fuzzy sets made simple. *IEEE Transactions on Fuzzy Systems*, *10*(2), 117–127.

Mizumoto, M., & Tanaka, K. (1976). Some properties of fuzzy sets of type 2. *Information and Control*, *31*(4), 312–340.

Mizumoto, M., & Tanaka, K. (1981). Fuzzy sets and type 2 under algebraic product and algebraic sum. *Fuzzy Sets and Systems*, *5*(3), 277–290.

Mu, Z., Zeng, S., & Liu, Q. (2018). Some interval-valued intuitionistic fuzzy Zhenyuan aggregation operators and their application to multi-attribute decision making. *International Journal of Uncertainty, Fuzziness and Knowledge-Based Systems*, *26*(04), 633–653.

Ou, Y., Yi, L., Zou, B., & Zheng, P. (2018). The linguistic intuitionistic fuzzy set TOPSIS method for linguistic multi-criteria decision makings. *International Journal of Computational Intelligence Systems*, *11*(1), 120.

Park, D. G., Kwun, Y. C., Park, J. H., & Park, I. Y. (2009). Correlation coefficient of interval-valued intuitionistic fuzzy sets and its application to multiple attribute group decision making problems. *Mathematical and Computer Modelling*, *50*(9–10), 1279–1293.

Pramanik, S., Jana, D. K., Mondal, S. K., & Maiti, M. (2015). A fixed-charge transportation problem in two-stage supply chain network in Gaussian type-2 fuzzy environments. *Information Sciences*, *325*, 190–214.

Rodriguez, R. M., Martinez, L., & Herrera, F. (2011). Hesitant fuzzy linguistic term sets for decision making. *IEEE Transactions on Fuzzy Systems*, *20*(1), 109–119.

Riaz, M., Salabun, W., Athar Farid, H. M., Ali, N., & Wątróbski, J. (2020). A robust q-rung orthopair fuzzy information aggregation using Einstein operations with application to sustainable energy planning decision management. *Energies*, *13*(9), 2155.

Riaz, M., Athar Farid, H. M., Kalsoom, H., Pamučar, D., & Chu, Y. M. (2020a). A robust q-rung orthopair fuzzy Einstein prioritized aggregation operators with application towards MCGDM. *Symmetry*, *12*(6), 1058.

Shi, L., & Ye, J. (2018). Dombi aggregation operators of neutrosophic cubic sets for multiple attribute decision-making. *algorithms*, *11*(3), 29.

Shi, L., & Yuan, Y. (2019). Hybrid weighted arithmetic and geometric aggregation operator of neutrosophic cubic sets for MADM. *Symmetry*, *11*(2), 278.

Smarandache, F. (1999). A unifying field in logics: Neutrosophic logic. In Florentin Smarandache *Philosophy* (pp. 1–141). American Research Press.

Smarandache, F. (2005). *A unifying field in logics: Neutrosophic logic. Neutrosophy, neutrosophic set, neutrosophic probability: Neutrosophic logic. Neutrosophy, neutrosophic set, neutrosophic probability*. Infinite Study.

Takac, Z. (2014). Aggregation of fuzzy truth values. *Information Sciences*, *271*, 1–13.

Tu, A., Ye, J., & Wang, B. (2018). Multiple attribute decision-making method using similarity measures of neutrosophic cubic sets. *Symmetry*, *10*(6), 215.

Wang, W., Liu, X., & Qin, Y. (2012). Interval-valued intuitionistic fuzzy aggregation operators. *Journal of Systems Engineering and Electronics*, *23*(4), 574–580.

Wang, H., Ju, Y., & Liu, P. (2019). Multi-attribute group decision-making methods based on q-rung orthopair fuzzy linguistic sets. *International Journal of Intelligent Systems*, *34*(6), 1129–1157.

Xu, Z., & Chen, J. (2007, August). On geometric aggregation over interval-valued intuitionistic fuzzy information. In *Fourth international conference on fuzzy systems and knowledge discovery (FSKD 2007)* (Vol. 2, pp. 466–471). IEEE.

Xu, Z. (2007). Methods for aggregating interval-valued intuitionistic fuzzy information and their application to decision making. *Control and decision*, *22*(2), 215–219.

Xu, Z., & Gou, X. (2017). An overview of interval-valued intuitionistic fuzzy information aggregations and applications. *Granular Computing*, *2*, 13–39.

Yager, R. R. (2016). Generalized orthopair fuzzy sets. *IEEE Transactions on Fuzzy Systems*, *25*(5), 1222–1230.

Ye, J. (2018). Operations and aggregation method of neutrosophic cubic numbers for multiple attribute decision-making. *Soft Computing*, *22*(22), 7435–7444.

Yue, N., Xie, J., & Chen, S. (2020). Some new basic operations of probabilistic linguistic term sets and their application in multi-criteria decision making. *Soft Computing*, *24*, 12131–12148.

Zadeh, L. A. (1965). Fuzzy sets. *Information and Control*, *8*(3), 338–353.

Zadeh, L. A. (1999). Fuzzy sets as a basis for a theory of possibility. *Fuzzy sets and Systems*, *100*, 9–34.

Zhang, H. (2014). Linguistic intuitionistic fuzzy sets and application in MAGDM. *Journal of Applied Mathematics*, *2014*, 1–11.

Zhang, H. Y., Peng, H. G., Wang, J., & Wang, J. Q. (2017). An extended outranking approach for multi-criteria decision-making problems with linguistic intuitionistic fuzzy numbers. *Applied Soft Computing*, *59*, 462–474.

Zhang, Z. (2018). Geometric Bonferroni means of interval-valued intuitionistic fuzzy numbers and their application to multiple attribute group decision making. *Neural Computing and Applications*, *29*, 1139–1154.

Zimmermann, H. J. (1991). Possibility theory, probability theory, and fuzzy set theory. In *Fuzzy set theory—and its applications*, Springer, Dordrecht. https://doi.org/10.1007/978-94-015-7949

Chapter 2

Preliminaries

Gagandeep Kaur

2.1 INTRODUCTION

In this chapter, the aspect of basic preliminaries of this book has been discussed. Fuzzy sets have been widely accepted as the tool for quantifying the uncertainties occurring in real-life scenarios. In standard crisp set theory, the elements belonging to a set are decided on the basis of a characteristic function which is either 0 or 1. In fuzzy set theory, the belongingness of an element in a set is decided by the membership function. The most commonly used range of membership functions is [0,1]. Membership function associated with a fuzzy set A defined over a universal set X is denoted as μ_A, as $\mu_A : X \longrightarrow [0,1]$.

There are certain properties associated with the membership functions, such as continuous, monotonic, and bounded.

The cutting-edge research being done in the present scenario of fuzzy set theory involves the study of information measures and aggregation operators. These two aspects are covered under various fuzzy environments such as intuitionistic fuzzy sets, interval-valued fuzzy sets, and orthopair fuzzy sets. To follow this, firstly, the description about the information measures is given below.

2.2 INFORMATION MEASURES

An information measure gives the amount of uncertainty enclosed in a fuzzy number. In other words, it helps in quantifying the amount of uncertainty in a fuzzy set. There are a number of distance measures from which measures such as distance, similarity, and entropy are the prominent ones. For it, let $\phi(X)$ be a collection of fuzzy sets over universal set $X = \{x_1, x_2, \ldots, x_n\}$.

a. **Distance measure:** Let A, B, and C be three fuzzy sets, then the distance measure d satisfies the following properties:
   (P1) $0 \leq d(A,B) \leq 1$;

DOI: 10.1201/9781003497219-2

(**P2**) $d(A,B) = d(B,A)$;
(**P3**) $d(A,B) = 0$ if $A=B$;
(**P4**) If $A \leq B \leq C$, then $d(A,B) \leq d(A,C)$ and $d(B,C) \leq d(A,C)$.

b. **Similarity measure:** Let A, B, and C be three fuzzy sets, then the distance measure d satisfies the following properties:
(**P1**) $0 \leq S(A,B) \leq 1$;
(**P2**) $S(A,B) = S(B,A)$;
(**P3**) $S(A,B) = 1$ if $A=B$;
(**P4**) If $A \leq B \leq C$, then $S(A,B) \geq S(A,C)$ and $S(B,C) \geq S(A,C)$.

c. **Correlation measure:** For two fuzzy sets A and B, the correlation coefficient satisfies the following properties:
(**P1**) $0 \leq K(A,B) \leq 1$;
(**P2**) $K(A,B) = K(B,A)$;
(**P3**) $K(A,B) = 1$ if $A=B$.

Similarly, there are a number of information measures which are being used in the fuzzy set theory. Subjected to the different environments, the information measures must satisfy the above-stated basic axioms. Various information measures classified under different environments have been given in the subsequent chapters.

2.3 AGGREGATION OPERATORS

Aggregation operators are used to present the collective amount of uncertain information held by different fuzzy sets by aggregating them under different mathematical notions. For it, there are two basic norm operations stated as t-norm and t-conorm. The associated properties of t-norm and t-conorm are given below:

(i) A t-norm (fuzzy intersection) 'T' is a binary operation on [0,1], i.e.,

$$T : [0,1] \times [0,1] \longrightarrow [0,1]$$

defined as $(A \cap B)(x) = T(A(x), B(x)), \forall x \in X$,
where A and B are arbitrary fuzzy sets. The mapping preserves the following axioms for all $a,b,c \in [0,1]$.

a) $T(a,1) = a$ (Boundary condition)
b) If $b < c$, then $T(a,b) < T(a,c)$ (Monotonicity)
c) $T(a,T(b,c)) = T(T(a,b),c)$ (Associativity)
d) $T(a,b) = T(b,a)$ (Commutativity)

(ii) Subsequently, t-conorm (fuzzy union) 'S' is also a binary operator on [0,1] given by

$$S: [0,1] \times [0,1] \longrightarrow [0,1]$$

defined by $(A \cup B)(x) = S(A(x), B(x)), \forall x \in X$.

Alike the t-norm, it also satisfies the conditions of boundary, monotonicity, associativity, and commutativity. Both the norms are inter-related to each other and their mutual relation is given as $S(a,b) = 1-T(1-a,\ 1-b)$.

A class of fuzzy intersection (t-norm) is obtained if t-norm also satisfies the additional axioms, i.e.,

e) T is a continuous function. (Continuity)
f) $T(a,a) < a$ (Sub-idempotency)
g) If $a_1 < a_2$ and $b_1 < b_2$, then $T(a_1,\ b_1) = T(a_2,\ b_2)$ (Strict monotonicity)

In a similar manner, for t-conorm, the sub-idempotency is replaced by $S(a,a) > a$ and is called superidempotency. A continuous t-norm which satisfies the sub-idempotency, i.e., $T(a,a) < a$ is called an Archimedean t-norm. If it also satisfies strict monotonicity, then it is called strict Archimedean t-norm. On the other hand, a continuous t-conorm that satisfies the super-idempotency, i.e., $S(a,a) > a$ is called an Archimedean t-conorm. If it is strict, then it is called strict Archimedean t-conorm.

Further, strict Archimedean t-norm and t-conorm can be expressed as continuous function y:[0,1) ⟶[0,1], respectively, for $a,b \in [0,1]$ as $T(a,b) = y^{-1}\ (y(a)+y(b))$ and $S(a,b) = \mathrm{u}^{-1}\ (u(a)+u(b))$, where y (or u) is decreasing (or increasing) function with $y(1) = 0$, $u(0) = 0$, and $y(a) = u(1-a)$. However, some standard union and intersection forms for $a,b \in [0,1]$ are defined as:

a) Standard intersection and union: $T(a,b) = min(a,b)$; $S(a,b) = min(a,b)$.
b) Algebraic product and algebraic sum: $T(a,b) = ab$; $S(a,b) = a+b-ab$.
c) Bounded difference and sum: $T(a,b) = max(0,a+b-1)$; $S(a,b) = min(1,\ a+b)$.
d) Drastic intersection and union:

$$T(a,b)=\begin{cases} a;\text{if } b=1 \\ b;\text{if } a=1 \\ 0;\text{ otherwise} \end{cases} \quad ; \quad S(a,b)=\begin{cases} a;\text{if } b=0 \\ b;\text{if } a=0 \\ 1;\text{ otherwise} \end{cases}$$

e) Yager class of t-norm and t-conorm:

$$T(a,b)=1-\min\left\{1,\left[(1-a)^{\xi}+(1-b)^{\xi}\right]^{\frac{1}{\xi}}\right\}$$

$$S(a,b)=\min\{1,(a^{\xi}+b^{\xi})^{\frac{1}{\xi}}\},$$

where $\xi > 0$.

Apart from that, some other t-norms and t-conorms with their generator functions are given in Table 2.1

Operator	$T(a,b)$	Generator
Yager (1980)	$1-\min\left\{1,\left[(1-a)^{\omega}+(1-b)^{\omega}\right]^{\frac{1}{\omega}}\right\}$	$\omega>0$
Dubois (1980)	$\dfrac{ab}{\max(a,b,\alpha)}$	$\alpha\in[0,1]$
Weber (1983)	$max\left(0,\dfrac{a+b+\lambda ab-1}{1+\lambda}\right)$	$\lambda>1$
Yu (1985)	$max\left[0,(1+\lambda)(a+b-1)-\lambda ab\right]$	$\lambda>-1$
Dombi (1982)	$\left\{1+\left[\left(\frac{1}{a}-1\right)^{\lambda}+\left(\frac{1}{b}-1\right)^{\lambda}\right]^{-\frac{1}{\lambda}}\right\}^{-1}$	$\lambda>0$
Frank (1979)	$1-log_{a}\left[1-\dfrac{(\lambda^{1-a}-1)(\lambda^{1-b}-1)}{\lambda-1}\right]$	$\lambda>0,\lambda\neq1$
Hamacher (1978)	$\dfrac{a+b+(r-2)ab}{r+(r-1)ab}$	$r>0$
Schweizer (1963)	$1-\left\{max\left(0,(1-a)^{p}+(1-b)^{p}-1\right)\right\}^{\frac{1}{p}}$	$p\neq0$

2.4 FUZZY SET THEORY IN DECISION-MAKING

Consider the set of alternatives $A=\{A_1, A_2, \ldots, A_m\}$ be the set of alternatives which are to be evaluated under 'n' criteria denoted by $C=\{C_1, C_2, \ldots, C_n\}$ and the experts give their preferences in the form of fuzzy numbers α_{ij}; $i = 1, 2, \ldots, m$; $j = 1, 2, \ldots, n$ given as:

$$D=\begin{bmatrix}\alpha_{11} & \alpha_{12} & \cdots & \alpha_{1n}\\ \alpha_{21} & \alpha_{22} & \cdots & \alpha_{2n}\\ \vdots & \vdots & \ddots & \vdots\\ \alpha_{m1} & \alpha_{m2} & \cdots & \alpha_{mn}\end{bmatrix}$$

DM matrix to be input:

Step 1: Tabulate the values in the form of alternatives and criteria.
Step 2: Normalize the rating values from cost to benefit type or vice-versa according to the nature of the problem.
Step 3: Aggregate the preference values related to alternatives into collective values using the suitable tool/technique.
Step 4: Utilize the appropriate defuzzification method to compute the crisp value related to each alternative.
Step 5: Rank the alternatives on the basis of defuzzified values and select the best alternative(s).

REFERENCES

Dombi, J. (1982). A general class of fuzzy operators, the De Morgan class of fuzzy operators and fuzziness measures induced by fuzzy operators. *Fuzzy Sets and Systems*, *8*(2), 149–163.

Dubois, D. (1980). *Fuzzy Sets and Systems: Theory and Applications*. New York: Academic Press.

Frank, M. J. (1979). On the simultaneous associativity of F(x,y) and x + y - F(x, y). *Aequationes Mathematicae*, *19*(2–3), 194–226.

Hamacher, H. (1978). Sensitivity analysis in fuzzy linear programming. *Fuzzy Sets and Systems*, *1*(4), 269–281.

Schweizer, B. (1963). Associative functions and abstract semigroups. *Publicationes Mathematicae Debrecen*, *10*, 69–81.

Weber, S. (1983). A general concept of fuzzy connectives, negations and implications based on r-norms and f-conorms. *Fuzzy Sets and Systems*, *11*(2), 115–134.

Yager, R. R. (1980). On a general class of fuzzy connectives. *Fuzzy Sets and Systems*, *4*(3), 235–242.

Yu, Y. D. (1985). Triangular norms and TNF-sigma algebia. *Fuzzy Sets and Systems*, *16*(3), 251–264.

Chapter 3

Multi-attributed decision-making using enhanced possibility degree measure of intuitionistic fuzzy set

Pooja Yadav, Vikash Patel, and Neeraj Lather

3.1 INTRODUCTION

Multi-attribute decision-making (MADM) has several uses in the technical and social sciences. MADM includes choosing the best alternative (Wang et al.,2021) by considering various factors. Finding the best framework for evaluating the performance of alternatives with regard to these criteria is one of the main issues for decision-makers in this process of MADM. In order to deal with uncertainty or vagueness, Zadeh (1965) created the concept of fuzzy set (FS) to handle these challenges. Intuitionistic fuzzy set (IFS) is a notion that was first developed by Atanassov (1986), and Atanassov and Gargov (1999) further extended it to interval-valued intuitionistic fuzzy set (IVIFS). IFSs are based on membership degree, non-membership degree, and the degree of hesitancy, whereas FS is defined by only a membership degree.

Extensive research has been done in the area of MADM. Effective MADM management requires two key considerations: First to create a suitable function that combines the varied preferences of decision-makers into a single set of preferences. Designing acceptable metrics for ranking the available options is the second aspect. In the context of MADM, Wei and Tang (2010) presented the possibility of degree measure (PDM) for intuitionistic fuzzy numbers (IFNs). Garg and Kumar (2019) developed an improved PDM to overcome the Wei and Tang's PDM (Wei and Tang, 2010). Kumar and Chen (2021) introduced the improved intuitionistic fuzzy Einstein weighted averaging (IIFEWA) aggregation operator (AO) for MADM approaches.

Over the past few decades, numerous researchers, including Kumar and Gupta (2023), Ma et al. (2020), Garg and Kaur (2020), and Mishra et al. (2019), have proposed various MADM methods based on the framework of fuzzy sets (FSs) and intuitionistic fuzzy sets (IFSs). The above-mentioned studies reveal that researchers have employed different measures such as accuracy, scores, or possibility measures to rank various intuitionistic fuzzy numbers (IFNs) (Zeng et al., 2019;Liu et al., 2020; and Dhankhar and Kumar, 2023). In our study, we observed that if we assume the same

DOI: 10.1201/9781003497219-3

membership degree for two IFNs in Garg and Kumar's (2019), PDM for IFNs provides an inaccurate ranking order (RO). To address this issue, we introduced a novel PDM for IFNs to overcome the limitations of Garg and Kumar's (2019) PDM for IFNs. Consequently, to overcome the drawbacks of Garg and Kumar's (2019) existing MADM approach, it becomes necessary to develop a new MADM approach within the IFNs' framework.

In this chapter, we have proposed the enhanced possibility degree measure (EPDM) for ranking of IFNs. This newly proposed EPDM for IFNs addresses the limitations of Garg and Kumar's (2019) PDM for IFNs. Garg and Kumar's (2019) PDM in certain situations fails to differentiate the RO of alternatives. The aim of this chapter is divided as follows: In Section 3.2, a brief introduction of fundamental concepts is presented, which are relevant to this chapter. In Section 3.3, we proposed an EPDM for the IFS environment to overcome the flaws of existing PDM of the IFS environment. The solution for MADM problem based on proposed EPDM is given in Section 3.4. Section 3.5 concludes the chapter.

3.2 PRELIMINARIES

Definition 3.1 (Atanassov, 1986) In universal set X, an IFS I_F is represented by

$$I_F = \{\langle x, \alpha(x), \beta(x)\rangle \mid x \in X\}, \tag{3.1}$$

Where $\alpha(x)$ and $\beta(x)$ are membership degree and non-membership degree of the element x to I_F, respectively, where $x \in X$, $0 \leq \alpha(x) + \beta(x) \leq 1$, $x \in X$, and hesitance degree is

$\pi(x) = 1 - \alpha(x) - \beta(x)$ of x to I_F, where $0 \leq \pi(x) \leq 1$, $x \in X$.

In the context of the IFS, Garg and Kumar (2019) commonly referred to pair $\langle\alpha, \beta\rangle$ an IFN in the IFS $I_F = \{\langle x, \alpha(x), \beta(x)\rangle \mid x \in X\}$.

Definition 3.2 (Atanassov, 1986) For comparing two IFNs $\chi_1 = \langle\alpha_1, \beta_1\rangle$ and $\chi_2 = \langle\alpha_2, \beta_2\rangle$, the operational guidelines are presented as follows:

(i) $\chi_1 \geq \chi_2 \Leftrightarrow \alpha_1 \geq \alpha_2$ and $\beta_1 \leq \beta_2$;
(ii) $\chi_1 = \chi_2 \Leftrightarrow \alpha_1 = \alpha_2$ and $\beta_1 = \beta_2$.

Definition 3.3 (Kumar and Chen, 2021) The IIFEWA AO for aggregating $\chi_1 = \langle\alpha_1, \beta_1\rangle, \chi_2 = \langle\alpha_2, \beta_2\rangle, \ldots, \chi_m = \langle\alpha_m, \beta_m\rangle$, is presented as follows:

$$\text{IIFEWA}(\chi_1, \chi_2, \ldots, \chi_m) = \left\langle \frac{\prod_{t=1}^{m}(1-\alpha_t)^{w_t} - \left[1 - \frac{1}{\varepsilon}\left(1 - \prod_{t=1}^{m}(1-\varepsilon\alpha_t)^{w_t}\right)\right]}{\prod_{t=1}^{m}(1-\alpha_t)^{w_t} + \left[1 - \frac{1}{\varepsilon}\left(1 - \prod_{t=1}^{m}(1-\varepsilon\alpha_t)^{w_t}\right)\right]}, \frac{2\left(1 - \frac{1}{\varepsilon}\left(1 - \prod_{t=1}^{m}(1-\varepsilon(1-\beta_t))^{w_t}\right)\right)}{\prod_{t=1}^{m}(2-\beta_t)^{w_t} + \left(1 - \frac{1}{\varepsilon}\left(1 - \prod_{t=1}^{m}(1-\varepsilon(1-\beta_t))^{w_t}\right)\right)} \right\rangle, \tag{3.2}$$

where the IFN χ_t has weight w_t such that $w_t \geq 0$ and $\sum_{t=1}^{m} w_t = 1$.

The ranking principle related to the PDM as described by Garg and Kumar (2019) is summarized below together with the existing PDM as it was introduced by Garg and Kumar in 2019.

Definition 3.4 For two IFNs $\chi_1 = \langle \alpha_1, \beta_1 \rangle$ and $\chi_2 = \langle \alpha_2, \beta_2 \rangle$, the existing PDM $\rho'(\chi_1 \geq \chi_2)$ of $\chi_1 \geq \chi_2$ is defined as follows:

(i) If either $\pi_1 \neq 0$ or $\pi_2 \neq 0$ then

$$\rho'(\chi_1 \succeq \chi_2) = min\left(max\left(\frac{1+\alpha_1 - 2\alpha_2 - \beta_2}{\pi_1 + \pi_2}, 0\right), 1\right). \tag{3.3}$$

(ii) If both $\pi_1 = \pi_2 = 0$ then

$$\rho'(\chi_1 \succeq \chi_2) = \begin{cases} 1 & : \alpha_1 > \alpha_2 \\ 0 & : \alpha_1 < \alpha_2 \\ 0.5 & : \alpha_1 = \alpha_2. \end{cases}$$

By employing Eq. (3.3), we get the possibility degree matrix (PDMx) $M' = \left[\rho'_{ti}\right]_{n \times n} = \left[\rho'(\chi_t \succeq \chi_i)\right]_{n \times n}$ for ordering n IFNs $\chi_1, \chi_2, \ldots, \chi_n$ as follows:

$$M' = \begin{pmatrix} \rho'_{11} & \rho'_{12} & \cdots & \rho'_{1n} \\ \rho'_{21} & \rho'_{22} & \cdots & \rho'_{2n} \\ \vdots & \vdots & \ddots & \vdots \\ \rho'_{n1} & \rho'_{n2} & \cdots & \rho'_{nn} \end{pmatrix}. \tag{3.4}$$

Following that, compute the ranking value (RV) φ_t for IFNs χ_t as

$$\varphi'_t = \frac{1}{n(n-1)}\left(\sum_{i=1}^{n} \rho'_{ti} + \frac{n}{2} - 1\right). \tag{3.5}$$

Choose the best IFNs by placing the ranking values (RVs) in decreasing order.

Example 3.1 Consider two IFNs $\chi_1 = \langle 0.3, 0.4 \rangle$ and $\chi_2 = \langle 0.3, 0.2 \rangle$. For comparing χ_1 and χ_2, utilize the existing PDM as defined in Eq. (3.3), we have $\rho'(\chi_1 \geq \chi_2) = 0.625$ and $\rho'(\chi_2 \geq \chi_1) = 0.375$.

Therefore, the PDMx is obtained by utilizing Eq. (3.4) in the following manner: $M' = \begin{pmatrix} 0.5000 & 0.625 \\ 0.375 & 0.5000 \end{pmatrix}$.

Hence, by utilizing Eq. (3.5), we get the RVs $\varphi_1 = 0.5625$ and $\varphi_2 = 0.4375$ of the IFNs χ_1 and χ_2, respectively. Since $\varphi_1 > \varphi_2$, therefore $\chi_1 > \chi_2$.

Additionally, we discovered that $\alpha_1 = \alpha_2 = 0.3$ and $\beta_1 = 0.4 > 0.2 = \beta_2$ indicating that the membership grades (MGs) of χ_1 and χ_2 are identical, and non-membership grades (NMG) of χ_2 are less than the non-membership grades (NMG) of χ_1. Consequently, in accordance with Definition 3.2, we obtain $\chi_2 > \chi_1$. Consequently, the existing PDM ρ' from the work of Garg and Kumar (2019) does not accurately determine the correct ranking order (RO) of the IFNs χ_1 and χ_2.

Hence, we need to propose a new PDM for IFNs.

3.3 ENHANCED POSSIBILITY DEGREE MEASURE (EPDM)

In this segment, we introduced an EPDM for the evaluation of IFN rankings.

Definition 3.5 For any two IFNs $\chi_1 = \langle\alpha_1,\beta_1\rangle$ and $\chi_2 = \langle\alpha_2,\beta_2\rangle$, the proposed EPDM $\rho(\chi_1 \geq \chi_2)$ of $\chi_1 \geq \chi_2$ is given as follows:

(i) If either $\pi_1 \neq 0$ or $\pi_2 \neq 0$ then

$$\rho(\chi_1 \succeq \chi_2) = min\left(max\left(\frac{1+\alpha_1-2\alpha_2-\beta_1+2\alpha_2\beta_2}{\pi_1+\pi_2+2\alpha_1\beta_1+2\alpha_2\beta_2},0\right),1\right). \quad (3.6)$$

(ii) If both $\pi_1 = \pi_2 = 0$ then

$$\rho(\chi_1 \succeq \chi_2) = \begin{cases} 1 & : \alpha_1 > \alpha_2 \\ 0 & : \alpha_1 < \alpha_2 \\ 0.5 & : \alpha_1 = \alpha_2. \end{cases} \quad (3.7)$$

Theorem 3.1 Consider χ_1 and χ_2 be any two IFNs, then

(a) $0 \leq \rho(\chi_1 \geq \chi_2) \leq 1$;
(b) $\rho(\chi_1 \geq \chi_2) = 0.5$ if $\chi_1 = \chi_2$;
(c) $\rho(\chi_1 \geq \chi_2) + \rho(\chi_2 \geq \chi_1) = 1$.

Proof:

(a) We do this by assuming

$$\kappa = \frac{1+\alpha_1-2\alpha_2-\beta_1+2\alpha_2\beta_2}{\pi_1+\pi_2+2\alpha_1\beta_1+2\alpha_2\beta_2}.$$

Now, there are the following three situations:

(1) If $\kappa \geq 1$ then
$\rho\,(\chi_1 \geq \chi_2) = \min(\max(\kappa,0),1) = \min(\kappa,1) = 1.$

(2) If $0 < \kappa < 1$ then
$\rho\,(\chi_1 \geq \chi_2) = \min(\max(\kappa,0),1) = \min(\kappa,1) = \kappa.$

(3) If $\kappa \leq 0$ then
$\rho\,(\chi_1 \geq \chi_2) = \min(\max(\kappa,0),1) = \min(0,1) = 0.$

Based on the three scenarios described above, it can be deduced that $0 \leq \rho(\chi_1 \geq \chi_2) \leq 1$.

(b) Let $\chi_1 = \langle\alpha_1,\beta_1\rangle$, $\chi_2 = \langle\alpha_2, \beta_2\rangle$ be two IFNs. If $\chi_1 = \chi_2$, which implies that $\alpha_1 = \alpha_2$ and $\beta_1 = \beta_2$. Then, Eq. (3.6) becomes

$$\rho(\chi_1 \succeq \chi_2) = \min\left(\max\left(\frac{1+\alpha_1-2\alpha_2-\beta_1+2\alpha_2\beta_2}{\pi_1+\pi_2+2\alpha_1\beta_1+2\alpha_2\beta_2},0\right),1\right)$$

$$= \min\left(\max\left(\frac{\pi_1+2\alpha_1\beta_1}{2\pi_1+4\alpha_1\beta_1},0\right),1\right) = \min(\max(0.5,0),1) = 0.5.$$

(c) Let

$$\kappa = \frac{1+\alpha_1-2\alpha_2-\beta_1+2\alpha_2\beta_2}{\pi_1+\pi_2+2\alpha_1\beta_1+2\alpha_2\beta_2},$$

$$\epsilon = \frac{1+\alpha_2-2\alpha_1-\beta_2+2\alpha_1\beta_1}{\pi_1+\pi_2+2\alpha_1\beta_1+2\alpha_2\beta_2}.$$

We have

$$\kappa+\epsilon = \frac{1+\alpha_1-2\alpha_2-\beta_1+2\alpha_2\beta_2+1+\alpha_2-2\alpha_1-\beta_2+2\alpha_1\beta_1}{\pi_1+\pi_2+2\alpha_1\beta_1+2\alpha_2\beta_2}$$

$$= \frac{\pi_1+\pi_2+2\alpha_2\beta_2+2\alpha_1\beta_1}{\pi_1+\pi_2+2\alpha_1\beta_1+2\alpha_2\beta_2} = 1.$$

(a) If $\kappa \leq 0$ and $\epsilon \geq 1$ then
$\rho(\chi_1 \geq \chi_2) + \rho(\chi_2 \geq \chi_1) = \min(\max(\kappa,0),1) + \min(\max(\epsilon,0),1) = \min(0,1) + \min(\epsilon,1) = 1.$

(b) If $0 < \kappa, \epsilon < 1$ then
$\rho(\chi_1 \geq \chi_2) + \rho(\chi_2 \geq \chi_1) = \min(\max(\kappa,0),1) + \min(\max(\epsilon,0),1) = \min(\kappa,1) + \min(\epsilon,1) = \kappa + \epsilon = 1.$

(c) If $\kappa \geq 1$ and $\epsilon \leq 0$ then
$\rho(\chi_1 \geq \chi_2) + \rho(\chi_2 \geq \chi_1) = \min(\max(\kappa,0),1) + \min(\max(\epsilon,0),1) = \min(\kappa,1) + \min(0,1) = 1.$

Theorem 3.2 For any two IFNs $\chi_1 = \langle\alpha_1,\beta_1\rangle$ and $\chi_2 = \langle\alpha_2,\beta_2\rangle$, the proposed EPDM $\rho(\chi_1 \geq \chi_2)$ satisfies the following characteristics:

(i) $\rho(\chi_1 \geq \chi_2) = 1$ if $\alpha_1 - \alpha_2 \geq \pi_2/2 + \alpha_1\beta_1$;
(ii) $\rho(\chi_1 \geq \chi_2) = 0$ if $\alpha_2 - \alpha_1 \geq \pi_1/2 + \alpha_2\beta_2$.

Proof: For two IFNs $\chi_1 = \langle\alpha_1,\beta_1\rangle$ and $\chi_2 = \langle\alpha_2,\beta_2\rangle$, we have

(i) Let $\alpha_1 - \alpha_2 \geq \pi_2/2 + \alpha_1\beta_1$, then we have

$$\frac{1+\alpha_1-2\alpha_2-\beta_1+2\alpha_2\beta_2}{\pi_1+\pi_2+2\alpha_1\beta_1+2\alpha_2\beta_2} = \frac{1+2\alpha_1-\alpha_1-2\alpha_2-\beta_1+2\alpha_2\beta_2}{\pi_1+\pi_2+2\alpha_1\beta_1+2\alpha_2\beta_2}$$

$$\geq \frac{\pi_1+\pi_2+2\alpha_1\beta_1+2\alpha_2\beta_2}{\pi_1+\pi_2+2\alpha_1\beta_1+2\alpha_2\beta_2} = 1.$$

Therefore, $\min\left(\max\left(\frac{1+\alpha_1-2\alpha_2-\beta_1+2\alpha_2\beta_2}{\pi_1+\pi_2+2\alpha_1\beta_1+2\alpha_2\beta_2},0\right),1\right) = 1.$

Hence, $\rho(\chi_1 \geq \chi_2) = 1$.

(ii) Let $\alpha_2 - \alpha_1 \geq \pi_1/2 + \alpha_2\beta_2$, then we have

$$\frac{1+\alpha_1-2\alpha_2-\beta_1+2\alpha_2\beta_2}{\pi_1+\pi_2+2\alpha_1\beta_1+2\alpha_2\beta_2} = \frac{1+2\alpha_1-\alpha_1-2\alpha_2-\beta_1+2\alpha_2\beta_2}{\pi_1+\pi_2+2\alpha_1\beta_1+2\alpha_2\beta_2}$$

$$\geq \frac{\pi_1-\pi_1-2\alpha_2\beta_2+2\alpha_2\beta_2}{\pi_1+\pi_2+2\alpha_1\beta_1+2\alpha_2\beta_2} = 0.$$

Therefore, $\min\left(\max\left(\frac{1+\alpha_2-2\alpha_1-\beta_2+2\alpha_1\beta_1}{\pi_1+\pi_2+2\alpha_1\beta_1+2\alpha_2\beta_2},0\right),1\right) = 0.$

Hence, $\rho(\chi_1 \geq \chi_2) = 0$.

Nevertheless, we formulate the possibility degree matrix (PDMx)

$M = [\rho_{ti}]_{nxn} = [\rho(\chi_t \geq \chi_i)]_{nxn}$, where $t, i = 1,2,\ldots,n$ to rank n IFNs $\chi_1, \chi_2,\ldots, \chi_n$ by employing Eq. (3.6) in the following manner:

$$M = \begin{pmatrix} \rho_{11} & \rho_{12} & \cdots & \rho_{1n} \\ \rho_{21} & \rho_{22} & \cdots & \rho_{2n} \\ \vdots & \vdots & \ddots & \vdots \\ \rho_{n1} & \rho_{n2} & \cdots & \rho_{nn} \end{pmatrix}. \tag{3.8}$$

The RV for IFNs is calculated in this section as follows:

$$\varphi_t = \frac{1}{n(n-1)}\left(\sum_{i=1}^{n} \rho_{ti} + \frac{n}{2} - 1\right). \tag{3.9}$$

Consequently, arrange the RVs in decreasing order φ_t, $t = 1,2,\ldots,n$ and pick the best IFN χ_t.

Example 3.2 Consider identical IFNs $\chi_1 = \langle 0.3,0.4 \rangle$ and $\chi_2 = \langle 0.3,0.2 \rangle$ as provided in Example 3.1 and use the proposed EPDM for ranking them. For this, we apply Eq. (3.6) to get the proposed EPDMs $\rho\ (\chi_1 \geq \chi_2) = 0.36$ and $\rho\ (\chi_1 \geq \chi_2) = 0.64$ of $\chi_1 \geq \chi_2$ and $\chi_2 \geq \chi_1$, respectively.
Thus, by utilizing Eq. (3.8), we compute the PDMx as

$$M = \begin{bmatrix} 0.5 & 0.36 \\ 0.64 & 0.5 \end{bmatrix}.$$

The ranking values φ_1 and φ_2 of the IFNs χ_1 and χ_2 are determined by employing Eq. (3.9), respectively, we have $\varphi_1 = 0.43$ and $\varphi_2 = 0.57$.
Since $\varphi_2 > \varphi_1$, therefore $\chi_2 > \chi_1$. Therefore, the proposed EPDM ρ can effectively address the limitations of the existing PDM ρ' (Garg and Kumar, 2019), as explained in Section 3.2.

3.4 PROPOSED MADM METHOD BASED ON THE PROPOSED EPDM OF IFNS

In this section, we design a new MADM approach for IFNs using the proposed EPDM technique for IFNs.

Given $C_1, C_2, \ldots, C_n$ and $O_1, O_2, \ldots, O_m$ are set of attributes and alternatives respectively, having $w_1, w_2, \ldots, w_n$ are their respective weights such that $\sum_{t=1}^{n} w_t = 1$ and $w_t > 0$. Decision-maker (DMk) evaluates the alternative O_k with respect to the $C_t (t = 1,2,\ldots, n)$ attributes using $\tilde{\chi}_{kt} = \langle \tilde{\alpha}_{kt}, \tilde{\beta}_{kt} \rangle, t = 1,2,\ldots,n$ and $k = 1,2,\ldots,m$IFNs.

Step 1: Arrange the DMk assessment using the decision matrix (DMx) $\tilde{D} = (\tilde{\chi}_{kt})_{m \times n}$ as outlined below:

$$\tilde{D} = \begin{array}{c} \\ O_1 \\ O_2 \\ \vdots \\ O_m \end{array} \begin{array}{c} \begin{array}{cccc} C_1 & C_2 & \cdots & C_n \end{array} \\ \begin{pmatrix} \tilde{\chi}_{11} & \tilde{\chi}_{12} & \cdots & \tilde{\chi}_{1n} \\ \tilde{\chi}_{21} & \tilde{\chi}_{22} & \cdots & \tilde{\chi}_{2n} \\ \vdots & \vdots & \ddots & \vdots \\ \tilde{\chi}_{m1} & \tilde{\chi}_{m2} & \cdots & \tilde{\chi}_{mn} \end{pmatrix} \end{array}.$$

Step 2: Convert the DMx $\tilde{D} = (\tilde{\chi}_{kt})_{m\times n} = \left(\left\langle \tilde{\alpha}_{kt}, \tilde{\beta}_{kt} \right\rangle\right)_{m\times n}$ into normalized DMx (NDMx) $D = (\chi_{kt})_{m\times n} = \langle \alpha_{kt}, \beta_{kt} \rangle$ as follows:

$$\tilde{\chi}_{kt} = \begin{cases} \left\langle \tilde{\alpha}_{kt}, \tilde{\beta}_{kt} \right\rangle: \text{ if } C_t \text{ is a benefit-type attribute} \\ \left\langle \tilde{\beta}_{kt}, \tilde{\alpha}_{kt} \right\rangle: \text{ if } C_t \text{ is a cost-type attribute.} \end{cases} \tag{3.10}$$

Step 3: Calculate the overall IFN$\chi_k = \langle \alpha_k, \beta_k \rangle$ of the alternative O_k, $k = 1,2,\ldots,m$ by utilizing Eq. (3.2), as demonstrated below:

$$\chi_k = \left\langle \begin{array}{c} \dfrac{\prod_{t=1}^{n}(1-\alpha_{kt})^{w_t} - \left[1-\dfrac{1}{\varepsilon}\left(1-\prod_{t=1}^{n}(1-\varepsilon\alpha_{kt})^{w_t}\right)\right]}{\prod_{t=1}^{n}(1-\alpha_{kt})^{w_t} + \left[1-\dfrac{1}{\varepsilon}\left(1-\prod_{t=1}^{n}(1-\varepsilon\alpha_{kt})^{w_t}\right)\right]}, \\ \dfrac{2\left(1-\dfrac{1}{\varepsilon}\left(1-\prod_{t=1}^{n}\left(1-\varepsilon(1-\beta_{kt})\right)^{w_t}\right)\right)}{\prod_{t=1}^{n}(2-\beta_{kt})^{w_t} + \left(1-\dfrac{1}{\varepsilon}\left(1-\prod_{t=1}^{n}\left(1-\varepsilon(1-\beta_{kt})\right)^{w_t}\right)\right)} \end{array} \right\rangle. \tag{3.11}$$

Step 4: By using Definition 3.5, PDMx $M = [\rho_{ki}]_{m\times m}$, k, $i = 1,2,\ldots,m$ as follows:

$$M = \begin{pmatrix} \rho_{11} & \rho_{12} & \cdots & \rho_{1m} \\ \rho_{21} & \rho_{22} & \cdots & \rho_{2m} \\ \vdots & \vdots & \ddots & \vdots \\ \rho_{k1} & \rho_{k2} & \cdots & \rho_{km} \end{pmatrix}. \tag{3.12}$$

Step 5: Calculate the RVs of the alternatives $\varphi_1,\varphi_2,\ldots,\varphi_m$ by using Eq. (3.9) as follows:

$$\varphi_k = \frac{1}{m(m-1)}\left(\sum_{i=1}^{m} \rho_{ki} + \frac{m}{2} - 1\right). \tag{3.13}$$

Step 6: Sort the RVs $\varphi_1,\varphi_2,\ldots,\varphi_m$ of alternatives $O_1,O_2,\ldots,O_m$ and get the RO in decreasing order.

Example 3.3 (Garg and Kumar, 2019) Government intends to select a contractor for any construction project from amidst the available contractors: "Firm A" (O_1), "Firm B" (O_2), "Firm C" (O_3), and "Firm D" (O_4). The government established five attributes for this task: "Corporate cost" (C_1) "completion time of the Corporate" (C_2), "technical proficiency" (C_3), "financial condition" (C_4), and "Corporate history" (C_5) with the weights $w_1 = 0.3$, $w_2 = 0.25$, $w_3 = 0.1$, $w_4 = 0.15$, and $w_5 = 0.2$. The primary objective of this problem is to choose the most suitable firm from the entire pool.

To address this issue, we employ the proposed method in the following manner:

Step 1: The DMk acquires the DMx$D = (\chi_{kt})_{4\times 5}$ by assessing the alternatives with respect to the attributes using the IFNs, as illustrated below:

$$D = \begin{array}{c} \\ O_1 \\ O_2 \\ O_3 \\ O_4 \end{array}\begin{array}{c} \begin{array}{ccccc} C_1 & C_2 & C_3 & C_4 & C_5 \end{array} \\ \begin{pmatrix} \langle 0.3,0.6\rangle & \langle 0.5,0.4\rangle & \langle 0.7,0.2\rangle & \langle 0.5,0.2\rangle & \langle 0.7,0.1\rangle \\ \langle 0.5,0.3\rangle & \langle 0.6,0.2\rangle & \langle 0.5,0.4\rangle & \langle 0.6,0.3\rangle & \langle 0.4,0.2\rangle \\ \langle 0.5,0.4\rangle & \langle 0.7,0.2\rangle & \langle 0.8,0.1\rangle & \langle 0.6,0.2\rangle & \langle 0.5,0.3\rangle \\ \langle 0.6,0.2\rangle & \langle 0.4,0.3\rangle & \langle 0.7,0.2\rangle & \langle 0.4,0.4\rangle & \langle 0.2,0.8\rangle \end{pmatrix} \end{array}$$

Step 2: Attributes C_1 and C_2 are considered cost type and the others are benefit type, the proposed method calculates the NDMx using Eq. (3.10) and demonstrates it as:

$$D = \begin{array}{c} \\ O_1 \\ O_2 \\ O_3 \\ O_4 \end{array}\begin{array}{c} \begin{array}{ccccc} C_1 & C_2 & C_3 & C_4 & C_5 \end{array} \\ \begin{pmatrix} \langle 0.6,0.3\rangle & \langle 0.4,0.5\rangle & \langle 0.7,0.2\rangle & \langle 0.5,0.2\rangle & \langle 0.7,0.1\rangle \\ \langle 0.3,0.5\rangle & \langle 0.2,0.6\rangle & \langle 0.5,0.4\rangle & \langle 0.6,0.3\rangle & \langle 0.4,0.2\rangle \\ \langle 0.4,0.5\rangle & \langle 0.2,0.7\rangle & \langle 0.8,0.1\rangle & \langle 0.6,0.2\rangle & \langle 0.5,0.3\rangle \\ \langle 0.2,0.6\rangle & \langle 0.3,0.4\rangle & \langle 0.7,0.2\rangle & \langle 0.4,0.4\rangle & \langle 0.2,0.8\rangle \end{pmatrix} \end{array}$$

Step 3: By utilizing Eq. (3.11), the proposed MADM methodology gets the overall IFN χ_k for the alternative O_k, where k = 1,2,3,4, ε = 0.99, χ_1 = ‹0.4947,0.3667›, χ_2 = ‹0.5401,0.2363›, χ_3 = ‹0.5988,0.2730›, and χ_4 = ‹0.5849,0.4151›.

Step 4: The PDMx is obtained by the proposed utilizing Eq. (3.12) and is presented as follows:

$$M = \begin{pmatrix} 0.5000 & 0.3415 & 0.2691 & 0.4497 \\ 0.6585 & 0.5000 & 0.4305 & 0.6097 \\ 0.7309 & 0.5695 & 0.5000 & 0.6820 \\ 0.5503 & 0.3903 & 0.3180 & 0.5000 \end{pmatrix}.$$

Step 5: By applying Eq. (3.13), the proposed MADM approach calculates the RVs φ_1 = 0.2134, φ_2 = 0.2666, φ_3 = 0.2902, and φ_4 = 0.2299 for the alternatives O_1, O_2, O_3, and O_4, respectively.

Step 6: Since $\varphi_3 > \varphi_2 > \varphi_4 > \varphi_1$, the RO of the alternative is $O_3 > O_2 > O_4 > O_1$. Consequently, O_3 is the optimal choice for this MADM problem.

3.5 CONCLUSION

In this chapter, we have proposed an enhanced possibility degree measure (EPDM) to effective ranking of intuitionistic fuzzy numbers (IFNs). The hypothesized EPDM between the two IFNs suggested that there was a chance that one IFN could be larger than the other. To achieve this goal, we initially identified limitations within the existing PDM of IFNs, particularly in specific scenarios.

In response to these limitations, we proposed the EPDM as a solution to address the shortcomings observed by Garg and Kumar (2019) possibility degree measure (PDM) for IFNs, which provided an inaccurate RO of different IFNs in some certain cases. Moreover, based on the newly introduced EPDM, we presented the concept of multi-attribute decision-making (MADM) problems. This approach aided in computing alternative ranking values and identifying the optimal alternative for the MADM problem with IFS environment. In the future, we could also extend the results of this chapter to some other uncertain and IFS environment.

REFERENCES

Atanassov, K., &Gargov, G. (1999). Interval valued intuitionistic fuzzy sets. *Fuzzy Sets and Systems, 31*(3), 343–349.

Atanassov, K. T. (1986). Intuitionistic fuzzy sets. *Fuzzy sets and Systems*, *20*(1), 87–96.

Dhankhar, C., &Kumar, K. (2023). Multi-attribute decision-making based on the advanced possibility degree measure of intuitionistic fuzzy numbers. *Granular Computing*, *8*(3), 467–478.

Garg, H., &Kaur, G. (2020). Novel distance measures for cubic intuitionistic fuzzy sets and their applications to pattern recognitions and medical diagnosis. *Granular Computing*, *5*, 169–184.

Garg, H., &Kumar, K. (2019). Improved possibility degree method for ranking intuitionistic fuzzy numbers and their application in multiattribute decision-making. *Granular Computing*, *4*, 237–247.

Kumar, K., &Chen, S. M. (2021). Multiattribute decision making based on the improved intuitionistic fuzzy Einstein weighted averaging operator of intuitionistic fuzzy values. *Information Sciences*, *568*, 369–383.

Kumar, M., &Gupta, S. K. (2023). Multicriteria decision-making based on the confidence level Q-rung orthopair normal fuzzy aggregation operator. *Granular Computing*, *8*(1), 77–96.

Liu, H., Tu, J., &Sun, C. (2020). Improved possibility degree method for intuitionistic fuzzy multi-attribute decision making and application in aircraft cockpit display ergonomic evaluation. *IEEE Access*, *8*, 202540–202554.

Ma, Z. M., &Xu, Z. S. (2020). Computation of generalized linguistic term sets based on fuzzy logical algebras for multi-attribute decision making. *Granular Computing*, *5*, 17–28.

Mishra, A. R., Singh, R. K., &Motwani, D. (2019). Multi-criteria assessment of cellular mobile telephone service providers using intuitionistic fuzzy WASPAS method with similarity measures. *Granular Computing*, *4*, 511–529.

Wang, H., Liu, Y., Liu, F., &Lin, J. (2021). Multiple attribute decision-making method based upon intuitionistic fuzzy partitioned dual Maclaurin symmetric mean operators. *International Journal of Computational Intelligence Systems*, *14*, 1–20.

Wei, C. P., &Tang, X.(2010). Possibility degree method for ranking intuitionistic fuzzy numbers. In *2010 IEEE/WIC/ACM International Conference on Web Intelligence and Intelligent Agent Technology*,*3*, 142–145. IEEE.

Zadeh, L. A.(1965). Fuzzy sets. *Information and Control*, *8*(3), 338–353.

Zeng, S., Chen, S. M., &Kuo, L. W. (2019). Multiattribute decision making based on novel score function of intuitionistic fuzzy values and modified VIKOR method. *Information Sciences*, *488*, 76–92.

Chapter 4

A new score function for interval-valued intuitionistic fuzzy set and its application to MADM problems with partial weight information

Reeta Bhardwaj, Naveen Mani, Lokesh Singh, and Amit Sharma

4.1 INTRODUCTION

In the real world, the most probable multi-attribute decision-making (MADM) problems, such as staff selection, supplier selection, contractor selection for building construction, etc., depend on many aspects simultaneously. These aspects make the problem more complex and uncertain for decision-makers to give the assessments towards the attributes, and decision-makers cannot provide their assessments in the form of crisp numbers. To remove such types of difficulties of the decision-makers, Zadeh (1965) introduced the concept of the fuzzy sets (FSs), and after that extensions of it like intuitionistic FS (IFS) Atanassov (1986) and interval-valued IFS (IVIFS) Atanassov (1989) have been proposed as powerful tools to handle the uncertainty. Recently, these theories have been applied more by researchers such as Garg (2017), Garg and Kumar (2018), Garg and Kumar (2019), Garg et al. (2017), Gupta et al. (2018), and Liu and Li (2016) to solve real-world MADM issues. Liu (2013) defined the Hamacher aggregation operators (AOs) for handling the MADM issues for IFNs information with the entropy weights. Other than these, researchers paid more attention to IVIFS theory for giving the assessment. Like, Garg and Kumar (2020a), Kumar and Garg (2018) presented a TOPSIS method for handling MADM issues under the IVIFNs environment based on the set pair analysis theory. Garg and Kumar (2020b) presented the possibility degree method for MADM issues under the IVIFNs environment.

Attribute's weight plays a crucial role during the Decision making (DM) process because fluctuation in the weights can change the ranking of the alternatives. In some MADM problems, the decision-maker provides complete information about the attribute's weight. For example, Park et al. (2011) established an optimization model based on the TOPSIS approach to obtain attribute's weight during solving the MAGDM problems under IVIFNs environment. Chen and Huang (2017) presented a DM approach by using the Xu's (2007b) score function into LP methodology for solving the MADM issues under the IVIFNs environment. Wang and Chen (2017)

DOI: 10.1201/9781003497219-4

and Wang and Chen (2018) utilized the score function in LP to obtain the weight vector. All these methods use the score functions into LP methodology for obtaining weight vectors. But, we found in this study that existing score and accuracy functions have lots of drawbacks. Therefore, these methods are not reliable for obtaining weight vectors.

After obtaining the attribute's weights, the main important task in MADM problems is to aggregate the multi-attribute information into a collective one. For instance, Xu and Yager (2006) presented geometric aggregation operator (AO) while Xu (2007a) presented some averaging AO for aggregating the different IFNs. Xu (2007b) defined the interval-valued intuitionistic fuzzy weighted averaging (IVIFWA) AO. Xu and Chen (2007) presented interval-valued intuitionistic fuzzy weighted geometric (IVIFWG) AO for aggregating the interval-valued IFNs (IVIFNs). After that, Liu (2013) defined the AOs by using the Hamacher t-norm to fuse IVIFNs information. Further, Garg et al. (2017) addressed some shortcomings of existing AOs. Liu (2013) defined Hamacher interactive AOs for IVIF information. Chen and Tsai (2016) analyzed by taking some numerical examples that the existing AOs such as Wang and Liu (2013a), Wang and Liu (2013b), Xu (2007), and Xu and Chen (2007 have some drawbacks. To overcome these drawbacks, Chen and Tsai (2016) defined a novel multiplication operation for IVIFNs and defined AOs based on it. Afterwards, by studying these AOs, we found that these AOs have some limitations. For example, if we take only one value [0,0] as membership/non-membership for any alternative, then geometric/averaging AOs directly set the collective matrix to zero for that alternative. Therefore, we need to improve this limitation.

In the last step of solving the MADM issues, ranking of the alternatives plays an important role. For this, Chen and Tan (1994) introduced the concept of score and accuracy function for IFS. Afterwards, Xu (2007) extended the idea of score and accuracy functions for IVIFNs. Wang and Chen (2017) defined a score function and a DM approach for handling the MADM issues. However, Wang and Chen (2018) found some drawbacks to the score function defined by Wang and Chen (2017), and they proposed a new score function with the application in MADM problems.

In certain cases, however, the existing score and accuracy function, and DM techniques based on the score and accuracy functions do not provide adequate details on alternatives. Therefore, in this article, we have presented some counterexamples to show the drawbacks of the existing score and accuracy functions of the IVIFNs. Afterwards, we have developed a new score function to compare the IVIFNs and overcome the drawbacks of the existing score and accuracy functions of IVIFNs. Apart from this, we have developed a DM approach by using LP methodology and proposed a score function for solving the MADM issues under the IVIFNs environment in which the weight information of attributes is partially known.

Therefore, based on the above analysis under the IVIFNs environment, the main targets of this paper are summarized as:

a). Some counterexamples are presented to show the drawbacks of the existing score and accuracy functions.
b). A new score function has been developed that overcomes the flaws of the existing score and accuracy functions.
c). A DM approach based on the LP methodology and proposed score function has been developed for IVIFNs information.
d). To evaluate the developed approach, a real-life illustrative has been taken, and results are compared with the existing strategies to show the preferences of the proposed approach.
e). At the last, two MADM examples are considered to show the drawbacks of the existing DM approaches (Wang and Chen 2017; Wang and Chen 2018) and advantages of the proposed DM approach.

To accomplish the above targets, this chapter is sorted out as follows: Section 4.2, briefly introduces the concepts of IVIFS and its existing score and accuracy functions. Section 4.3 presents the proposed score function that overcomes the flaws of existing score and accuracy functions. In Section 4.4, we have developed a DM approach based on the proposed score and LP methodology under the IVIFNs environment. Two examples are discussed to illustrate the effectiveness of the developed DM approach, and results are compared with the results of existing methods. Finally, Section 4.5 concludes the paper.

4.2 PRELIMINARIES

In this segment, a few fundamental concepts related to the IVIFS theory with existing score and accuracy function are characterized as follows:

Definition 4.1 (Atanassov, 1989) An IVIFS A is defined as

$$A = \{\langle x,[\tau_A(x),\eta_A(x)],[\theta_A(x),\upsilon_A(x)]\rangle \mid x \in X\}, \tag{4.1}$$

where $\tau_A(x),\eta_A(x),\theta_A(x),\upsilon_A(x) \in [0,1]$ represents membership and non-membership degrees of x to A, respectively, such that $0 \le \eta_A(x)+\upsilon_A(x) \le 1$ holds for $\forall\, x$. Usually, the pair $\langle[\tau,\eta],[\theta,\upsilon]\rangle$ is called an IVIFN.

Definition 4.2 (Atanassov, 1989) Let $A = \langle[\tau,\eta],[\theta,\upsilon]\rangle$, $A_1 = \langle[\tau_1,\eta_1],[\theta_1,\upsilon_1]\rangle$ and $A_2 = \langle[\tau_2,\eta_2],[\theta_2,\upsilon_2]\rangle$ be any three IVIFNs, then

(i) $A_1 \subseteq A_2$ if $\tau_1 \le \tau_2, \eta_1 \le \eta_2, \theta_1 \ge \theta_2$ and $\upsilon_1 \ge \upsilon_2$.
(ii) $A_1 = A_2$ iff $A_1 \subseteq A_2$ and $A_1 \supseteq A_2$.
(iii) $A^c = \langle[\theta,\upsilon],[\tau,\eta]\rangle$.

Definition 4.3 (Wang & Chen, 2017) For an IVIFN $A = \langle[\tau,\eta],[\theta,\upsilon]\rangle$, the score function is defined as follows:

$$S_{wc}(A) = \frac{\tau + \eta + \sqrt{\eta\upsilon}(1-\tau-\theta) + \sqrt{\tau\theta}(1-\eta-\upsilon)}{2}. \quad (4.2)$$

Definition 4.4 (Wang & Chen, 2018) For an IVIFN $A = \langle[\tau,\eta],[\theta,\upsilon]\rangle$, the Wang and Chen's score function is defined as follows:

$$S_{NWC}(A) = \frac{(\tau+\eta)(\tau+\theta) - (\theta+\upsilon)(\eta+\upsilon)}{2}. \quad (4.3)$$

Definition 4.5 (Wang & Chen, 2018) For an IVIFN $A = \langle[\tau,\eta],[\theta,\upsilon]\rangle$, the Wang and Chen's accuracy function is defined as follows:

$$H_{NWC}(A) = \frac{(1-\tau+\eta)(1-\tau-\theta) - (1-\theta+\upsilon)(1-\eta-\upsilon)}{2}. \quad (4.4)$$

A few illustrations are given to show that the existing score and accuracy functions are unable to rank the alternative as follows:

Example 4.1 Here, we consider the two IVIFNs $A_1 = \langle[0.0,0.0],[0.10,0.10]\rangle$ and $A_2 = \langle[0.0,0.0],[0.2,0.3]\rangle$ as alternatives. To rank these alternatives, by utilizing Eq. (4.2), we get $S_{WC}(A_1) = S_{WC}(A_2) = 0$. Hence, the score function given in Eq. (4.2) is unable to rank A_1 and A_2 in this case.

Example 4.2 Let $A_1 = \langle[0.3,0.3],[0.3,0.3]\rangle$ and $A_2 = \langle[0.2,0.2],[0.2,0.2]\rangle$ be two IVIFNs as alternatives. To rank these alternatives, by utilizing Eq. (4.3) and Eq. (4.4), we get $S_{NWC}(A_1) = S_{NWC}(A_2)=0$ and $H_{NWC}(A_1) = H_{NWC}(A_2)=0$. Hence, the score function and accuracy function given in Eq. (4.3) and Eq. (4.4) are unable to rank A_1 and A_2 in this case.

From the obtained results, it is clear that the score and accuracy functions $S_{WC}(.)$, $S_{NWC}(.)$, and $H_{NWC}(.)$ cannot rank the alternatives in the above cases.

4.3 A NEW SCORE FUNCTION FOR IVIFN

From the results of Examples 4.1 and 4.2, it is clear that the existing score and accuracy functions are invalid for ranking the IVIFNs in some situations. Therefore, there is a need for a new score function that overcomes the flaws of the existing score and accuracy functions. To handle this, we develop a new score function for ranking the IVIFNs.

Definition 4.6 For an IVIFN $A = \langle[\tau,\eta],[\theta,\upsilon]\rangle$, we proposed a new score function as

$$KS(A) = \begin{cases} \dfrac{\tau(2-\eta-\upsilon)+\eta(2-\tau-\theta)-\theta-\upsilon}{2} & \text{if } [\tau,\eta] \neq [\theta,\upsilon], \\ \dfrac{\tau(2-\eta-\upsilon)+\eta(2-\tau-\theta)}{2} & \text{if } [\tau,\eta] = [\theta,\upsilon]. \end{cases} \tag{4.5}$$

The larger the value of $KS(A)$, the larger the IVIFN A.
The proposed score function satisfies the following properties:

> **Property 4.1** For an IVIFN $A = \langle[\tau,\eta],[\theta,\upsilon]\rangle$, $KS(A) \in [-1,1]$.
> **Property 4.2** For an IVIFN $A = \langle[1,1],[0,0]\rangle$, $KS(A) = 1$.
> **Property 4.3** For an IVIFN $A = \langle[0,0],[1,1]\rangle$, $KS(A) = -1$.

Theorem 4.1 For any two IVIFNs A_1 and A_2 with proposed score function $KS(A_1)$ and $KS(A_2)$ respectively. If $A_1 \neq A_2$ then $KS(A_1) \neq KS(A_2)$.

Proof. Let $A_1 = \langle[\tau_1,\eta_1],[\theta_1,\upsilon_1]\rangle$ and $A_2 = \langle[\tau_2,\eta_2],[\theta_2,\upsilon_2]\rangle$ be two IVIFNs such that $A_1 \neq A_2$. First, we assume that $A_1 \prec A_2$, i.e.. $\tau_1 < \tau_2, \eta_1 < \eta_2, \theta_1 < \theta_2$ and $\upsilon_1 < \upsilon_2$, which implies $\tau_1 - \tau_2 < 0, \eta_1 - \eta_2 < 0, \theta_2 - \theta_1 < 0$, and $\upsilon_2 - \upsilon_1 < 0$. Since

$$KS(A_1) = \frac{\tau_1(2-\eta_1-\upsilon_1)+\eta_1(2-\tau_1-\theta_1)-\theta_1-\upsilon_1}{2},$$

$$KS(A_2) = \frac{\tau_2(2-\eta_2-\upsilon_2)+\eta_2(2-\tau_2-\theta_2)-\theta_2-\upsilon_2}{2}.$$

Now, we have

$$\begin{aligned} 2(KS(A_1) - KS(A_2)) & \\ &= \tau_1(2-\eta_1-\upsilon_1)-\tau_2(2-\eta_2-\upsilon_2)+\eta_1(2-\tau_1-\theta_1)-\eta_2(2-\tau_2-\theta_2) \\ &\quad +(\theta_2-\theta_1)+(\upsilon_2-\upsilon_1) \\ &< \tau_2(2-\eta_1-\upsilon_1)-\tau_2(2-\eta_2-\upsilon_2)+\eta_2(2-\tau_1-\theta_1)-\eta_2(2-\tau_2-\theta_2) \\ &= \tau_2(\eta_2+\upsilon_2-\eta_1-\upsilon_1)+\eta_2(\tau_2+\theta_2-\tau_1-\theta_1) \\ &= \tau_2(\eta_2-\eta_1)+\tau_2(\upsilon_2-\upsilon_1)+\eta_2(\tau_2-\tau_1)+\eta_2(\theta_2-\theta_1) \\ &< \tau_2(\eta_2-\eta_1)+\eta_2(\tau_2-\tau_1) \neq 0 \end{aligned}$$

Therefore, if $A_1 \prec A_2$ then $KS(A_1) \neq KS(A_2)$. Similarly, results hold for $A_1 \succ A_2$. Hence, if $A_1 \neq A_2$ then $KS(A_1) \neq KS(A_2)$ while equality holds iff $\tau_1 = \tau_2, \eta_2 = \eta_1, \theta_1 = \theta_2$ and $\upsilon_1 = \upsilon_2$

Example 4.3 If we apply the function $KS(.)$ on the consider data in Example 4.1 to rank the alternatives, then we get $KS(A_1) = -0.1000$ and $KS(A_2) = -0.2500$. Thus, $A_1 \succ A_2$.
Example 4.4 Here, we reconsider the data given in Example 4.2. After that, we utilized the proposed ranking principle to rank the alternatives and get $KS(A_1)$=0.4200 and $KS(A_1)$=0.3200. Thus, $A_1 \succ A_2$.

From the result of the above examples, it is clear that the proposed score function successfully overcome the drawbacks of the existing score and accuracy functions. Keeping the advantages of the proposed score function, in the next section, we will develop a DM approach for handling MADM issues in the IVIFNs environment.

4.4 MADM APPROACH BASED ON THE PROPOSED SCORE FUNCTION

Consider an MADM issue in which there are m alternatives $A = \{A_1, A_2, \ldots, .A_m\}$ which are assessed under the set $\mathcal{G} = \{\mathcal{G}_1, \mathcal{G}_2, \ldots, \mathcal{G}_n\}$ of different attributes, and the weights of the attributes are partially known. Experts prefer to evaluate the A_k alternatives under the $\mathcal{G}_t$ attributes in terms of the IVIFN $\tilde{\gamma}_{kt} = \langle[\tilde{\tau}_{kt}, \tilde{\eta}_{kt}], [\tilde{\theta}_{kt}, \tilde{\upsilon}_{kt}]$, and arrange the collective information in the form of the decision matrix $\tilde{R} = \left(\tilde{\gamma}_{kt}\right)_{m \times n}$ as

$$\tilde{R} = \begin{array}{c} \\ A_1 \\ A_2 \\ \vdots \\ A_m \end{array} \begin{array}{c} \begin{array}{cccc} \mathcal{G}_1 & \mathcal{G}_2 & \cdots & \mathcal{G}_n \end{array} \\ \begin{pmatrix} \tilde{\gamma}_{11} & \tilde{\gamma}_{12} & \cdots & \tilde{\gamma}_{1n} \\ \tilde{\gamma}_{21} & \tilde{\gamma}_{22} & \cdots & \tilde{\gamma}_{2n} \\ \vdots & \vdots & \ddots & \vdots \\ \tilde{\gamma}_{m1} & \tilde{\gamma}_{m2} & \cdots & \tilde{\gamma}_{mn} \end{pmatrix} \end{array}.$$

Then, we summarized the following steps of the developed approach as follows:

(Step 1): Decision matrix $\tilde{R}$ is change over into its normalized matrix, indicated by $R = (\gamma_{kt})_{m \times n}$, to remove the impact of physical dimension, as follows:

$$\gamma_{kt} = \begin{cases} \langle [\tilde{\tau}_{kt}, \tilde{\eta}_{kt}], [\tilde{\theta}_{kt}, \tilde{\upsilon}_{kt}] \rangle : \text{if } \mathcal{G}_t \text{ is benefit-type attribute} \\ \langle [\tilde{\theta}_{kt}, \tilde{\upsilon}_{kt}], [\tilde{\tau}_{kt}, \tilde{\eta}_{kt}] \rangle : \text{if } \mathcal{G}_t \text{ is cost-type attribute} \end{cases} \tag{4.6}$$

(Step 2): In this step, we obtained the score matrix $KS = (KS(\gamma_{kt}))_{m\times n}$ by utilizing the proposed score $KS(\gamma_{kt})$ for each IVIFN γ_{kt} as defined in Definition 4.6

(Step 3): Based on the score matrix KS, we construct an LP problem (LPP) to obtain the attribute's weight when they are partially known as:

$$\max \sum_{k=1}^{m}\sum_{t=1}^{n} KS(\gamma_{kt})\omega_t \text{ s.t.}$$

$$\omega_1^l \le \omega_1 \le \omega_1^u,$$
$$\omega_2^l \le \omega_2 \le \omega_2^u,$$
$$\vdots$$
$$\omega_n^l \le \omega_n \le \omega_n^u,$$
$$\sum_{t=1}^{n} \omega_t = 1.$$

(Step 4): Calculate the overall performance $WKS(A_k)$ for each alternative A_k, k=1, 2, …, m as follows:

$$WKS(A_k) = \sum_{t=1}^{n} KS(\gamma_{kt})\omega_t \tag{4.7}$$

(Step 5): Rank the alternatives A_k according to the decreasing order of $WKS(A_k)$.

Example 4.5 We consider an MADM problem with three alternative A_k, k=1, 2, 3, under the three attributes $\mathcal{G}_t, t = 1,2,3,$ with optimal weights of attributes $\omega_1, \omega_2, \omega_3$ such that $0.10 \le \omega_1 \le 0.80, 0.20 \le \omega_2 \le 0.85, 0.25 \le \omega_3 \le 0.85$ and $\omega_1 + \omega_2 + \omega_3 = 1$. The assessments of the alternatives are given by the decision-maker(s) in terms of IVIFNs and summarized in the decision matrix $\tilde{R} = (\gamma_{kt})_{3\times 3}$ as follows:

$$\tilde{R} = \begin{array}{c} \\ A_1 \\ A_2 \\ A_3 \end{array} \begin{array}{c} \begin{array}{ccc} \mathcal{G}_1 & \mathcal{G}_2 & \mathcal{G}_3 \end{array} \\ \begin{pmatrix} \langle [0.2,0.6],[0.2,0.4] \rangle & \langle [0.3,0.4],[0.2,0.3] \rangle & \langle [0.2,0.2],[0.2,0.2] \rangle \\ \langle [0.3,0.5],[0.1,0.5] \rangle & \langle [0.3,0.5],[0.1,0.3] \rangle & \langle [0.2,0.3],[0.1,0.2] \rangle \\ \langle [0.4,0.4],[0.0,0.6] \rangle & \langle [0.3,0.3],[0.3,0.3] \rangle & \langle [0.1,0.1],[0.1,0.1] \rangle \end{pmatrix} \end{array}.$$

On the above data, if we apply the existing DM approach given by Wang and Chen (2018) to rank the alternatives, then we have the following steps:
If we utilized the proposed DM approach on the above considered data, then we have the following steps:

(Step 1): No need of a normalizing process.
(Step 2): Calculate the score matrix $KS = (KS(\gamma_{kt}))_{3\times3}$ and obtained results are summarized as

$$KS = \begin{pmatrix} 0.2800 & 0.2450 & 0.3200 \\ 0.2500 & 0.3800 & 0.2550 \\ 0.2200 & 0.4200 & 0.1800 \end{pmatrix}.$$

(Step 3): To obtain the attribute's weight, we form an LPP as:

$$max\ 0.7500\omega_1 + 1.0450\omega_2 + 0.7550\omega_3 \text{ s.t.}$$

$$0.10 \le \omega_1 \le 0.80,$$

$$0.20 \le \omega_2 \le 0.85,$$

$$0.25 \le \omega_3 \le 0.85$$

$$\omega_1 + \omega_2 + \omega_3 = 1$$

On solving the above LPP, we get $\omega_1 = 0.10, \omega_2 = 0.65$, and $\omega_3 = 0.25$.
(Step 4): By utilizing Eq. (4.7), we get the overall performance of each alternative A_k, k=1,2,3, as follows:

$$WKS(A_1)=0.2673,\ WKS(A_2)=0.3357,\ WKS(A_3)=0.3400.$$

(Step 5): Since $WKS(A_3) > WKS(A_2) > WKS(A_1)$, therefore the ranking order of the alternatives A_1, A_2, and A_3 is $A_3 \succ A_2 \succ A_1$. Hence, A_3 is the best alternative for the task.

If we apply the existing DM approach given by Wang and Chen (2018) to rank the alternatives A_1, A_2, and A_3, then we get $A_3 \succ A_1{=}A_2$. Hence, the MADM approach proposed by Wang and Chen (2018) cannot distinguish the ranking order between the alternative A_1 and A_2.

Example 4.6 Here, we take an MADM problem from Wang and Chen (2017) which has four alternatives A_k, k=1, 2, 3, 4 under the three attributes $\mathcal{G}_t, t = 1,2,3$. The attributes ω_1, ω_2, and ω_3 are partially known such that $0.10 \le \omega_1 \le 0.80, 0.20 \le \omega_2 \le 0.85, 0.25 \le \omega_3 \le 0.85$ and $\omega_1 + \omega_2 + \omega_3 = 1$. The decision-maker gives the preferences in terms of IVIFNs and are summarized in decision matrix $\widetilde{R} = (\gamma_{kt})_{4\times3}$ as

$$\tilde{R} = \begin{array}{c} A_1 \\ \\ A_2 \\ \\ A_3 \\ \\ A_4 \end{array} \overset{\begin{array}{ccc} \mathcal{G}_1 & \mathcal{G}_2 & \mathcal{G}_3 \end{array}}{\begin{pmatrix} \left\langle [0.40,0.50], [0.30,0.40] \right\rangle & \left\langle [0.40,0.60], [0.20,0.40] \right\rangle & \left\langle [0.10,0.30], [0.50,0.60] \right\rangle \\ \left\langle [0.53,0.70], [0.05,0.10] \right\rangle & \left\langle [0.60,0.63], [0.16,0.30] \right\rangle & \left\langle [0.49,0.70], [0.10,0.20] \right\rangle \\ \left\langle [0.30,0.60], [0.30,0.40] \right\rangle & \left\langle [0.50,0.60], [0.30,0.40] \right\rangle & \left\langle [0.50,0.60], [0.10,0.30] \right\rangle \\ \left\langle [0.70,0.80], [0.10,0.20] \right\rangle & \left\langle [0.60,0.70], [0.10,0.30] \right\rangle & \left\langle [0.30,0.40], [0.10,0.20] \right\rangle \end{pmatrix}}.$$

If we utilize the proposed DM approach on the above considered data, then we have the following steps:

(Step 1): No need of a normalizing process.

(Step 2): Calculate the score matrix $KS = (KS(\gamma_{kt}))_{3\times 3}$ and obtained results are summarized as

$$KS = \begin{pmatrix} 0.0975 & 0.1600 & -0.1425 \\ 0.3700 & 0.2408 & 0.3065 \\ 0.1100 & 0.1300 & 0.2475 \\ 0.3400 & 0.2775 & 0.1900 \end{pmatrix}.$$

(Step 3): To obtain the attribute's weight, we form an LPP as:

$$\begin{aligned} & max\, 0.9175\omega_1 + 0.8083\omega_2 + 0.6015\omega_3 \text{ s.t.} \\ & 0.10 \le \omega_1 \le 0.80, \\ & 0.20 \le \omega_2 \le 0.85, \\ & 0.25 \le \omega_3 \le 0.85 \\ & \omega_1 + \omega_2 + \omega_3 = 1 \end{aligned}$$

On solving the above LPP, we get $\omega_1 = 0.55, \omega_2 = 0.20$, and $\omega_3 = 0.25$.

(Step 4): By utilizing Eq. (4.7), we get the overall performance of each alternative A_k, k=1, 2, 3, 4 as follows:

$WKS(A_1)$=0.0500, $WKS(A_2)$=0.3283,
$WKS(A_3)$=0.1484, $WKS(A_4)$=0.2900.

(Step 5): Since $WKS(A_2)>WKS(A_4)>WKS(A_3)>WKS(A_1)$, therefore the ranking order of the alternatives A_1, A_2, A_3 and A_4 is $A_2 \succ A_4 \succ A_3 \succ A_1$. Hence, A_2 is the best alternative for the task.

If we apply the existing DM approach given by Wang and Chen (2017) to rank the alternatives A_1, A_2, A_3, and A_4, then we get $A_2 = A_4 \succ A_3 \succ A_1$. Hence, the MADM approach proposed by Wang and Chen (2017) cannot distinguish the ranking order between the alternative A_2 and A_4.

The results show that the proposed DM approach is more effective and suitable for handling MADM issues compared with the other existing DM approaches (Wang & Chen, 2017; Wang and Chen, 2018) when the weights of attributes are partially known.

4.5 CONCLUSION

In this article, we have proposed a new score function for comparing two IVIFNs that overcomes the flaws of the other existing score functions defined by the authors Wang and Chen (2017) and Wang and Chen (2018). A few numerical examples are presented to show the advantages of the proposed score functions, and some properties of it have also been discussed. Afterwards, we have developed an MADM approach by using the proposed score function and LP methodology for the IVIFNs environment. At last, a real-life case of MADM issue is given to demonstrate the effectiveness of the developed DM approach, and results are compared with other existing approaches. Lastly, two MADM issues are considered to show the advantages of the proposed DM approach with the existing DM approaches (Wang and Chen 2017; Wang and Chen 2018). From computed results and the comparative study, it has been concluded that the proposed DM approach is sensible and practicable and gives another way for handling MADM issues beneath the IVIFNs environment. In the future, we will apply the proposed DM approach to the other field.

REFERENCES

Atanassov, K. T. (1986). Intuitionistic fuzzy sets. *Fuzzy Sets and Systems*, *20*(1), 87–96.

Atanassov, K. T., & Gargov, G. (1989) Interval valued intuitionistic fuzzy sets. *Fuzzy Sets and System*, *31*(3), 343–349.

Chen, S. M., & Huang, Z. C. (2017). Multiattribute decision making based on interval-valued intuitionistic fuzzy values and linear programming methodology. *Information Sciences*, *381*, 341–351.

Chen, S. M., & Tan, J. M. (1994). Handling multicriteria fuzzy decision-making problems based on vague set theory. *Fuzzy Sets and Systems*, *67*(2), 163–172.

Chen, S. M., & Tsai, W. H. (2016). Multiple attribute decision making based on novel interval-valued intuitionistic fuzzy geometric averaging operators. *Information Sciences*, *367*, 1045–1065.

Garg, H. (2017). Novel intuitionistic fuzzy decision making method based on an improved operation laws and its application. *Engineering Applications of Artificial Intelligence*, *60*, 164–174.

Garg, H., & Kumar, K. (2018). A novel correlation coefficient of intuitionistic fuzzysets based on the connection number of set pair analysis and its application. *Scientia Iranica*, *25*(4), 2373–2388.

Garg, H., & Kumar, K. (2019). Linguistic interval-valued Atanassov intuitionistic fuzzy sets and their applications to group decision making problems. *IEEE Transactions on Fuzzy Systems*, 27(12), 2302–2311.

Garg, H., & Kumar, K. (2020a). A novel exponential distance and its based TOPSIS method for interval-valued intuitionistic fuzzy sets using connection number of SPA theory. *Artificial Intelligence Review*, *53*, 595–624.

Garg, H., & Kumar, K. (2020b). A novel possibility measure to interval-valued intuitionistic fuzzy set using connection number of set pair analysis and its applications. *Neural Computing and Applications*, *32*, 3337–3348.

Garg, H., Agarwal, N., & Tripathi, A. (2017). Some improved interactive aggregation operators under interval-valued intuitionistic fuzzy environment and its application to decision making process. *Scientia Iranica*, *24*(5), 2581–2604.

Gupta, P., Mehlawat, M. K., Grover, N., & Pedrycz, W. (2018). Multi-attribute group decision making based on extended TOPSIS method under interval-valued intuitionistic fuzzy environment. *Applied Soft Computing*, *69*, 554–567.

Kumar, K., & Garg, H. (2018). TOPSIS method based on the connection number of set pair analysis under interval-valued intuitionistic fuzzy set environment. *Computational and Applied Mathematics*, *37*, 1319–1329.

Liu, J. C., & Li, D. F. (2016). Corrections to "TOPSIS-based nonlinear-programming methodology for multi-attribute decision making with interval-valued intuitionistic fuzzy sets" [Apr 10 299-311]. *IEEE Transactions on Fuzzy Systems*, *26*(1), 391–391.

Liu, P. (2013). Some Hamacher aggregation operators based on the interval-valued intuitionistic fuzzy numbers and their application to group decision making. *IEEE Transactions on Fuzzy Systems*, *22*(1), 83–97.

Park, J. H., Park, I. Y., Kwun, Y. C., & Tan, X. (2011). Extension of the TOPSIS method for decision making problems under interval-valued intuitionistic fuzzy environment. *Applied Mathematical Modelling*, *35*(5), 2544–2556.

Wang, C. Y., & Chen, S. M. (2017). An improved multiattribute decision making method based on new score function of interval-valued intuitionistic fuzzy values and linear programming methodology. *Information Sciences*, *411*, 176–184.

Wang, C. Y., & Chen, S. M. (2018). A new multiple attribute decision making method based on linear programming methodology and novel score function and novel accuracy function of interval-valued intuitionistic fuzzy values. *Information Sciences*, *438*, 145–155.

Wang, W., & Liu, X. (2013a). Interval-valued intuitionistic fuzzy hybrid weighted averaging operator based on Einstein operation and its application to decision making. *Journal of Intelligent & Fuzzy Systems*, *25*(2), 279–290.

Wang, W., & Liu, X. (2013b). The multi-attribute decision making method based on interval-valued intuitionistic fuzzy Einstein hybrid weighted geometric operator. *Computers & Mathematics with Applications*, *66*(10), 1845–1856.
Xu, Z. (2007a). Intuitionistic fuzzy aggregation operators. *IEEE Transactions on Fuzzy Systems*, *15*(6), 1179–1187.
Xu, Z. (2007b). Methods for aggregating interval-valued intuitionistic fuzzy information and their application to decision making. *Control and Decision*, *22*(2), 215–219.
Xu, Z., & Chen, J. (2007, August). On geometric aggregation over interval-valued intuitionistic fuzzy information. In *Fourth International Conference on Fuzzy Systems and Knowledge Discovery (FSKD 2007)* (Vol. 2, pp. 466–471). IEEE.
Xu, Z., & Yager, R. R. (2006). Some geometric aggregation operators based on intuitionistic fuzzy sets. *International Journal of General Systems*, *35*(4), 417–433.
Zadeh, L. A. (1965). Fuzzy sets. *Information and Control*, *8*(3), 338–353.

Chapter 5

Neutrosophic cubic set-based operators for library ranking system

Rohit Khatri and Gagandeep Kaur

5.1 INTRODUCTION

Decision-making (DM) holds a significant part in day-to-day activities as well as in complicated problem-resolving situations. Under variable situations, there are a number of pertinent issues such as the availability of more than one feasible alternative, more than one classified criteria, information entities or data units, and a DM technique. However, one more dimension in the present scenarios is found when the unprocessed information entities are incomplete, vague, or uncertain in nature. These situations are not efficiently handled by the crisp-set theory; thus, the theory of fuzzy sets (FSs) (Zadeh 1965) or their extended version in the form of intuitionistic FSs (IFSs) (Atanassov 1986) as well as interval-valued intuitionistic fuzzy sets (Atanassov 1989) was devised to deal with such imprecise values. To model real-life situations more closely, a dimension of indeterminacy was added to the primitive IFS and neutrosophic sets were introduced by F. Smarandache (1999). Based on the neutrosophic sets, some more advanced forms such as interval-valued neutrosophic sets (IVNSs) (H. Wang 2005) and single-valued neutrosophic sets (SVNSs) (H. Wang 2010) were introduced.

Aggregation operators (AOs) play a significant role in aggregating the data values. Some averaging AOs for intuitionistic fuzzy numbers (IFNs) have been defined by Xu (2007). However, Wang, Liu, and Qin (2012) presented some operators to aggregate different interval-valued intuitionistic fuzzy (IVIF) numbers (IVIFNs). Zhao (2010) gave generalized AOs for intuitionistic fuzzy sets. Many decision-making approaches were designed on the basis of Einstein AOs on interval-valued intuitionistic fuzzy sets such as Wang and Liu (2013) proposed a multi-criteria decision-making (MCDM) method based on interval-valued Einstein hybrid weighted geometric operators. Consequently, many scholars have shown interest in proposing more AOs such as Hamacher AOs (Liu 2014), Frank norm operators (Nancy and Garg 2016), Bonferroni mean operator (Garg and Arora 2018), and so on. These operators gave the facility to set different parameter values and they can closely represent real-life problems in an efficient manner. Researchers Ye (2016), Ji (2018), Liu and Wang (2016), Liu and Luo (2017),

DOI: 10.1201/9781003497219-5

and Liu and Wang (2014) also focused on implementing these AOs to solve various DM problems.

It is noticeable that Bonferroni mean can reflect the interrelationship between all the permuted pairs of the input numbers while the MSM has the ability to capture the interrelationships between the multi-input arguments. These advantages of MSM make it more generalized, rational, and robust as compared to other AOs. In light of this, many researchers put forth the MSM operators under various environments, such as Qin and Liu (2014) have proposed the MSM operator for aggregating the IFSs. Liu and Qin (2017) have worked under the linguistic IFS environment to process the information using MSM operators. Moreover, Wang, Yang, and Li 2018 have captured the single-valued neutrosophic linguistic numbers while Liu and Zhang (2018) gave trapezoidal neutrosophic numbers to formulate the respective MSM operators. Moved by the striking work proposed by these researchers, we have worked into the direction of proposing the MSM operators by a hybrid version of more than one fuzzy environment.

However, it is analyzed that in practical DM situations, sometimes it becomes a complex job to express the available information by framing it into one extended FS environment only. There is a need for the hybridization of more than one fuzzy environment to efficiently handle the unprocessed pieces of uncertain information. So, Jun, Kim, and Yang (2012) corroborated a hybrid environment called cubic set (CS) by merging IVNS and SVNS together. This creation of CS was limited to the consideration of truth values only. The aspect of falsity along with the truth values was added to form cubic intuitionistic fuzzy sets (CIFSs) as given by authors (Kaur and Garg 2018a; Kaur and Garg 2018b). The CIFSs take into account only the membership as well as non-membership intervals and do not stress on the indeterminacy of the data entities, but neutrosophic cubic sets (NCSs) cover the indeterminacy aspect too. Suppose that an experimentalist has to perform some experiment E and before performing it, i.e., at time T_1, he is sure that the outcome measure may range between [0.40,0.50], unsure about it between [0.10,0.20], and indeterminate for about [0.5,0.10] and after completion of the experiment at time T_1, he obtained a crisp value that agrees the pre-assumed intervals. Say at this point, he obtained such values that agree 50% to the interval [0.40,0.50], 10% to [0.10, 0.20], and 5% indeterminate to the interval [0.5,0.10], so he may formulate an NCS by merging these two time constraints as (([0.40,0.50],[0.10,0.20],[0.5,0.10]), (0.50,0.10,0.5)). This concept of NCS was put forth by Ali, Deli, and Smarandache (2016) in which indeterminacy was also considered. Further, Jun, Smarandache, and Kim (2017) investigated the NCSs for the P-union and P-intersection and discussed their all related properties. Afterwards, many researchers worked on establishing various DM techniques. For instance, Ye (2018) developed some average aggregation operators for processing the NCSs. Shi and Ye (2018) used Dombi t-norm and t-conorm to develop the aggregation operators and applied them to the DM scenario. Apart from the aggregation operators, various information measures have also been developed under the NCS environment such as Lu and Ye (2017) worked on cosine similarity measures to process the DM scenario. Further, Pramanik, Dalapati, et al. (2017) implemented the TODIM (Tomada de decisao interativa e multicritévio) strategy to the NCSs and outlined the desirable rankings of various alternatives in the DM problem.

Based on these studies, it has been found that NCS is a powerful tool to handle uncertainties and the MSM operator has the capability of capturing the interrelationships among the data entities. Thus, motivated by the advantages of both of them, in the present chapter, we focus on formulating MSM for NCS. The primary objectives of this chapter are listed as follows:

(i) To propose an NCMSM operator by combining the features of MSM as well as NCSs and to check them for various desirable properties.
(ii) To establish the concept of NCWMSM based on the NCSs and to analyze some special reductions to another primitive operator by varying the value of the parameters involved.
(iii) To establish an MCDM based on the proposed approach and to apply it to real-life DM problems through an illustrative example.
(iv) To compare the proposed approach with the existing environments in order to validate their superiority.

This chapter is designed in the way that Section 5.2 consists of the preliminary concepts associated with the concept of NCS as well as MSM operator. Section 5.3 gives the outlook of the proposed NCMSM operator and some of its desirable properties. In Section 5.4, the concept is extended to the weighted considerations of data values and NCWMSM is proposed. A DM approach based on the proposed operators is given in Section 5.5 with real-life applications as well as comparison analyses. Lastly, Section 5.6 concludes the whole proposed work.

5.2 PRELIMINARIES

This section introduces the related basic concepts. For this, let us assume X as a universe of discourse, with a generic element x in it, then the following are some basic definitions based on it:

Definition 5.1 (Smarandache 1998) A neutrosophic set (NS) α in X is defined as

$$\alpha = \{(x, \zeta(x), \rho(x), \vartheta(x) \mid x \in X)\} \tag{5.1}$$

where the truth, indeterminacy, and falsity degrees are characterized by $\zeta(x), \rho(x)$ and $\vartheta(x)$, respectively, such that $\zeta, \rho, \vartheta \to (0^-, 1^+)$ and satisfy the conditions $0^- \leq \zeta(x) + \rho(x) + \vartheta(x) \leq 3^+$.

To enhance the concept of neutrosophic set and to apply it to real-life and engineering problems, Wang (2010) provided the following definition.

Definition 5.2 (Jun, Smarandache and Kim 2017) A neutrosophic cubic set (NCS) α in X is defined as

$$\alpha = (x, \mathcal{A}(x), \lambda(x)) \tag{5.2}$$

where $\mathcal{A}(x) = \{(x, [\zeta^L(x), \zeta^U(x)], [\rho^L(x), \rho^U(x)], [\vartheta^L(x), \vartheta^U(x)])\}$ is an IVNS and $\lambda(x) = \{(x, \zeta(x), \rho(x), \vartheta(x) \mid x \in X)\}$ is an SVNS in X.

For the sake of simplicity, an element in NCS is represented as $\alpha = (([\zeta^L, \zeta^U], [\rho^L, \rho^U], [\vartheta^L, \vartheta^U]), (\zeta, \rho, \vartheta))$ and is called as neutrosophic cubic number (NCN), where $[\zeta^L, \zeta^U], [\rho^L, \rho^U], [\vartheta^L, \vartheta^U] \subseteq [0,1]$ and $0 \leq \zeta, \rho, \vartheta \leq 1$ satisfying the conditions $0 \leq \zeta^U + \rho^U + \vartheta^U \leq 3$ and $0 \leq \zeta + \rho + \vartheta \leq 3$.

Definition 5.3 (Shi and Ye 2018) Let $\alpha = (([\zeta^L, \zeta^U], [\rho^L, \rho^U], [\vartheta^L, \vartheta^U]), (\zeta, \rho, \vartheta))$ be an NCN, then the score is defined as:

$$S(\alpha) = \frac{4 + \zeta^L - \rho^L - \vartheta^L + \zeta^U - \rho^U - \vartheta^U + 2 + \zeta - \rho - \vartheta}{9} \tag{5.3}$$

Moreover, the accuracy and certainty of an NCN are defined as:

$$A(\alpha) = \frac{\zeta^L - \rho^L + \zeta^U - \rho^U + \zeta - \rho}{3} \tag{5.4}$$

$$C(a) = \frac{\zeta^L + \zeta^U + \zeta}{3} \tag{5.5}$$

Here, $0 \leq S(\alpha), A(\alpha), C(\alpha) \leq 1$.

Definition 5.4 Let $\alpha_1 = (([\zeta_1^L, \zeta_1^U], [\vartheta_1^L, \rho_2^U], [\vartheta_1^L, \vartheta_1^U]), (\zeta_1, \rho_1, \vartheta_1))$ and $\alpha_2 = (([\zeta_2^L, \zeta_2^U], [\rho_2^L, \rho_2^U], [\vartheta_2^L, \vartheta_2^U]), (\zeta_2, \rho_2, \vartheta_2))$ be two NCNs, then the comparison rules for NCNs are defined as:

(i) If $S(\alpha_1) < S(\alpha_2)$, then $\alpha_1 < \alpha_2$.
(ii) If $S(\alpha_1) = S(\alpha_2)$ and $A(\alpha_1) < A(\alpha_2)$, then $\alpha_1 < \alpha_2$.
(iii) If $S(\alpha_1) = S(\alpha_2)$, $A(\alpha_1) = A(\alpha_2)$ and $C(\alpha_1) < C(\alpha_2)$, then $\alpha_1 < \alpha_2$.
(iv) If $S(\alpha_1) = S(\alpha_2)$, $A(\alpha_1) = A(\alpha_2)$ and $C(\alpha_1) = C(\alpha_2)$, then $\alpha_1 = \alpha_2$.

Definition 5.5(Ye 2018) Let $\alpha_1 = \left(\left(\left[\zeta_1^L,\zeta_1^U\right],\left[\rho_1^L,\rho_1^U\right],\left[\vartheta_1^L,\vartheta_1^U\right]\right),\left(\zeta_1,\rho_1,\vartheta_1\right)\right)$ and $\alpha_2 = \left(\left(\left[\zeta_2^L,\zeta_2^U\right],\left[\rho_2^L,\rho_2^U\right],\left[\vartheta_2^L,\vartheta_2^U\right]\right),\left(\zeta_2,\rho_2,\vartheta_2\right)\right)$ be two NCNs and $\xi > 0$ be a real number, then the operational laws are defined as follows:

(i) $$\alpha_1 \oplus \alpha_2 = \left(\begin{array}{c}\left(\begin{array}{c}\left[\zeta_1^L+\zeta_2^L-\zeta_1^L\zeta_2^L,\zeta_1^U+\zeta_2^U-\zeta_1^U\zeta_2^U\right],\\ \left[\rho_1^L\rho_2^L,\rho_1^U\rho_2^U\right],\left[\vartheta_1^L\vartheta_2^L,\vartheta_1^U\vartheta_2^U\right]\end{array}\right),\\ \left(\zeta_1+\zeta_2-\zeta_1\zeta_2,\rho_1\rho_2,\vartheta_1\vartheta_2\right)\end{array}\right)$$

(ii) $$\alpha_1 \otimes \alpha_2 = \left(\begin{array}{c}\left(\begin{array}{c}\left[\zeta_1^L\zeta_2^L,\zeta_1^U\zeta_2^U\right],\left[\rho_1^L+\rho_2^L-\rho_1^L\rho_2^L,\rho_1^U+\rho_2^U-\rho_1^U\rho_2^U\right],\\ \left[\vartheta_1^L+\vartheta_2^L-\vartheta_1^L\vartheta_2^L,\vartheta_1^U+\vartheta_2^U-\vartheta_1^U\vartheta_2^U\right]\end{array}\right),\\ \left(\zeta_1\zeta_2,\rho_1+\rho_2-\rho_1\rho_2,\vartheta_1+\vartheta_2-\vartheta_1\vartheta_2\right)\end{array}\right)$$

(iii) $$\xi\alpha_1 = \left(\begin{array}{l}\left(\left[1-\left(1-\zeta_1^L\right)^\xi,1-\left(1-\zeta_1^U\right)^\xi\right],\left[\left(\rho_1^L\right)^\xi,\left(\rho_1^U\right)^\xi\right],\left[\left(\vartheta_1^L\right)^\xi,\left(\vartheta_1^U\right)^\xi\right]\right),\\ \left(1-\left(1-\zeta_1\right)^\xi,\left(\rho_1\right)^\xi,\left(\vartheta_1\right)^\xi\right)\end{array}\right)$$

(iv) $$\left(\alpha_1\right)^\xi = \left(\begin{array}{c}\left(\begin{array}{c}\left[\left(\zeta_1^L\right)^\xi,\left(\zeta_1^U\right)^\xi\right],\left[1-\left(1-\rho_1^L\right)^\xi,1-\left(1-\rho_1^U\right)^\xi\right],\\ \left[1-\left(1-\vartheta_1^L\right)^\xi,1-\left(1-\vartheta_1^U\right)^\xi\right]\end{array}\right),\\ \left(\left(\zeta_1\right)^\xi,1-\left(1-\rho_1\right)^\xi,1-\left(1-\vartheta_1\right)^\xi\right)\end{array}\right)$$

Definition 5.6 (Maclaurin 1729) Let a_i (i = 1, 2, …, n) be a collection of non-negative real numbers and k=1, 2, …, n. If

$$MSM^{(k)}\left(a_1,a_2,\ldots,a_n\right) = \left(\frac{\sum_{\substack{1\le i_1<\ldots\\ <i_k\le n}}\prod_{j=1}^{k} a_{i_j}}{\binom{n}{k}}\right)^{\left(\frac{1}{k}\right)} \tag{5.6}$$

then MSM$^{(k)}$ is called the Maclaurin symmetric mean (MSM), where $(i_1, i_2, \ldots, i_k)$ traversal of all the k-tuple combination of $(1,2,\ldots,n)$, $\binom{n}{k}$ is the binomial coefficient.

5.3 NEUTROSOPHIC CUBIC MACLAURIN SYMMETRIC MEAN (NCMSM) OPERATOR

In this section, we shall extend the concept of Maclaurin symmetric mean to the NC environment and will investigate some of their desirable properties. For this, we consider a family of NCNs α_i; $(i=1, 2, \ldots, n)$ denoted by Ω.

Definition 5.7 For a family of NCNs $\alpha_i = \left(\left(\left[\zeta_i^L,\zeta_i^U\right],\left[\rho_i^L,\rho_i^U\right],\left[\vartheta_i^L,\vartheta_i^U\right]\right), \left(\zeta_i,\rho_i,\vartheta_i\right)\right),(i=1,2,\ldots,n)$, and for $(k = 1, 2, \ldots, n)$, an NCMSM$^{(k)}$ operator is defined as

$$NCMSM^{(k)}\left(\alpha_1,\alpha_2,\ldots,\alpha_n\right) = \left(\frac{\oplus_{\substack{1\le i_1<\\ \ldots<i_k\le n}} \left(\otimes^n_{\{j=1\}}\alpha_{i_j}\right)}{\binom{n}{k}}\right)^{1/k} \tag{5.7}$$

Theorem 5.1 For a family of NCNs $\alpha_i = \left(\left(\left[\zeta_i^L,\zeta_i^U\right],\left[\rho_i^L,\rho_i^U\right],\left[\vartheta_i^L,\vartheta_i^U\right]\right), \alpha_i = \left(\zeta_i,\rho_i,\vartheta_i\right)\right),(i=1,2,\ldots,n)$, and $(k=1,2,\ldots,n)$, the aggregated value using the NCMSM operator is still an NCN and is given by

$$NCMSM^{(k)}\left(\alpha_1,\alpha_2,\ldots,\alpha_n\right) = \left(\begin{array}{c}\left(\begin{array}{c}\left[f\left(\zeta_{i_j}^L\right),f\left(\zeta_{i_j}^U\right)\right],\left[g\left(\rho_{i_j}^L\right),g\left(\rho_{i_j}^U\right)\right],\\ \left[g\left(\vartheta_{i_j}^L\right),g\left(\vartheta_{i_j}^U\right)\right]\end{array}\right),\\ \left(f\left(\zeta_{i_j}\right),g\left(\rho_{i_j}\right),g\left(\vartheta_{i_j}\right)\right)\end{array}\right) \tag{5.8}$$

where $f\left(x_{i_j}\right) = \left(1-\left(\prod_{\substack{1\le i_1<\ldots\\ <i_k\le n}}\left(1-\prod_{j=1}^{k}x_{i_j}\right)\right)^{\frac{1}{\binom{n}{k}}}\right)^{\frac{1}{k}}$

and $g\left(x_{i_j}\right)=1-\left(1-\left(\prod_{\substack{1\le i_1<\dots\\<i_k\le n}}\left(1-\prod_{j=1}^{k}\left(1-x_{i_j}\right)\right)\right)^{\frac{1}{\binom{n}{k}}}\right)^{\frac{1}{k}}$

Proof. Using the operational laws as stated in Definition 5.5, we have

$$\overset{k}{\underset{j=1}{\otimes}}\alpha_{i_j}=\left(\begin{array}{c}\left(\begin{array}{c}\left[\prod_{j=1}^{k}\zeta_{i_j}^{L},\prod_{j=1}^{k}\zeta_{i_j}^{U}\right],\left[1-\prod_{j=1}^{k}\left(1-\rho_{i_j}^{L}\right),1-\prod_{j=1}^{k}\left(1-\rho_{i_j}^{U}\right)\right],\\ \left[1-\prod_{j=1}^{k}\left(1-\vartheta_{i_j}^{L}\right),1-\prod_{j=1}^{k}\left(1-\vartheta_{i_j}^{U}\right)\right]\end{array}\right),\\ \prod_{j=1}^{k}\zeta_{i_j},1-\prod_{j=1}^{k}\left(1-\rho_{i_j}\right),1-\prod_{j=1}^{k}\left(1-\vartheta_{i_j}\right)\end{array}\right) \quad (5.9)$$

and $\underset{((1\le i_1<\dots<i_k\le n))}{\oplus}\left(\overset{k}{\underset{j=1}{\otimes}}\alpha_{(i_j)}\right)=\Bigg(\Bigg[1-\prod_{((1\le i_1<\dots<i_k\le n))}\left(1-\prod_{j=1}^{k}\zeta_{(i_j)}^{L}\right),1$

$-\prod_{((1\le i_1<\dots<i_k\le n))}\left(1-\prod_{j=1}^{k}\zeta_{(i_j)}^{U}\right)\Bigg],\Bigg[\prod_{((1\le i_1<\dots<i_k\le n))}\left(1-\prod_{j=1}^{k}\rho_{(i_j)}^{L}\right),$

$\prod_{((1\le i_1<\dots<i_k\le n))}\left(1-\prod_{j=1}^{k}\rho_{(i_j)}^{U}\right)\Bigg],$

$\left[\prod_{((1\le i_1<\dots<i_k\le n))}\left(1-\prod_{j=1}^{k}\vartheta_{(i_j)}^{L}\right),\prod_{((1\le i_1<\dots<i_k\le n))}\left(1-\prod_{j=1}^{k}\vartheta_{(i_j)}^{U}\right)\right]$

$\prod_{((1\le i_1<\dots<i_k\le n))}\left(1-\prod_{j=1}^{k}\zeta_{(i_j)}\right),\prod_{((1\le i_1<\dots<i_k\le n))}\left(1-\prod_{j=1}^{k}\rho_{(i_j)}\right)$

$\prod_{((1\le i_1<\dots<i_k\le n))}\left(1-\prod_{j=1}^{k}\vartheta_{(i_j)}\right)$

then we obtain

$$\frac{1}{\binom{n}{k}}\left(\underset{\substack{1\le i_1<\dots\\<i_k\le n}}{\oplus}\left(\otimes_{j=1}^{k}\alpha_{i_j}\right)\right)=\left(\begin{pmatrix}\left[f_1\left(\zeta_{i_j}^{L}\right),f_1\left(\zeta_{i_j}^{U}\right)\right],\left[g_1\left(\rho_{i_j}^{L}\right),g_1\left(\rho_{i_j}^{U}\right)\right],\\ \left[g_1\left(\vartheta_{i_j}^{L}\right),g_1\left(\vartheta_{i_j}^{U}\right)\right]\end{pmatrix},\\ \left(f_1\left(\zeta_{i_j}\right),g_1\left(\rho_{i_j}\right),g_1\left(\vartheta_{i_j}\right)\right)\right)$$

where

$$f_1\left(x_{i_j}\right)=1-\left(\prod_{\substack{1\le i_1<\dots\\<i_k\le n}}\left(1-\prod_{j=1}^{k}x_{i_j}\right)\right)^{\frac{1}{\binom{n}{k}}}\text{ and }g_1\left(x_{i_j}\right)=\left(\prod_{\substack{1\le i_1<\dots\\<i_k\le n}}\left(1-\prod_{j=1}^{k}\left(1-x_{i_j}\right)\right)\right)^{\frac{1}{\binom{n}{k}}}$$

Therefore,

$$NCMSM^{(k)}\left(\alpha_1,\alpha_2,\dots,\alpha_n\right)=\left(\begin{pmatrix}\left[f_1\left(\zeta_{i_j}^{L}\right),f_1\left(\zeta_{i_j}^{U}\right)\right],\left[g_1\left(\rho_{i_j}^{L}\right),g_1\left(\rho_{i_j}^{U}\right)\right],\\ \left[g_1\left(\vartheta_{i_j}^{L}\right),g_1\left(\vartheta_{i_j}^{U}\right)\right]\end{pmatrix},\\ \left(f_1\left(\zeta_{i_j}\right),g_1\left(\rho_{i_j}\right),g_1\left(\vartheta_{i_j}\right)\right)\right)\tag{5.10}$$

Where $f_1\left(x_{i_j}\right)=1-\left(\prod_{\substack{1\le i_1<\dots\\<i_k\le n}}\left(1-\prod_{j=1}^{k}x_{i_j}\right)\right)^{\frac{1}{\binom{n}{k}}}$ and $g_1\left(x_{i_j}\right)=\left(\prod_{\substack{1\le i_1<\dots\\<i_k\le n}}\left(1-\prod_{j=1}^{k}\left(1-x_{i_j}\right)\right)\right)^{\frac{1}{\binom{n}{k}}}$.

In addition, since, $0\le\zeta_{i_j}^{U}\le 1$, implies that $0\le\prod_{j=1}^{k}\zeta_{i_j}^{U}\le 1$. Thus, we have $0\le 1-\prod_{j=1}^{k}\zeta_{i_j}^{U}\le 1$ which leads to $0\le\prod_{\substack{1\le i_1<\dots\\<i_k\le n}}\left(1-\prod_{j=1}^{k}\zeta_{i_j}^{U}\right)\le 1$. It follows that $0\le f\left(\zeta_{i_j}^{U}\right)\le 1$. Also, $0\le\rho_{i_j}^{U}\le 1$, which implies that $0\le\prod_{j=1}^{k}\left(1-\rho_{i_j}^{U}\right)\le 1$. Further, $0\le g\left(\rho_{i_j}^{U}\right)\le 1$ and $0\le g\left(\vartheta_{i_j}^{U}\right)\le 1$. In a similar manner, $0\le f\left(\zeta_{i_j}\right),g\left(\rho_{i_j}\right),g\left(\vartheta_{i_j}\right)\le 1$. Thus, $0\le f\left(\zeta_{i_j}^{U}\right)+g\left(\rho_{i_j}^{U}\right)+g\left(\vartheta_{i_j}^{U}\right)\le 3$ and $0\le f\left(\zeta_{i_j}\right)+g\left(\rho_{i_j}\right)+g\left(\vartheta_{i_j}\right)\le 3\cdot$

Property 5.1 If α_i, (i=1, 2, …, n) are all equal, i.e., $\alpha_i = \alpha$ $=\left(\left(\left[\zeta^L,\zeta^U\right],\left[\rho^L,\rho^U\right],\left[\vartheta^L,\vartheta^U\right]\right),(\zeta,\rho,\vartheta)\right)$, then

$$\text{NCMSM}^{(k)}(\alpha,\alpha,\ldots,\alpha)=\alpha$$

Proof. Let $\alpha_i=\left(\left(\left[\zeta_i^L,\zeta_i^U\right],\left[\rho_i^L,\rho_i^U\right],\left[\vartheta_i^L,\vartheta_i^U\right]\right),(\zeta_i,\rho_i,\vartheta_i)\right)$ and

$\alpha=\left(\left(\left[\zeta^L,\zeta^U\right],\left[\rho^L,\rho^U\right],\left[\vartheta^L,\vartheta^U\right]\right),(\zeta,\rho,\vartheta)\right)$ be NCNs such that $\alpha_i=\alpha$ for all i and j. Then

$$f\left(\zeta_{i_j}^L\right)=\left(1-\left(\prod_{\substack{1\le i_1<\ldots\\<i_k\le n}}\left(1-\prod_{j=1}^{k}\zeta_{i_j}^L\right)\right)^{\frac{1}{\binom{n}{k}}}\right)^{\frac{1}{k}}=\left(1-\left(\prod_{\substack{1\le i_1<\ldots\\<i_k\le n}}\left(1-\prod_{j=1}^{k}\zeta^L\right)\right)^{\frac{1}{\binom{n}{k}}}\right)^{\frac{1}{k}}$$

$$=\left(1-\left(\prod_{\substack{1\le i_1<\ldots\\<i_k\le n}}\left(1-\left(\zeta^L\right)^k\right)\right)^{\frac{1}{\binom{n}{k}}}\right)^{\frac{1}{k}}=\left(1-\left(\left(\left(1-\left(\zeta^L\right)^k\right)\right)^{\binom{n}{k}}\right)^{\frac{1}{\binom{n}{k}}}\right)^{\frac{1}{k}}=\zeta^L$$

Similarly, $g\left(\vartheta_{i_j}^L\right)=g\left(\vartheta^L\right)$; thus, from Theorem 5.1, we have

$$\text{NCMSM}^{(k)}(\alpha_1,\alpha_2,\ldots,\alpha_n)=\begin{pmatrix}\left(\left[f\left(\zeta_{i_j}^L\right),f\left(\zeta_{i_j}^U\right)\right],\left[g\left(\rho_{i_j}^L\right),g\left(\rho_{i_j}^U\right)\right],\left[g\left(\vartheta_{i_j}^L\right),g\left(\vartheta_{i_j}^U\right)\right]\right),\\ \left(f\left(\zeta_{i_j}\right),g\left(\rho_{i_j}\right),g\left(\vartheta_{i_j}\right)\right)\end{pmatrix}$$

$$=\left(\left(\left[\zeta^L,\zeta^U\right],\left[\rho^L,\rho^U\right],\left[\vartheta^L,\vartheta^U\right]\right),(\zeta,\rho,\vartheta)\right)=\alpha$$

which completes the proof.

Property 5.2 (Monotonicity) Let $\alpha_i=\left(\left(\left[\zeta_{\alpha_i}^L,\zeta_{\alpha_i}^U\right],\left[\rho_{\alpha_i}^L,\rho_{\alpha_i}^U\right],\left[\vartheta_{\alpha_i}^L,\vartheta_{\alpha_i}^U\right]\right),\right.$ $\left.\left(\zeta_{\alpha_i},\rho_{\alpha_i},\vartheta_{\alpha_i}\right)\right)$ and $\beta_i=\left(\left(\left[\zeta_{\beta_i}^L,\zeta_{\beta_i}^U\right],\left[\rho_{\beta_i}^L,\rho_{\beta_i}^U\right],\left[\vartheta_{\beta_i}^L,\vartheta_{\beta_i}^U\right]\right),\left(\zeta_{\beta_i},\rho_{\beta_i},\vartheta_{\beta_i}\right)\right)$, ($i$ = 1,2, …, n), be two collections of NCNs. If $\left[\zeta_{\alpha_i}^L,\zeta_{\alpha_i}^U\right]\subseteq\left[\zeta_{\beta_i}^L,\zeta_{\beta_i}^U\right],\left[\rho_{\alpha_i}^L,\rho_{\alpha_i}^U\right]$ $\supseteq\left[\rho_{\beta_i}^L,\rho_{\beta_i}^U\right],\left[\vartheta_{\alpha_i}^L,\vartheta_{\alpha_i}^U\right]\supseteq\left[\vartheta_{\beta_i}^L,\vartheta_{\beta_i}^U\right],\zeta_{\alpha_i}\le\zeta_{\beta_i},\rho_{\alpha_i}\ge\rho_{\beta_i}$ and $\vartheta_{\alpha_i}\ge\vartheta_{\beta_i}$ for all i, $\text{NCMSM}^{(k)}(\alpha_1,\alpha_2,\ldots,\alpha_n)\le\text{NCMSM}^{(k)}(\beta_1,\beta_2,\ldots,\beta_n)$

Proof. Since $k \geq 1$, and $\left[\zeta_{\alpha_i}^L, \zeta_{\alpha_i}^U\right] \subseteq \left[\zeta_{\beta_i}^L, \zeta_{\beta_i}^U\right], \left[\rho_{\alpha_i}^L, \rho_{\alpha_i}^U\right] \supseteq \left[\rho_{\beta_i}^L, \rho_{\beta_i}^U\right], \left[\vartheta_{\alpha_i}^L, \vartheta_{\alpha_i}^U\right] \supseteq \left[\vartheta_{\beta_i}^L, \vartheta_{\beta_i}^U\right], \zeta_{\alpha_i} \leq \zeta_{\beta_i}, \rho_{\alpha_i} \geq \rho_{\beta_i}$ and $\vartheta_{\alpha_i} \geq \vartheta_{\beta_i}$, then for all ($I$ = 1, 2,..., n) and for (j=1, 2, ..., k) we obtain

$$\prod_{j=1}^{k} \zeta_{\alpha_{ij}}^L \leq \prod_{j=1}^{k} \zeta_{\beta_{ij}}^L \Rightarrow 1 - \prod_{j=1}^{k} \zeta_{\alpha_{ij}}^L \geq 1 - \prod_{j=1}^{k} \zeta_{\beta_{ij}}^L \Rightarrow \prod_{\substack{1 \leq i_1 < \ldots \\ < i_k \leq n}} \left(1 - \prod_{j=1}^{k} \zeta_{\alpha_{ij}}^L\right)$$

$$\geq \prod_{\substack{1 \leq i_1 < \ldots \\ < i_k \leq n}} \left(1 - \prod_{j=1}^{k} \zeta_{\beta_{ij}}^L\right) \Rightarrow \left(1 - \left(\prod_{\substack{1 \leq i_1 < \ldots \\ < i_k \geq n}} \left(1 - \prod_{j=1}^{k} \zeta_{\alpha_{ij}}^L\right)\right)^{\binom{n}{k}}\right)^{\frac{1}{k}}$$

$$\leq \left(1 - \left(\prod_{\substack{1 \leq i_1 < \ldots \\ < i_k \leq n}} \left(1 - \prod_{j=1}^{k} \zeta_{\beta_{ij}}^L\right)\right)^{\binom{n}{k}}\right)^{\frac{1}{k}}$$

Similarly, we can have the expression for ζ^U as well as ζ. Also, we have

$$\rho_{\alpha_{ij}}^L \geq \rho_{\beta_{ij}}^L \Rightarrow \left(1 - \rho_{\alpha_{ij}}^L\right) \leq \left(1 - \rho_{\beta_{ij}}^L\right) \Rightarrow 1 - \prod_{j=1}^{k} \left(1 - \rho_{\alpha_{ij}}^L\right) \geq 1 - \prod_{j=1}^{k} \left(1 - \rho_{\beta_{ij}}^L\right)$$

$$\Rightarrow \left(\prod_{\substack{1 \leq i_1 < \ldots \\ < i_k \leq n}} \left(1 - \prod_{j=1}^{k} \left(1 - \rho_{\alpha_{ij}}^L\right)\right)\right)^{\frac{1}{\binom{n}{k}}} \geq \left(\prod_{\substack{1 \leq i_1 < \ldots \\ < i_k \leq n}} \left(1 - \prod_{j=1}^{k} \left(1 - \rho_{\beta_{ij}}^L\right)\right)\right)^{\frac{1}{\binom{n}{k}}}$$

$$\Rightarrow \left(1 - \left(\prod_{\substack{1 \leq i_1 < \ldots \\ < i_k \leq n}} \left(1 - \prod_{j=1}^{k} \left(1 - \rho_{\alpha_{ij}}^L\right)\right)\right)^{\frac{1}{\binom{n}{k}}}\right)^{\frac{1}{k}} \leq \left(1 - \left(\prod_{\substack{1 \leq i_1 < \ldots \\ < i_k \leq n}} \left(1 - \prod_{j=1}^{k} \left(1 - \rho_{\beta_{ij}}^L\right)\right)\right)^{\frac{1}{\binom{n}{k}}}\right)^{\frac{1}{k}}$$

$$\Rightarrow 1 - \left(1 - \left(\prod_{\substack{1 \leq i_1 < \ldots \\ < i_k \leq n}} \left(1 - \prod_{j=1}^{k} \left(1 - \rho_{\alpha_{ij}}^L\right)\right)\right)^{\frac{1}{\binom{n}{k}}}\right)^{\frac{1}{k}} \geq 1 - \left(1 - \left(\prod_{\substack{1 \leq i_1 < \ldots \\ < i_k \leq n}} \left(1 - \prod_{j=1}^{k} \left(1 - \rho_{\beta_{ij}}^L\right)\right)\right)^{\frac{1}{\binom{n}{k}}}\right)^{\frac{1}{k}}$$

Similarly, we can have expressions for ρ^U, ϑ^L, ϑ^U as well as for ρ and for ϑ. Therefore, by comparison rules as defined in Definition 5.4, we obtain

$$\text{NCMSM}^{(k)}\,(\alpha_1, \alpha_2, \ldots, \alpha_n) \le \text{NCMSM}^{(k)}\,(\beta_1, \beta_2, \ldots, \beta_n)$$

which completes the proof of the property.

Property 5.3 (Boundedness) Let $\alpha_i = \left(\left(\left[\zeta_i^L, \zeta_i^U\right], \left[\rho_i^L, \rho_i^U\right], \left[\vartheta_i^L, \vartheta_i^U\right]\right), \left(\zeta_i, \rho_i, \vartheta_i\right)\right)$, $(i = 1, 2, \ldots, n)$, be a collection of NCNs and let

$$\alpha^- = \left(\begin{array}{c}\left(\left[min_i\zeta_i^L, min_i\zeta_i^U\right], \left[max_i\rho_i^L, max_i\rho_i^U\right], \left[max_i\vartheta_i^L, max_i\vartheta_i^U\right]\right.\\ \left(min_i\zeta_i, max_i\rho_i, max_i\vartheta_i\right)\end{array}\right. \text{ and}$$

$$\alpha^+ = \left(\begin{array}{c}\left(\left[max_i\zeta_i^L, max_i\zeta_i^U\right], \left[min_i\rho_i^L, min_i\rho_i^U\right], \left[min_i\vartheta_i^L, min_i\vartheta_i^U\right]\right),\\ \left(max_i\zeta_i, min_i\rho_i, min_i\vartheta_i\right)\end{array}\right) \text{ then}$$

$$\alpha^- \le \text{NCMSM}^{(k)}\,(\alpha_1, \alpha_2, \ldots, \alpha_n) \le \alpha^+.$$

Proof. Based on Properties 1 and 2, we have

$$\text{NCMSM}^{(k)}\,(\alpha_1, \alpha_2, \ldots, \alpha_n) \ge \text{NCNMSM}^{(k)}\,(\alpha^-, \alpha^-, \ldots, \alpha^-) = \alpha^-$$

and

$$\text{NCMSM}^{(k)}\,(\alpha_1, \alpha_2, \ldots, \alpha_n) \le \text{NCNMSM}^{(k)}\,(\alpha^+, \alpha^+, \ldots, \alpha^+) = \alpha^+$$

which completes the proof.

5.4 NEUTROSOPHIC CUBIC WEIGHTED MACLAURIN SYMMETRIC MEAN

From Section 5.3, it is evident that the NCMSM does not consider the importance of the values to be aggregated. Nevertheless, in real-life practical situations, the elements generally possess some priority values or the weight vector. To handle such situations, we introduce the definition of neutrosophic cubic weighted Maclaurin symmetric mean (NCWMSM) as follows:

Definition 5.8 Let $\alpha_i = \left(\left(\left[\zeta_i^L, \zeta_i^U\right], \left[\rho_i^L, \rho_i^U\right], \left[\vartheta_i^L, \vartheta_i^U\right]\right), \left(\zeta_i, \rho_i, \vartheta_i\right)\right)$ ($i = 1, 2, \ldots, n$) be a collection of NCNs, $\omega = (\omega_1, \omega_2, \ldots, \omega_n)^T$ is the weight vector of α_i, such that $\omega_i \in (0,1)$ and $\sum_{i=1}^{n} \omega_i = 1$. If

$$\mathrm{NCWMSM}_{\omega}^{(k)}\left(\alpha_1, \alpha_2, \ldots, \alpha_n\right) = \left(\frac{\underset{1 \le i_1 < \ldots < i_k \le n}{\oplus} \left(\overset{k}{\underset{j=1}{\otimes}} \omega_{i_j} \alpha_{i_j} \right)}{\binom{n}{k}} \right)^{\frac{1}{k}}$$

then the NCWMSM is called the neutrosophic cubic weighted Maclaurin symmetric mean.

By using the operational laws, we can derive the following theorem.

Theorem 5.2 Let $1 \le k \le n$ ($k \in Z$) and $\alpha_i = \left(\left(\left[\zeta_i^L, \zeta_i^U\right], \left[\rho_i^L, \rho_i^U\right], \left[\vartheta_i^L, \vartheta_i^U\right]\right), \left(\zeta_i, \rho_i, \vartheta_i\right)\right)$ ($i = 1, 2, \ldots, n$) be a collection of NCNs. Then, the aggregated values, by using the NCWMSM, are also an NCN and are given as

$$\mathrm{NCWMSM}_{\omega}^{(k)}\left(\alpha_1, \alpha_2, \ldots, \alpha_n\right) = \begin{pmatrix} \left(\left[f_\omega\left(\zeta_{i_j}^L\right), f_\omega\left(\zeta_{i_j}^U\right)\right], \left[g_\omega\left(\rho_{i_j}^L\right), g_\omega\left(\rho_{i_j}^U\right)\right], \left[g_\omega\left(\vartheta_{i_j}^L\right), g_\omega\left(\vartheta_{i_j}^U\right)\right]\right), \\ \left(f_\omega\left(\zeta_{i_j}\right), g_\omega\left(\rho_{i_j}\right), g_\omega\left(\vartheta_{i_j}\right)\right) \end{pmatrix}$$

$$\text{where } f_\omega\left(x_{i_j}\right) = \left(1 - \left(\prod_{\substack{1 \le i_1 < \ldots \\ < i_k \le n}} \left(1 - \prod_{j=1}^{k} \left(1 - \left(1 - x_{i_j}\right)^{\omega_{i_j}} \right) \right) \right)^{\frac{1}{\binom{n}{k}}} \right)^{\frac{1}{k}}$$

$$\text{and } g_\omega\left(x_{i_j}\right) = 1 - \left(1 - \left(\prod_{\substack{1 \le i_1 < \ldots \\ < i_k \le n}} \left(1 - \prod_{j=1}^{k} \left(1 - \left(x_{i_j}\right)^{\omega_{i_j}} \right) \right) \right)^{\frac{1}{\binom{n}{k}}} \right)^{\frac{1}{k}} \tag{5.11}$$

Proof. Similar to Theorem 5.1.

5.5 DECISION-MAKING APPROACH

The decision-making approach provides us with the facility of aligning our study to the real world. It is a useful approach to find the best alternative under a set of some feasible criterion. Assume the set of alternatives to be $A = A_1, A_2, \ldots, A_m$ which are to be analyzed by the expert(s) in accordance with several criteria $C = C_1, C_2, \ldots, C_n$. The following steps are adopted to find the best fit among the available alternatives.

1. Collect the information rating of alternatives corresponding to criteria and summarize in the form of NCNs $\alpha_{ij} = \left(\left(\left[\zeta_{ij}^L, \zeta_{ij}^U\right], \left[\rho_{ij}^L, \rho_{ij}^U\right], \left[\vartheta_{ij}^L, \vartheta_{ij}^U\right]\right), \left(\zeta_{ij}, \rho_{ij}, \vartheta_{ij}\right)\right)$: $(i = 1, 2, \ldots, m)$; $(j = 1, 2, \ldots, n)$. These rating values are expressed as a decision matrix D:

$$D = \begin{bmatrix} \alpha_{11} & \alpha_{12} & \cdots & \alpha_{1n} \\ \alpha_{21} & \alpha_{22} & \cdots & \alpha_{2n} \\ \vdots & \vdots & \ddots & \vdots \\ \alpha_{m1} & \alpha_{m2} & \cdots & \alpha_{mn} \end{bmatrix}$$

2. Aggregated values are computed corresponding to each alternative A_i $(i = 1, 2, \ldots, m)$ making use of Eq. (5.11) to obtain the collective value p_i corresponding to each alternative A_i, $(i=1,2,\ldots,m)$.
3. Compute the score values as given in Eq. (5.5).
4. Rank the alternatives on the basis of their order in accordance with the score values.

5.5.1 Illustrative example

This portion gives an account of the practical applicability of the DM approach highlighted above.

Management holds a very strong position in the coordination of resources and services in any of the concerned areas, whether it is a small classroom or a big metropolitan city. Management is the core need for the successful functioning of all the parts of the system. Similar, is its need in the library system. Often, the regional, institutional, or state libraries need for proper coordination of the staff, library personals, etc., to carry on the smooth functioning of the library as a controlled unit. This case study focuses on the intense feedback analysis obtained from the members of libraries situated at four different locations, viz. A_1, A_2, A_3, and A_4. The library management has collected feedback from the members at two different stages. Firstly, the feedback is obtained from them (in the form of IVNN) before becoming the member and the second feedback is obtained (in the form of SVNN) after registration for the membership card. The major feature of

Table 5.1 Preference values

	C_1	C_2	C_3	C_4
A_1	(([0.20,0.30], [0.15,0.20], [0.30,0.35]), (0.25,0.16,0.32)	(([0.45,0.50], [0.02,0.10], [0.30,0.40]), (0.47,0.05,0.35))	(([0.20,0.25], [0.15,0.25], [0.10,0.20]), (0.24,0.20,0.15))	(([0.10,0.60], [0.30,0.40], [0.50,0.80]), (0.50,0.50,0.70))
A_2	(([0.30,0.40], [0.16,0.20], [0.30,0.35]), (0.35,0.18,0.32))	(([0.60,0.70], [0.30,0.60], [0.30,0.70]), (0.70,0.50,0.40))	(([0.30,0.70], [0.70,0.80], [0.60,0.70]), (0.40,0.20,0.65))	(([0.20,0.50], [0.40,0.90], [0.50,0.80]), (0.30,0.50,0.60))
A_3	(([0.40,0.60], [0.50,0.70], [0.10,0.20]), (0.45,0.52,0.15))	(([0.30,0.90], [0.20,0.70], [0.30,0.50]), (0.40,0.22,0.35))	(([0.40,0.60], [0.20,0.30], [0.15,0.20]), (0.45,0.22,0.18))	(([0.25,0.35], [0.65,0.70], [0.40,0.50]), (0.28,0.68,0.43))
A_4	(([0.50,0.60], [0.20,0.40], [0.30,0.50]), (0.54,0.23,0.37))	(([0.30,0.90], [0.40,0.60], [0.50,0.80]), (0.37,0.42,0.69))	(([0.70,0.80], [0.20,0.30], [0.10,0.15]), (0.77,0.22,0.14))	(([0.30,0.40], [0.20,0.25], [0.10,0.15]), (0.35,0.22,0.14))

these two feedbacks is that the first feedback shows the person's expectation level towards the library and the second feedback shows the person's satisfaction level from the services provided. So, basically the whole analysis gives insights regarding the expectation versus satisfaction of the library members, and to achieve this simultaneous analysis all the obtained values are merged together to form NCNs as shown in Table 5.1. The Human Resource Manager (HRM) analyzed the filled feedback form and categorized the feedbacks roughly into four major criteria, namely C_1: "Library environment," C_2: "Content access," C_3: "Cleanliness and maintenance," and C_4: "Flexibility." The questionnaire consisted of questions eliciting the characteristic criteria based on numerable factors, which are highlighted as follows:

(i) **Library environment:** The library environment plays an important role in determining the best library among the available ones. This mainly comprises the ambience provided by the library management to the members. It was seen that the required amount of seating facilities contributed in a major manner towards the library atmosphere and convenience to the members. Secondly, it was analyzed that the libraries having adequate silence and less disturbance were preferred by people. On the other hand, discipline maintenance was also demanded to a large extent. So, based on these instances, the library environment was seen to have three further aspects: seating arrangement, silence, and discipline maintained.

(ii) **Content access:** The content available in a library is the main asset to the members. The characteristic of the availability of content was further categorized into different aspects. Some of the members were found keen towards the availability of the latest book editions while some people were more aligned towards the available access to the latest paid journals. Also, many of the members wanted the library to be updated regularly with the daily news chapter or with the regular editions of magazines. So, the content access was found to be comprising three sub-factors, namely availability of books, access to paid journals, and availability to the daily news chapter as well as magazines.

(iii) **Cleanliness and maintenance:** A place where a person has to devote some of the hours from their schedule has to be properly clean and well-maintained. The same is the case with a library. It was found that people were easily attracted to the libraries which are neat and clean. From the analysis, it was found that people have different genres under this criterion too. Some were concerned towards the proper sanitation facilities while some were more influenced by the regular cleaning of the available infrastructure such as desks. It was also found that people are attracted by the proper ordering of the books kept on the library shelves so that they can be found easily at their earliest convenience. So, these criteria are subdivided into three sub-factors, namely proper sanitation facilities and regular cleaning of desks and books kept in a proper manner.

(iv) **Flexibility:** It was found that people prefer those libraries more which have not a very rigid work frame such that they should have enough time for returning the books, should not be having costly membership fees, etc. Also, it was found that the cooperation of the staff members in making the people comfortable in all the queries is also much preferred when they talk about the flexibility of the staff members. So, the flexibility criterion is found to have three sub-categories: sufficient days for returning books, pocket-friendly membership cards, and cooperative staff.

These four criteria have further categorized elements which are found in the feedback forms obtained from the members. These weighted criteria are taken as $\omega = (0.28, 0.14, 0.21, 0.37)^T$. The elements and their weight contribution are highlighted in the Figure 5.1. The aim is to rank the libraries according to the quality of the service provided by them and to determine which library is most preferable by people.

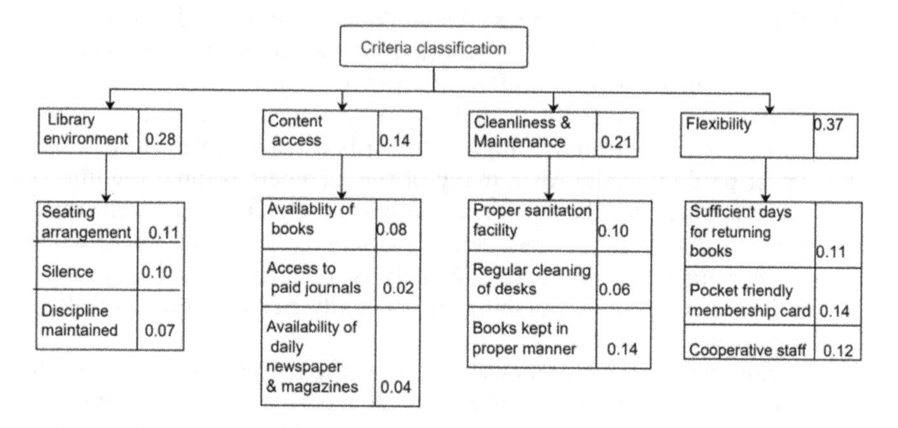

Figure 5.1 Criteria classification with weight information.

The following are the steps of the proposed approach implemented as:

1. The information collected regarding the alternatives A_i, $(i = 1, 2, ..., 4)$ corresponding to the said criteria C_j, $(j = 1, 2, ..., 4)$ are tabulated in Table 5.1.
2. Aggregate the preference values using Eq. (5.11):

$$p_1 = \text{NCMSM}^{(2)}(\alpha_{11}, \alpha_{12}, \alpha_{13}, \alpha_{14})$$
$$\begin{pmatrix} ([0.0555, 0.1230], [0.6199, 0.7062], [0.7411, 0.8211]), \\ (0.1042, 0.6875, 0.7899) \end{pmatrix}$$

Similarly,

$$p_2 = \begin{pmatrix} ([0.0911, 0.1831], [0.7813, 0.8973], [0.8116, 0.8966]), \\ (0.1230, 0.7597, 0.8419) \end{pmatrix}$$

$$p_3 = \begin{pmatrix} ([0.0946, 0.2043], [0.7990, 0.8829], [0.6939, 0.7637]), \\ (0.1126, 0.8112, 0.7249) \end{pmatrix}$$

$$p_4 = \begin{pmatrix} ([0.1383, 0.2390], [0.7013, 0.7790], [0.6841, 0.7569]), \\ (0.1626, 0.7177, 0.7260) \end{pmatrix}$$

3. Using Eq. (5.5), the score values are computed as: $S(p_1) = 0.2130$, $S(p_2) = 0.1565$, $S(p_3) = 0.1929$, and $S(p_4) = 0.2416$.
4. Based on the previous step, the alternatives are ranked as $A_4 \succ A_1 \succ A_3 \succ A_2$ (Table 5.2).

Table 5.2 Different values of *k*

Parameter value	A_1	A_2	A_3	A_4	*Ranking*
k = 1	0.2202	0.1683	0.1997	0.2614	$A_4 \succ A_1 \succ A_3 \succ A_2$
k = 2	0.2130	0.1565	0.1929	0.2416	$A_4 \succ A_1 \succ A_3 \succ A_2$
k = 3	0.2088	0.1502	0.1884	0.2274	$A_4 \succ A_1 \succ A_3 \succ A_2$
k = 4	0.2052	0.1445	0.1836	0.2088	$A_4 \succ A_1 \succ A_3 \succ A_2$

5.5.2 Comparison studies

In order to outline the superiority of our proposed approach, the comparison analysis is conducted on the basis of three different categories of neutrosophic environment, namely interval-valued neutrosophic, single-valued neutrosophic, and neutrosophic cubic fuzzy set environments. It is evident from the modelling of an NCN in the illustrative example that it captures the two-dimensional information ranging over two different aspects simultaneously in which the IVNN represents a person's expectation from the library and the SVNN represents the satisfaction level after the membership holding. So, we have compared the proposed approach with theories based on IVNN (Table 5.3).

It is seen that the NCN gets reduced to IVNN by not considering the satisfaction level analysis simultaneously with the expectation level analysis. Upon comparing the interval neutrosophic number weighted average (INNWA) operator with the interval neutrosophic number weighted geometric (INNWG) operator as in the existing approach (Zhang 2014), we get that the obtained ranking is $A_4 \succ A_1 \succ A_3 \succ A_2$, which is the same as that of our proposed approach. On the other hand, by analyzing the results using an interval-valued neutrosophic set generalized weighted average (IVNSGWA) operator (Aiwu 2015), we again obtain the same ranking as that of the proposed approach. This clearly shows the results of the approach along with

Table 5.3 Comparison with the existing approaches

Existing approach	*Operator*	$S(A_1)$	$S(A_2)$	$S(A_3)$	$S(A_4)$	*Ranking*
H. Zhang (2014)	INNWA	0.6162	0.5529	0.6025	0.6822	$A_4 \succ A_1 \succ A_3 \succ A_2$
H. Zhang (2014)	INNWG	0.5765	0.5019	0.5652	0.6384	$A_4 \succ A_1 \succ A_3 \succ A_2$
Z. Aiwu (2015)	IVNSGWA	0.6263	0.5702	0.6124	0.6896	$A_4 \succ A_1 \succ A_3 \succ A_2$
H. Garg (2018)	SVNPMM	0.9986	0.9997	0.9989	0.9999	$A_4 \succ A_1 \succ A_3 \succ A_2$
J. Ye (2014)	SNWA	0.6227	0.6135	0.6195	0.6618	$A_4 \succ A_1 \succ A_3 \succ A_2$
J. Chen (2017)	SVNDWAA	0.6418	0.5664	0.5942	0.7110	$A_4 \succ A_1 \succ A_3 \succ A_2$
Shi and Ye (2018)	NCDWAA	0.6382	0.5471	0.6051	0.7186	$A_4 \succ A_1 \succ A_3 \succ A_2$

the additional benefits of flexibility of considering satisfaction levels also. This facility is not in the IVNN-based theories.

Gradually, if we do not consider the expectation pattern but consider only the contentment levels, then we utilize theories based on SVNN. On comparing with the single-valued neutrosophic prioritized Muirhead mean (SVNPMM) operator (Garg 2018), it is found that the ranking is obtained as $A_4 \succ A_2 \succ A_3 \succ A_1$. Here, the conclusion about the most preferred library located at location (A_4) remains the same as that of our proposed approach but according to this theory, the least preferred one is the library at location (A_1) instead of that at location (A_2) as outlined by our methodology. This significant difference is because we have set the parameter value "p" in the existing operator as 1 and have not considered the expectation levels corresponding to which the satisfaction levels are analyzed. Thus, incomplete modelling of the situation by SVNN is the root cause of the difference in ranking. On the other hand, using the operators single-valued neutrosophic weighted average (SVNW) as well as single-valued neutrosophic dual weighted average aggregation (SVNDWAA) operator (Chen 2017) put forth by and respectively, the ranking is found similar to the proposed approach. This symmetry of the obtained rankings clearly depicts the capability of our approach to give the same outputs as that of the existing ones but with more detailed modelling of the uncertain and vague information.

For further analyzing the output in the NCN environment by considering both the expectation and satisfaction levels, we have compared the proposed method with that of the neutrosophic cubic Dombi weighted aggregation average operator as given by (Shi and Ye 2018), we have found the ranking to be $A_4 \succ A_1 \succ A_3 \succ A_2$ which is exactly the same as that of the proposed operator's outcomes. This clearly shows the alignment of our approach to all the IVNN, SVNN, and NCN-based theories.

5.6 CONCLUSION

The NCS theory captures the indeterminate judgements in the form of hybridization of IVNS and SVNS. To establish the concept of simultaneous representation of degrees of trueness and falsity along with indeterminacy, the notion of NCSs was created. Focusing on the advantages of the MSM operators as well as NCSs, we have defined weighted and non-weighted NCMSM aggregation operators and have investigated their desirable properties. Subsequently, to signify the applications of the proposed AOs, we have proposed a flexible decision-making approach. To justify the practical resilience, the proposed theory has been exemplified by a case study based on demand–supply analysis in an automobile manufacturing company. Finally, to analyze whether the formulated theory has overcome the limitations of the existing ones, an extensive comparative analysis has been conducted with the existing decision-making theories. From the comparative study, it has been analyzed that the proposed aggregation operators have dealt with the uncertainties in an efficient way. The future scope of work lies in applying the proposed approach to more DM problems such as supplier selection, pattern recognition problems, and medical diagnosis (Garg 2016).

REFERENCES

Aiwu, Z., D. Jianguo, and G. Hongjun. "Interval valued neutrosophic sets and multi-attribute decision-making based on generalized weighted aggregation operator." *Journal of Intelligent \& Fuzzy Systems* 29, no. 6 (2015): 2697–2706.

Ali, M., I. Deli, and F. Smarandache. "The theory of neutrosophic cubic sets and their applications in pattern recognition." *Journal of Intelligent & Fuzzy Systems* 30 (2016): 1957–1963.

Atanassov, K. T. "Intuitionistic fuzzy sets." *Fuzzy Sets and Systems* 20 (1986): 87–96.

Atanassov, K., and G. Gargov. "Interval valued intuitionistic fuzzy sets." *Fuzzy Sets and Systems* 3 (1989): 343–349.

Chen, J., and J. Ye. "Some single-valued neutrosophic dombi weighted aggregation operators for multiple attribute decision-making." *Symmetry* 9, no. 6 (2017): 82.

Garg, H. "Generalized intuitionistic fuzzy interactive geometric interaction operators using Einstein t-norm and t-conorm and their application to decision making." *Computers & Industrial Engineering* 101 (2016): 53–69.

Garg, H. "A new generalized pythagorean fuzzy information aggregation using Einstein operations and its application to decision making." *International Journal of Intelligent Systems* 31, no. 9 (2016): 886–920.

Garg, H., and R. Arora. "Bonferroni mean aggregation operators under intuitionistic fuzzy soft set environment and their applications to decision-making." *Journal of the Operational Research Society* 69, no. 11 (2018): 1711–1724.

Garg, H., and Nancy. "Multi-criteria decision-making method based on prioritized Muirhead mean aggregation operator under neutrosophic set environment." *Symmetry* 10, no. 7 (2018): 280.

Ji, P., J. Wang, and H. Zhang. "Frank prioritized bonferroni mean operator with single-valued neutrosophic sets and its application in selecting third-party logistics providers." *Neural Computing and Applications* 30, no. 3 (2018): 799–823.

Jun, Y. B., C. S. Kim, and K. O. Yang. "Cubic sets." *Annals of Fuzzy Mathematics and Informatics* 4, no. 1 (2012): 83–98.

Jun, Y. B., F. Smarandache, and C. S. Kim. "P-union and {P}-intersection of neutrosophic cubic sets." *Analele Universitatii "Ovidius" Constanta-Seria Matematica* 1 (2017): 99–115.

Kaur, G., and H. Garg. "Cubic intuitionistic fuzzy aggregation operators." *International Journal for Uncertainty Quantification* 8, no. 5 (2018a): 1–23.

Kaur, G., and H. Garg. "Multi-attribute decision-making based on Bonferroni mean operators under cubic intuitionistic fuzzy set environment." *Entropy* 20, no. 1 (2018b): 65.

Liu, P. "Some hamacher aggregation operators based on the interval-valued intuitionistic fuzzy numbers and their application to group decision making." *IEEE Transactions on Fuzzy Systems* 22, no. 1 (2014): 83–97.

Liu, C., and Y. Luo. "Power aggregation operators of simplified neutrosophic sets and their use in multi-attribute group decision making." *IEEE/CAA Journal of Automatica Sinica* 6, no. 2 (2017): 1–10.

Liu, P., and X. Qin. "Maclaurin symmetric mean operators of linguistic intuitionistic fuzzy numbers and their application to multiple-attribute decision-making." *Journal of Experimental & Theoretical Artificial Intelligence* 29, no. 6 (2017): 1173–1202.

Liu, P., and Y. Wang. "Interval neutrosophic prioritized owa operator and its application to multiple attribute decision making." *Journal of Systems Science and Complexity* 29, no. 3 (2016): 681–697.

Liu, P., and Y. Wang. "Multiple attribute decision-making method based on single-valued neutrosophic normalized weighted bonferroni mean." *Neural Computing and Applications* 25 (2014): 2001–2010.

Liu, P., and X. Zhang. "Some Maclaurin symmetric mean operators for single-valued trapezoidal neutrosophic numbers and their applications to group decision making." *International Journal of Fuzzy Systems* 20, no. 1 (2018): 45–61.

Lu, Z., and J. Ye. "Cosine measures of neutrosophic cubic sets for multiple attribute decision-making." *Symmetry* 9, no. 7 (2017): 121.

Maclaurin, C. "A second letter to {M}artin {F}olkes, {E}sq.; concerning the roots of equations, with the demonstartion of other rules in algebra, phil." *Transactions* 36 (1729): 59–96.

Nancy, and H.Garg. "Novel single-valued neutrosophic decision making operators under Frank norm operations and its application." *International Journal of Uncertainty Quantification* 6, no. 4 (2016): 361–375.

Pramanik, S., S. Dalapati, S. Alam, and T. K. Roy. "{NC}-{TODIM} based {MAGDM} under a neutrosophic cubic set environment." *Information* 8, no. 4 (2017): 149.

Pramanik, S., S. Dalapati, S. Alam, T. K. Roy, and F. Smarandache. *Neutrosophic cubic MCGDM method based on similarity measure* 16, no. 1 (2017): 44–56.

Qin, J., and X. Liu. "An approach to intuitionistic fuzzy multiple attribute decision making based on maclaurin symmetric mean operators." *Journal of Intelligent & Fuzzy Systems* 27, no. 5 (2014): 2177–2190.

Shi, L., and J. Ye. "Dombi aggregation operators of neutrosophic cubic sets for multiple attribute decision-making." *Algorithms* 11, no. 3 (2018): 29.

Smarandache, F. "A unifying field in logics: Neutrosophic logic." *Philosophy*. American Research Press, 1999: 1–141.

Smarandache, F. N. *Neutrosophic {P}robability, {S}et, and {L}ogic, {P}ro{Q}uest {I}nformation \& {L}earning*. Ann Arbor, 1998.

Wang, W., and X. Liu. "Interval-valued intuitionistic fuzzy hybrid weighted averaging operator based on Einstein operation and its application to decision making." *Journal of Intelligent & Fuzzy Systems* 25 (2013): 279–290.

Wang, W., and X. Liu. "The multi-attribute decision making method based on interval-valued intuitionistic fuzzy {E}instein hybrid weighted geometric operator." *Computers and Mathematics with Applications* 66 (2013): 1845–1856.

Wang, W., X. Liu, and Y. Qin. "Interval-valued intuitionistic fuzzy aggregation operators." *Journal of Systems Engineering and Electronics* 23, no. 4 (2012): 574–580.

Wang, H., F.Smarandache, Y. Q. Zhang, and R.Sunderrman. *Interval Neutrosophic Sets and Logic: Theory and Applications in Computing.* Phoenix, AZ: Hexis, 2005.

Wang, H., F. Smarandache, Y. Zhang, and R. Sunderraman. "Single valued neutrosophic sets." *Multispace and Multistructure* 4 (2010): 410–413.

Wang, J., Y. Yang, and L. Li. "Multi-criteria decision-making method based on single-valued neutrosophic linguistic Maclaurin symmetric mean operators." *Neural Computing and Applications* 30, no. 5 (2018): 1529–1547.

Xu, Z. "Intuitionistic fuzzy aggregation operators." *IEEE Transactions on Fuzzy* 15, no. 6 (2007): 1179–1187.

Ye, J. "A multicriteria decision-making method using aggregation operators for simplified neutrosophic sets." *Journal of Intelligent \& Fuzzy Systems* 26, no. 5 (2014): 2459–2466.

Ye, J. "Exponential operations and aggregation operators of interval neutrosophic sets and their decision making methods." *SpringerPlus* 5, no. 1 (2016): 1488–1506.

Ye, J. "Operations and aggregation method of neutrosophic cubic numbers for multiple attribute decision-making." *Soft Computing* 22 (2018): 7435–7444.

Zadeh, L. A. "Fuzzy sets." *Information and Control* 8 (1965): 338–353.

Zhang, H., J. Wang, and X. Chen. "Interval neutrosophic sets and their application in multicriteria decision making problems." *The Scientific World Journal* (2014): 1–5.

Zhao, H., Z. Xu, M. Ni, and S. Liu. "Generalized aggregation operators for intuitionistic fuzzy sets." *International Journal of Intelligent Systems* 25 (2010): 1–30.

Chapter 6

New distance measure for intuitionistic fuzzy set and its application in multi-attribute decision-making

Palash Dutta and Abhilash Kangsha Banik

6.1 INTRODUCTION: BACKGROUND AND DRIVING FORCES

Serial killers, a rare and captivating phenomenon, have drawn intense interest due to the horror they evoke and the distinctive personalities behind their violent acts (Hermann, Morrison, Sor, & Norman, 1983). Dr Hervey Cleckley's interviews detailed in "The Mask of Sanity" (Clecley, 1941) reveal how psychopaths conceal their disorder beneath a veneer of normalcy. Dr Robert Hare, a prominent figure, provided valuable insights with the Psychopathy Checklist (Hare, 2003a), marking traits like manipulation, superficiality, irresponsibility, callousness, lack of empathy, impulsivity, and violence (Gao & Raine, 2010; Hare, 2003a). About 1% of the population is considered clinically psychopathic (Fox & Levin, 2007), and Hare's assertion underscores the connection between psychopathy and criminal behaviour (Hare, 1996). The evolution of Dr Hare's Psychopathic Checklist into the Psychopathic Checklist-Revised (PCL-R) involved revisions, assessing personality traits with a score above 30 indicating psychopathy (Hare, 2003b; Norris, 2011). For a cost-effective psychopathy evaluation, the Hare Psychopathy Checklist: Screening Version (PCL-SV) was introduced (Hart, Cox, & Hare, 1995), mirroring PCL-R with scores ranging from 0 to 24. Diagnosing psychopathy introduces uncertainty due to a lack of reliable evidence. The widely used Hare PCL-R assesses various factors, including interpersonal relationships, affective responsiveness, impulsivity, and criminal behaviour. A Multi-Attribute Decision-Making (MADM) approach to the 20 checklist items facilitates a comprehensive assessment of psychopathy.

Addressing decision-making problems amid uncertainty poses challenges, prompting the development of various methodologies. Zadeh (1965) introduced Fuzzy Set (FS) Theory as an innovative approach, revolutionizing decision-making processes with its effectiveness in problem-solving. While FS proved successful, its reliance on membership degree (MD) limited its applicability. Atanosov (1986, 1999, 2012) expanded on this with the creation of Intuitionistic Fuzzy Set (IFS), incorporating non-membership

 DOI: 10.1201/9781003497219-6

degree (ND) alongside MD, where MD+ND≤1, and introducing hesitancy degree (HD). In IFS, each element is assigned MD, ND, and HD, providing a more comprehensive tool than FS. Zhang's (2014) research on Linguistic IFS (LIFS) further integrated linguistic values and IFS, offering a singular expression that combines qualitative and quantitative information. The introduction of IFS as a coping mechanism for uncertainty represents a highly inventive and versatile concept widely applied in various contexts.

IFS-based distance and similarity measures have proven useful in resolving decision-making issues with uncertainty. The distance measure (Burillo and Bustince 1996) devised with IFS has two components: MD and ND which Szmidt and Kacprzyk (2000) improvised by adding a third component, HD, and putting forth new distance measures based on it. A measurement of distance based on Jensen–Shannon divergence was proposed by Xiao (2019). In some recent studies, Chen and Deng (2020) expounded on the importance of the degree of hesitation and derived a distance measure from it. Mahanta and Panda (2021) proposed a distance measure between IFSs and applied it to the mask determination problem. By creating and using a similarity measure in the assessment of software quality, Nguyen and Chou (2021) led a fairly recent study. Garg and Rani (2021) defined a distance measure based on the square area's diagonal and the right triangle's point of intersection. Gohain et al. suggested several measures (Gohain, Dutta, Gogoi, & Chutia, 2021; Gohain, Chutia, & Dutta, 2022) that used both the difference of the maximum of the cross-evaluation factor and the difference of the minimum of the cross-evaluation factor. Using four alternative ideas of centres, including centroid, orthocentre, circumcentre, and incentre of the triangle, Garg and Rani (2022) established unique distance measures based on IFSs.

6.1.1 Motivation of the work

The existence of crime poses a challenge to societal solidity. Many killers exhibit psychopathic tendencies, but a diagnosable disorder of psychopathy necessitates thorough evaluations and research. The 20-item Psychopathic Checklist, PCL-R, developed by Hare, looks into the fundamental features and personal characteristics of psychopathy. But due to the presence of uncertainty in the form of vague information, it becomes puzzling for assessment. Further, just confining to a 3-point scale limits its applicability. Similar drawbacks could be seen in PCL-SV. This is the main reason why uncertainty arises in psychopathic analysis as a decision-making problem.

In order to handle such decision-making issues, fuzzy approaches are now relevant. In fuzzy procedures instead of crisp values, the membership values are considered, which can be more operative in handling complications concerning uncertainty. Nonetheless, with time it was understood that just considering membership degree is not adequate, and thus IFS was presented considering both membership and non-membership degrees.

Distance and similarity measures have been an imperative tool in unravelling decision-making problems. To measure distance/similarity between IFSs, many measures have been devised. The prevailing measures have shortcomings in some way or other. The use of the new distance measure for IFS in the PCL-R assessment can help in solving MADM problems by allowing for the combination of ratings expressed in linguistic forms of IFS. The proposed method can help decision-makers in the psychopathy assessment process by providing a new tool for measuring the distance between two IFSs, thereby enabling more accurate and reliable decision-making. Further, the vague information of PCL-R and PCL-SV ratings could be explained in linguistic forms. So, based on the motivations, the basic objective of our work is as follows:

i) Propose a novel distance measure based on IFS.
ii) Establish propositions based on the measure.
iii) Apply the proposed measure in a modified method for evaluating psychopathy using PCL-SV and PCL-R ratings.
iv) Demonstrate the applicability of the proposed measure through a case study involving a serial killer.

6.1.2 Structure of the work

The structure is arranged in the manner described below. The fundamental notion of psychopathy in offenders is discussed in this introduction section. Also, a broad outlook and the literature review of the existing studies have been given. The following outline is provided as follows. Preliminary definitions of FS, IFS and its operations, LIFS, distance and similarity measures, and more are covered in Section 6.2. The topic of existing distance measures is covered in Section 6.3 after that. Then a novel distance measure based on IFS has been proposed along with some properties. Drawbacks of the existing measures have been discussed showing the benefit of the proposed measure. In the next section, i.e., Section 6.4, a reformed method for solving psychopathic analysis has been proposed. Also, a case study of a serial killer is done. Then in Section 6.5, an appropriate conclusion is given. Finally, in Section 6.6, the future work/target is given

6.2 PRELIMINARIES

Some existing definitions are revised in this segment.

Definition 6.1 (Intuitionistic Fuzzy Set): (Atanassov, 1986) We consider $\mathcal{U}$ as a universal set. Then an IFS $\mathcal{I}$ is defined by $\mathcal{I} = \{\langle u_i, \mathcal{G}_{\mathcal{I}}(u_i), \mathcal{N}_{\mathcal{I}}(u_i)\rangle; u_i \in \mathcal{U}\}$, where $\mathcal{G}_{\mathcal{I}}(u_i) : \mathcal{U} \to [0,1]$ define the MD and $\mathcal{N}_{\mathcal{I}}(x) : \mathcal{U} \to [0,1]$ define the ND, such that $0 \le \mathcal{G}_{\mathcal{I}}(u_i) + \mathcal{N}_{\mathcal{I}}(u_i) \le 1$.

The HD is defined by $\mathcal{H}_{\mathcal{I}}(u_i) = 1 - [\mathcal{G}_{\mathcal{I}}(u_i) + \mathcal{N}_{\mathcal{I}}(u_i)]$ and $\mathcal{H}_{\mathcal{I}}(u_i) \in [0,1]$ such that $\mathcal{G}_{\mathcal{I}}(u_i) + \mathcal{N}_{\mathcal{I}}(u_i) + \mathcal{H}_{\mathcal{I}}(u_i) = 1$.

Definition 6.2 (Linguistic Intuitionistic Fuzzy Set): (Zhang, 2014) We consider $\mathcal{U}$ as a universal set and $s_L(u_i) \in S$. Then an LIFS L is defined by $L = \{\langle (u_i, s_L(u_i)), \mathcal{G}_L(u_i), \mathcal{N}_L(u_i) \rangle; u_i \in \mathcal{U}\}$, where $\mathcal{G}_L(u_i) : \mathcal{U} \to [0,1]$ define the MD and $\mathcal{N}_L(x) : \mathcal{U} \to [0,1]$ define the ND, such that $0 \le \mathcal{G}_L(u_i) + \mathcal{N}_L(u_i) \le 1$.

The HD is defined by $\mathcal{H}_L(u_i) = 1 - [\mathcal{G}_L(u_i) + \mathcal{N}_L(u_i)]$ and $\mathcal{H}_L(u_i) \in [0,1]$ such that $\mathcal{G}_L(u_i) + \mathcal{N}_L(u_i) + \mathcal{H}_L(u_i) = 1$.

Definition 6.3 (Chen & Deng, 2020) We consider, $\mathcal{P}, \mathcal{T}, \mathcal{W} \in$IFS ($\mathcal{U}$), where $\mathcal{U}$ is the universe of discourse. Then, D: IFS×IFS → [0,1] determines the distance between $\mathcal{P}$ and $\mathcal{T}$ if it fulfils the resulting axioms:

a. $0 \le D(\mathcal{P}, \mathcal{T}) \le 1$
b. $D(\mathcal{P}, \mathcal{T}) = 0$ iff $\mathcal{P} = \mathcal{T}$
c. $D(\mathcal{P}, \mathcal{T}) = D(\mathcal{T}, \mathcal{P})$
d. If $\mathcal{P}, \mathcal{T}, \mathcal{W} \in$IFS ($\mathcal{U}$) such that $\mathcal{P} \subseteq \mathcal{T} \subseteq \mathcal{W}$, then $D(\mathcal{P}, \mathcal{T}) \le D(\mathcal{P}, \mathcal{W})$ and $D(\mathcal{T}, \mathcal{W}) \le D(\mathcal{P}, \mathcal{W})$

6.3 DISTANCE MEASURE BASED ON IFS

The existing distance measures are discussed in the segment. Then, based on IFS, we provide an innovative distance measure. The different distance measures have been compared with our suggested measure in comparative research.

6.3.1 Existing distance measures

In Table 6.1, the existing distance measures are given.

Now, several drawbacks of the existing distance measure such as in Burillo and Bustince (1996), Szmidt and Kacprzyk (2000), Chen and Deng (2020), Xiao (2019), Mahanta and Panda (2021), Nguyen and Chou (2021), Garg and Rani (2021), Gohain et al. (2021, 2022), and Garg and Rani (2022) could be seen.

Considering the two profiles such that in Profile 1, A_1=(0.3,0.3), A_2=(0.4,0.4) and in Profile 2, B_1=(0.3,0.4), B_2=(0.4,0.3). From rational perspectives, Profile 1's IFSs appear more comparable to each other than Profile 2's as $B_1 = B_2^c$. Distance measures such as D_H, D_E, D_{NC}, and D_{MP} do not differentiate the IFSs, whereas the distance measures D_{SK1}, D_{SK2}, and D_X make the incorrect choice.

Table 6.1 Existing distance measures

Author	*Distance measure*
Burillo and Bustince (1996)	$D_H(\mathcal{P},\mathcal{T}) = [\frac{1}{2n}\sum_{i=1}^{n}(\lvert \mathcal{G}_{\mathcal{P}}(u_i) - \mathcal{G}_{\mathcal{T}}(u_i) \rvert + \lvert \mathcal{N}_{\mathcal{P}}(u_i) - \mathcal{N}_{\mathcal{T}}(u_i) \rvert)]$
	$D_E(\mathcal{P},\mathcal{T}) = [\frac{1}{2n}\sum_{i=1}^{n}\{(\mathcal{G}_{\mathcal{P}}(u_i) - \mathcal{G}_{\mathcal{T}}(u_i))^2 + (\mathcal{N}_{\mathcal{P}}(u_i) - \mathcal{N}_{\mathcal{T}}(u_i))^2\}]^{1/2}$
Szmidt and Kacprzyk (2000)	$D_{SK1}(\mathcal{P},\mathcal{T}) = [\frac{1}{2n}\sum_{i=1}^{n}(\lvert \mathcal{G}_{\mathcal{P}}(u_i) - \mathcal{G}_{\mathcal{T}}(u_i) \rvert + \lvert \mathcal{N}_{\mathcal{P}}(u_i) - \mathcal{N}_{\mathcal{T}}(u_i) \rvert$ $+ \lvert \mathcal{H}_{\mathcal{P}}(u_i) - \mathcal{H}_{\mathcal{T}}(u_i) \rvert)]$
	$D_{SK2}(\mathcal{P},\mathcal{T}) = [\frac{1}{2n}\sum_{i=1}^{n}\{(\mathcal{G}_{\mathcal{P}}(u_i) - \mathcal{G}_{\mathcal{T}}(u_i))^2 + (\mathcal{N}_{\mathcal{P}}(u_i) - \mathcal{N}_{\mathcal{T}}(u_i))^2$ $+ (\mathcal{H}_{\mathcal{P}}(u_i) - \mathcal{H}_{\mathcal{T}}(u_i))^2\}]^{1/2}$
Chen and Deng (2020)	$D^1_{CD}(\mathcal{P},\mathcal{T}) = \frac{1}{2n}\sum_{i=1}^{n}[(\lvert \mathcal{G}_{\mathcal{P}}(u_i) - \mathcal{G}_{\mathcal{T}}(u_i) \rvert + \lvert \mathcal{N}_{\mathcal{P}}(u_i) - \mathcal{N}_{\mathcal{T}}(u_i) \rvert)$ $\times (1 - \frac{1}{2}\lvert \mathcal{H}_{\mathcal{P}}(u_i) - \mathcal{H}_{\mathcal{T}}(u_i) \rvert)]$
	$D^2_{CD}(\mathcal{P},\mathcal{T}) = \frac{1}{2n}\sum_{i=1}^{n}[(\lvert \mathcal{G}_{\mathcal{P}}(u_i) - \mathcal{G}_{\mathcal{T}}(u_i) \rvert + \lvert \mathcal{N}_{\mathcal{P}}(u_i) - \mathcal{N}_{\mathcal{T}}(u_i) \rvert)$ $\times \cos(\frac{\pi}{6}\lvert \mathcal{H}_{\mathcal{P}}(u_i) - \mathcal{H}_{\mathcal{T}}(u_i) \rvert)]$
	$D^3_{CD}(\mathcal{P},\mathcal{T}) = [\frac{1}{2n}\sum_{i=1}^{n}((\mathcal{G}_{\mathcal{P}}(u_i) - \mathcal{G}_{\mathcal{T}}(u_i))^2 + (\mathcal{N}_{\mathcal{P}}(u_i) - \mathcal{N}_{\mathcal{T}}(u_i))^2)$ $\times (1 - \frac{1}{2}\lvert \mathcal{H}_{\mathcal{P}}(u_i) - \mathcal{H}_{\mathcal{T}}(u_i) \rvert)^2]^{1/2}$
Xiao (2019)	$D_X(\mathcal{P},\mathcal{T}) = \frac{1}{n}\sum_{i=1}^{n}\frac{1}{2}[\mathcal{G}_{\mathcal{P}}(u_i)\log\frac{\mathcal{G}_{\mathcal{P}}(u_i)}{\mathcal{G}_{\mathcal{P}}(u_i) + \mathcal{G}_{\mathcal{T}}(u_i)} + \mathcal{G}_{\mathcal{T}}(u_i)\log\frac{\mathcal{G}_{\mathcal{T}}(u_i)}{\mathcal{G}_{\mathcal{P}}(u_i) + \mathcal{G}_{\mathcal{T}}(u_i)}$ $+ \mathcal{N}_{\mathcal{P}}(u_i)\log\frac{\mathcal{N}_{\mathcal{P}}(u_i)}{\mathcal{N}_{\mathcal{P}}(u_i) + \mathcal{N}_{\mathcal{T}}(u_i)} + \mathcal{N}_{\mathcal{T}}(u_i)\log\frac{\mathcal{N}_{\mathcal{T}}(u_i)}{\mathcal{N}_{\mathcal{P}}(u_i) + \mathcal{N}_{\mathcal{T}}(u_i)}$ $+ \mathcal{H}_{\mathcal{P}}(u_i)\log\frac{\mathcal{H}_{\mathcal{P}}(u_i)}{\mathcal{H}_{\mathcal{P}}(u_i) + \mathcal{H}_{\mathcal{T}}(u_i)} + \mathcal{H}_{\mathcal{T}}(u_i)\log\frac{\mathcal{H}_{\mathcal{T}}(u_i)}{\mathcal{H}_{\mathcal{P}}(u_i) + \mathcal{H}_{\mathcal{T}}(u_i)}]$

Table 6.1 Existing distance measures (*Continued*)

Author	*Distance measure*
Mahanta and Panda (2021)	$D_{MP}(\mathcal{P},\mathcal{T}) = \frac{1}{n}\sum_{i=1}^{n} \frac{\lvert \mathcal{G}_\mathcal{P}(u_i)-\mathcal{G}_\mathcal{T}(u_i)\rvert + \lvert \mathcal{N}_\mathcal{P}(u_i)-\mathcal{N}_\mathcal{T}(u_i)}{\mathcal{G}_\mathcal{P}(u_i)+\mathcal{G}_\mathcal{T}(u_i)+\mathcal{N}_\mathcal{P}(u_i)+\mathcal{N}_\mathcal{T}(u_i)}$
Nguyen and Chou (2021)	$D_{NC}(\mathcal{P},\mathcal{T}) = 1-\sum_{i=1}^{n}\frac{1}{2}\left[\frac{e^{-\lvert\mathcal{G}_\mathcal{P}(u_i)-\mathcal{G}_\mathcal{T}(u_i)\rvert}-e^{-1}}{1-e^{-1}}+1-\lvert(\mathcal{N}_\mathcal{P}(u_i)-\mathcal{N}_\mathcal{T}(u_i))\rvert\right]$
Garg and Rani (2021)	$D_{GR}(\mathcal{P},\mathcal{T})=\frac{1}{2n}\sum_{i=1}^{n}\Big[\lvert(\mathcal{G}_\mathcal{P}(u_i)-\mathcal{G}_\mathcal{T}(u_i))\rvert$ $(\mathcal{N}_\mathcal{P}(u_i)-\mathcal{N}_\mathcal{T}(u_i))+(\mathcal{N}_\mathcal{P}(u_i)\mathcal{H}_\mathcal{T}(u_i)-\mathcal{N}_\mathcal{T}(u_i)\mathcal{H}_\mathcal{P}(u_i))$ $+\left\lvert\frac{+(\mathcal{H}_\mathcal{P}(u_i)-\mathcal{H}_\mathcal{T}(u_i))}{1-\mathcal{H}_\mathcal{P}(u_i)\mathcal{H}_\mathcal{T}(u_i)}\right\rvert\Big]$
Gohain et al. (2021)	$D_{\mathcal{G}1}(\mathcal{P},\mathcal{T})=\frac{1}{n}\sum_{i=1}^{n}\left[\frac{\lvert\mathcal{G}_\mathcal{P}(u_i)-\mathcal{G}_\mathcal{T}(u_i)\rvert+\lvert\mathcal{N}_\mathcal{P}(u_i)-\mathcal{N}_\mathcal{T}(u_i)\rvert}{(1-\mathcal{G}_\mathcal{P}(u_i))(1-\mathcal{G}_\mathcal{T}(u_i))+(1-\mathcal{N}_\mathcal{P}(u_i))(1-\mathcal{N}_\mathcal{T}(u_i))}\right.$ $+\frac{1}{4}\Big\{\lvert\min(\mathcal{G}_\mathcal{P}(u_i),\mathcal{N}_\mathcal{T}(u_i))-\min(\mathcal{G}_\mathcal{T}(u_i),\mathcal{N}_\mathcal{P}(u_i))\rvert$ $\left.+\lvert\max(\mathcal{G}_\mathcal{P}(u_i),\mathcal{N}_\mathcal{T}(u_i))-\max(\mathcal{G}_\mathcal{T}(u_i),\mathcal{N}_\mathcal{P}(u_i))\rvert\Big\}\right]$
Gohain et al. (2022)	$D_{\mathcal{G}2}(\mathcal{P},\mathcal{T})=\frac{5}{12n}\sum_{i=1}^{n}\left[\frac{\lvert G_p(u_i)-G_T(u_i)\rvert+\lvert N_p(u_i)-N_T(u_i)\rvert}{1+(1+G_p(u_i))(1+G_T(u_i))(1+N_p(u_i))(1+N_T(u_i))}\right.$ $+\lvert\min(G_p(u_i),N_T(u_i))-\min(G_T(u_i),N_P(u_i))\rvert$ $\left.+\lvert\max(G_p(u_i),N_T(u_i))-\max(G_T(u_i),N_P(u_i))\rvert\right]$
Garg and Rani (2022)	$D_{GRI}(\mathcal{P},\mathcal{T})=\left[\sum_{i=1}^{n}w_i\left\{\frac{\lvert(\mathcal{G}_\mathcal{P}(u_i)-\mathcal{G}_\mathcal{T}(u_i))-(\mathcal{N}_\mathcal{P}(u_i)-\mathcal{N}_\mathcal{T}(u_i))\rvert}{2}\right\}\right.$ $\times\left\{1-\frac{\mathcal{H}_\mathcal{P}(u_i)+\mathcal{H}_\mathcal{T}(u_i)}{2}\right\}$ $\left.+\sum_{i=1}^{n}w_i\left\{\frac{\lvert 2(\mathcal{N}_\mathcal{P}(u_i)-\mathcal{N}_\mathcal{T}(u_i))-(\mathcal{G}_\mathcal{P}(u_i)-\mathcal{G}_\mathcal{T}(u_i))\rvert}{3}\right\}\times\left\{\frac{\mathcal{H}_\mathcal{P}(u_i)+\mathcal{H}_\mathcal{T}(u_i)}{2}\right\}\right]$

Table 6.1 Existing distance measures (*Continued*)

Author	*Distance measure*

$$D_{GR2}(\mathcal{P},\mathcal{T}) = \left[\sum_{i=1}^{n} w_i \left\{ \frac{\left|(\mathcal{G}_{\mathcal{P}}(u_i) - \mathcal{G}_{\mathcal{T}}(u_i)) - (\mathcal{N}_{\mathcal{P}}(u_i) - \mathcal{N}_{\mathcal{T}}(u_i))\right|}{2} \right\} \right.$$
$$\times \left\{ 1 - \frac{\mathcal{H}_{\mathcal{P}}(u_i) + \mathcal{H}_{\mathcal{T}}(u_i)}{2} \right\}$$
$$\left. + \sum_{i=1}^{n} w_i \left\{ \frac{\left|3(\mathcal{N}_{\mathcal{P}}(u_i) - \mathcal{N}_{\mathcal{T}}(u_i)) - (\mathcal{G}_{\mathcal{P}}(u_i) - \mathcal{G}_{\mathcal{T}}(u_i))\right|}{4} \right\} \times \left\{ \frac{\mathcal{H}_{\mathcal{P}}(u_i) + \mathcal{H}_{\mathcal{T}}(u_i)}{2} \right\} \right]$$

$$D_{GR3}(\mathcal{P},\mathcal{T}) = \left[\sum_{i=1}^{n} w_i \left\{ \frac{\left|(\mathcal{G}_{\mathcal{P}}(u_i) - \mathcal{G}_{\mathcal{T}}(u_i)) - (\mathcal{N}_{\mathcal{P}}(u_i) - \mathcal{N}_{\mathcal{T}}(u_i))\right|}{2} \right\} \times \right.$$
$$\left\{ 1 - \frac{\mathcal{H}_{\mathcal{P}}(u_i) + \mathcal{H}_{\mathcal{T}}(u_i)}{2} \right\}$$
$$\left. + \sum_{i=1}^{n} w_i \left\{ \frac{\left|5(\mathcal{N}_{\mathcal{P}}(u_i) - \mathcal{N}_{\mathcal{T}}(u_i)) - 3(\mathcal{G}_{\mathcal{P}}(u_i) - \mathcal{G}_{\mathcal{T}}(u_i))\right|}{8} \right\} \times \left\{ \frac{\mathcal{H}_{\mathcal{P}}(u_i) + \mathcal{H}_{\mathcal{T}}(u_i)}{2} \right\} \right]$$

$$D_{GR4}(\mathcal{P},\mathcal{T}) = \left[\sum_{i=1}^{n} w_i \left\{ \frac{\left|(\mathcal{G}_{\mathcal{P}}(u_i) - \mathcal{G}_{\mathcal{T}}(u_i)) - (\mathcal{N}_{\mathcal{P}}(u_i) - \mathcal{N}_{\mathcal{T}}(u_i))\right|}{2} \right\} \right.$$
$$\times \left\{ 1 - \frac{\mathcal{H}_{\mathcal{P}}(u_i) + \mathcal{H}_{\mathcal{T}}(u_i)}{2} \right\}$$
$$\left. + \sum_{i=1}^{n} w_i \left\{ \frac{\left|\sqrt{5}(\mathcal{N}_{\mathcal{P}}(u_i) - \mathcal{N}_{\mathcal{T}}(u_i)) - (\mathcal{G}_{\mathcal{P}}(u_i) - \mathcal{G}_{\mathcal{T}}(u_i))\right|}{\sqrt{5}+1} \right\} \times \left\{ \frac{\mathcal{H}_{\mathcal{P}}(u_i) + \mathcal{H}_{\mathcal{T}}(u_i)}{2} \right\} \right]$$

Furthermore, when comparing Profiles 3 and 4, $A_3=(0.4,0.2)$, $A_4=(0.5,0.3)$, and $B_3=(0.4,0.2)$, $B_4=(0.5,0.2)$, it is evident that Profile 4's IFSs are more comparable to Profile 3's since the membership values of both profiles' IFSs follow an exact pattern, with only non-membership values differing from one another. Numerous distance measures such as D_{GR}, D_{DS}, D_{GR1}, D_{GR2}, D_{GR3}, and D_{GR4} make a wrong choice.

From the above study, we can see that the existing measures have numerous drawbacks which need to be gotten rid of. So, there is a need for new measures free from such shortcomings.

6.3.2 Novel distance measure

Now, in the previous part, we could see the drawbacks of the existing measures. One of the preliminary and sensible distance measures is the Hamming distance D_H proposed by Burillo and Bustince [4]. Based on it, a lot of modified forms of this distance measure have been developed such as D_{SK1}, D_{MP}, D_{CD1}, and many more have been developed. But each of these measures has limitations which are given above and a detailed analysis of which is given in the later sections. So, we aim to provide a novel distance measure in this part where we have added the product cross-evaluation factor to D_H.

The novel distance measure we present in this part is centred on IFS and is described as:

Definition 6.7 We consider, $\mathcal{P} = \{\langle u_i, \mathcal{G}_\mathcal{P}(u_i), \mathcal{N}_\mathcal{P}(u_i)\rangle; u_i \in \mathcal{U}\}$ and $\mathcal{T} = \{\langle u_i, \mathcal{G}_\mathcal{T}(u_i), \mathcal{N}_\mathcal{T}(u_i)\rangle;\ u_i \in \mathcal{U}\}$ are two IFSs defined in $\mathcal{U}$ where $\mathcal{U}$ is the universal set. The distance measure of $\mathcal{P}$ and $\mathcal{T}$ is defined by

$$D_A(\mathcal{P},\mathcal{T}) = \frac{1}{3n}\sum_{i=1}^{n}\Big[\left|\mathcal{G}_\mathcal{P}(u_i) - \mathcal{G}_\mathcal{T}(u_i)\right| + \left|\mathcal{N}_\mathcal{P}(u_i) - \mathcal{N}_\mathcal{T}(u_i)\right| + \left|\mathcal{G}_\mathcal{P}(u_i)\mathcal{N}_\mathcal{T}(u_i) - \mathcal{G}_\mathcal{T}(u_i)\mathcal{N}_\mathcal{P}(u_i)\right|\Big]$$

Theorem 6.1 $D_A(\mathcal{P},\mathcal{T})$ satisfies all the axioms of distance measure.

Proof: i) If $\mathcal{G}_\mathcal{P}(u_i), \mathcal{N}_\mathcal{P}(u_i), \mathcal{G}_\mathcal{T}(u_i), \mathcal{N}_\mathcal{P}(u_i), \in [0,$ then clearly $0 \le D_A(\mathcal{P},\mathcal{T}) \le 1$

ii) If $D_A(\mathcal{P},\mathcal{T}) = 0$, then

$$|\mathcal{G}_\mathcal{P}(u_i) - \mathcal{G}_\mathcal{T}(u_i)| + |\mathcal{N}_\mathcal{P}(u_i) - \mathcal{N}_\mathcal{T}(u_i)| + |\mathcal{G}_\mathcal{P}(u_i)\mathcal{N}_\mathcal{T}(u_i) - \mathcal{G}_\mathcal{T}(u_i)\mathcal{N}_\mathcal{P}(u_i)| = 0$$

$$\Rightarrow \mathcal{G}_\mathcal{P}(u_i) = \mathcal{G}_\mathcal{T}(u_i), \mathcal{N}_\mathcal{P}(u_i) = \mathcal{N}_\mathcal{T}(u_i) \Rightarrow \mathcal{P} = \mathcal{T}$$

Conversely, $\mathcal{P} = \mathcal{T}$, then $\mathcal{G}_\mathcal{P}(u_i) = \mathcal{G}_\mathcal{T}(u_i), \mathcal{N}_\mathcal{P}(u_i) = \mathcal{N}_\mathcal{T}(u_i)$.
Then, $D_A(\mathcal{P},\mathcal{T}) = 0$.
Hence, $D_A(\mathcal{P},\mathcal{T}) = 0$ iff $\mathcal{P} = \mathcal{T}$.

iii) Clearly, $D_A(\mathcal{P},\mathcal{T}) = D_A(\mathcal{T},\mathcal{P})$.

iv) If $\mathcal{P} \subseteq \mathcal{T} \subseteq \mathcal{W}$, then $\mathcal{G}_\mathcal{P}(u_i) \le \mathcal{G}_\mathcal{T}(u_i) \le \mathcal{G}_\mathcal{W}(u_i)$ and $\mathcal{N}_\mathcal{P}(u_i) \ge \mathcal{N}_\mathcal{T}(u_i) \ge \mathcal{N}_\mathcal{W}(u_i)$.

Then, $|\mathcal{G}_\mathcal{P}(x) - \mathcal{G}_\mathcal{W}(u_i)| \ge |\mathcal{G}_\mathcal{P}(u_i) - \mathcal{G}_\mathcal{T}(u_i)|, |\mathcal{N}_\mathcal{P}(u_i) - \mathcal{N}_\mathcal{W}(u_i)| \ge |\mathcal{N}_\mathcal{P}(u_i) - \mathcal{N}_\mathcal{T}(u_i)|$

and $|\mathcal{G}_\mathcal{P}(u_i)\mathcal{N}_\mathcal{W}(u_i) - \mathcal{G}_\mathcal{W}(u_i)\mathcal{N}_\mathcal{P}(u_i)| \ge |\mathcal{G}_\mathcal{P}(u_i)\mathcal{N}_\mathcal{T}(u_i) - \mathcal{G}_\mathcal{T}(u_i)\mathcal{N}_\mathcal{P}(u_i)|$

$$\Rightarrow |\mathcal{G}_\mathcal{P}(u_i) - \mathcal{G}_\mathcal{W}(u_i)| + |\mathcal{N}_\mathcal{P}(u_i) - \mathcal{N}_\mathcal{W}(u_i)| + |\mathcal{G}_\mathcal{P}(u_i)\mathcal{N}_\mathcal{W}(u_i) - \mathcal{G}_\mathcal{W}(u_i)\mathcal{N}_\mathcal{P}(u_i)|$$

$$\ge |\mathcal{G}_\mathcal{P}(u_i) - \mathcal{G}_\mathcal{T}(u_i)| + |\mathcal{N}_\mathcal{P}(u_i) - \mathcal{N}_\mathcal{T}(u_i)| + |\mathcal{G}_\mathcal{P}(u_i)\mathcal{N}_\mathcal{T}(u_i) - \mathcal{G}_\mathcal{T}(u_i)\mathcal{N}_\mathcal{P}(u_i)|$$

So, $D_A(\mathcal{P},\mathcal{W}) \ge D_A(\mathcal{P},\mathcal{T})$.

Similarly, $D_A(\mathcal{P},\mathcal{W}) \ge D_A(\mathcal{T},\mathcal{W})$.

Hence, $D_A(\mathcal{P},\mathcal{T})$ satisfies all the axioms of distance measure.

The distance formula described in **Definition 6.7** is proven in the aforementioned theorem's proof to be a distance measure.

Remark: The similarity measure of $\mathcal{P}$ and $\mathcal{T}$ is defined by $S_A(\mathcal{P},\mathcal{T}) = 1 - D_A(\mathcal{P},\mathcal{T})$.

Theorem 6.2 If $\mathcal{P} = \{a,b\}$ and $\mathcal{T} = \{c,d\}$, then $D_A(\mathcal{P}^c,\mathcal{T}) = \dfrac{|a-d|+|b-c|+|bd-ac|}{3} = D_A(\mathcal{P},\mathcal{T}^c)$.

Now, we define some basic propositions based on the proposed distance measure which are as follows:

Proposition 6.1 If $\mathcal{P} = \{a,b\}$ and $\mathcal{T} = \{b,a\}$, then

$$D_A(\mathcal{P},\mathcal{T}) = \frac{2|a-b|+|a^2-b^2|}{3}$$

Proposition 6.2 If $\mathcal{P} = \{a,a\}$, $\mathcal{T} = \{b,1-b\}$ and $\mathcal{W} = \{1-b,b\}$, then

$$D_A(\mathcal{P},\mathcal{T}) = \frac{|a-b|+|a+b-1|+|a-2ab|}{3} = D_A(\mathcal{P},\mathcal{W})$$

Proposition 6.3 If $\mathcal{P} = \{a,1-a\}, \mathcal{T} = \{b,1-b\}$, then

$$D_A(\mathcal{P},\mathcal{T}) = |a-b|$$

Proposition 6.4 If $\mathcal{P} = \{a,1-a\}$, $\mathcal{T} = \{1-a,a\}$, then

$$D_A(\mathcal{P},\mathcal{T}) = |2a-1|$$

6.3.3 Comparative study with examples

The following numerical examples and a comparison with the current approaches are being done to demonstrate the supremacy and effectiveness of the suggested distance measure. Here are some numerical examples.

Example 6.1 Consider the problem of arresting a criminal. Two offenders' evaluations are provided by the IFSs A = (0.5, 0.3) and B = (0.5, 0.2). As E = (0.4, 0.2) denote the study's expectation, the task is to locate or group the criminals according to their evaluation so that one of them is apprehended.

Table 6.2 Comparison of distance measures in Example 6.1

Distance	*D (E,A)*	*D (E, B)*	*Comment*
D_H	0.1	0.05	Distinguishable and correct
D_E	0.1	0.071	Distinguishable and correct
D_{SK1}	0.2	0.1	Distinguishable and correct
D_{SK2}	0.17	0.1	Distinguishable and correct
D_X	0.158	0.079	Distinguishable and correct
D_{NC}	0.125	0.075	Distinguishable and correct
D_{GR}	0.06	0.10	Distinguishable but incorrect
D_{CD1}	0.09	0.0475	Distinguishable and correct
D_{CD2}	0.0995	0.0499	Distinguishable and correct
D_{CD3}	0.09	0.0672	Distinguishable and correct
D_{MP}	0.1429	0.0769	Distinguishable and correct
D_{GC}	0.1038	0.0537	Distinguishable and correct
D_{GM}	0.1028	0.0520	Distinguishable and correct
D_{GR1}	0.01	0.0442	Distinguishable but incorrect
D_{GR2}	0.015	0.0412	Distinguishable but incorrect
D_{GR3}	0.0075	0.0456	Distinguishable but incorrect
D_{GR4}	0.0115	0.0433	Distinguishable but incorrect
D_A	0.073	0.04	Distinguishable and correct

Since the MDs of the IFSs in A and B are exactly in a similar pattern and the only differences can be found in ND, it stands to reason that the IFSs evaluation of offender B is less distanced from offender A. B and E's NDs are the same, while A and E's are different. From this vantage point, it makes sense and follows logic to claim that B is closer to E than A. Therefore, it seems sense that B is the intended offender rather than A. Many methods fell short of discovering this information.

Table 6.2 shows that distance measures like D_{GR}, D_{GR1}, D_{GR2}, D_{GR3}, and D_{GR4} create a seemingly illogical choice of the candidate. However, by choosing B over A, the suggested measures make a sensible choice.

Example 6.2 We look at a different issue with IFSs T = (0.3, 0.3), U = (0.4, 0.4), V = (0.3, 0.4), and W = (0.4,0.3). According to rational thinking, V and W are complements of each other, whereas T and U are comparable in nature. It becomes obvious that T and U are less apart than V and W.

It can be seen in Table 6.3, distance measures such as D_H, D_E, D_{NC}, D_{MP}, the distance values between Z and W are exactly the same as between T and U, whereas in D_{SK1}, D_{SK2}, and D_X, the distance values between V and W are less than T and U. But, in our proposed measure, distance values between V and W are more than T and U thus making a logical selection.

Table 6.3 Comparison of distance measures in Example 6.2

Distance	*D (T, U)*	*D (V,W)*	*Comment*
D_H	0.1	0.1	Indistinguishable
D_E	0.1	0.1	Indistinguishable
D_{SK1}	0.2	0.1	Distinguishable but incorrect
D_{SK2}	0.17	0.1	Distinguishable but incorrect
D_X	0.1554	0.0867	Distinguishable but incorrect
D_{NC}	0.1253	0.1253	Indistinguishable
D_{GR}	0.05	0.1214	Distinguishable and correct
D_{CD1}	0.09	0.1	Distinguishable and correct
D_{CD2}	0.0995	0.1	Distinguishable and correct
D_{CD3}	0.09	0.1	Distinguishable and correct
D_{MP}	0.1429	0.1429	Indistinguishable
D_{GC}	0.0446	0.0946	Distinguishable and correct
D_{GM}	0.0193	0.1027	Distinguishable and correct
D_{GR1}	0.01	0.1	Distinguishable and correct
D_{GR2}	0.015	0.1	Distinguishable and correct
D_{GR3}	0.0075	0.1	Distinguishable and correct
D_{GR4}	0.0115	0.1	Distinguishable and correct
D_A	0.0667	0.09	Distinguishable and correct

Example 6.3 We take into consideration a different issue with IFSs H = (0.4, 0.4), I = (0.4, 0.6), J = (0.3,0.3), and K= (0.6,0.4). It seems sense in general to assume that H and J are more comparable than H and I or H and K. It becomes apparent that the distance between H and J is closer than the distances between H and I and H and K.

In Table 6.4, for distance measures such as D_H, D_{SK1}, D_{CD1}, and D_{CD2}, the distance values could not distinguish the pairs whereas in D_{NC} and D_{MP}, the distance values could distinguish but were not logically correct. Further, D_X violated property 1 of the distance measure. But, in our proposed measure, the distance values between H and J are less than the distance values between H and I and between H and K, thus making a logical selection.

6.3.3 Discussion on the proposed measure

The above numerical and comparative study shows the drawbacks of the existing measures, whereas our proposed measure has none of these shortcomings.

In the literature, it is observed that some existing experiments verify the rationality and effectiveness of the existing similarity measures. Therefore, a comparative analysis between the proposed measure and existing approaches is conducted in this segment. The illustrative examples,

Table 6.4 Comparison of distance measures in Example 6.3

Distance	*D (H, I)*	*D (H, J)*	*D (H, K)*	*Comment*
D_H	0.1	0.1	0.1	Indistinguishable
D_E	0.141	0.1	0.141	Distinguishable and correct
D_{SK1}	0.2	0.2	0.2	Indistinguishable
D_{SK2}	0.2	0.17	0.2	Distinguishable and correct
D_X	N/A	0.1554	N/A	Property I violated
D_{NC}	0.1	0.1253	0.1434	Distinguishable but incorrect
D_{GR}	0.06	0.05	0.16	Distinguishable and correct
D_{CD1}	0.09	0.09	0.09	Indistinguishable
D_{CD2}	0.0995	0.0995	0.0995	Indistinguishable
D_{CD3}	0.1273	0.09	0.1273	Distinguishable and correct
D_{MP}	0.111	0.1429	0.111	Distinguishable but incorrect
D_{GC}	0.0885	0.0446	0.0955	Distinguishable and correct
D_{GM}	0.0988	0.0193	0.0988	Distinguishable and correct
D_{GR1}	0.1033	0.01	0.0967	Distinguishable and correct
D_{GR2}	0.1050	0.015	0.0950	Distinguishable and correct
D_{GR3}	0.1025	0.0075	0.0975	Distinguishable and correct
D_{GR4}	0.1038	0.0115	0.0962	Distinguishable and correct
D_A	0.0933	0.0667	0.0933	Distinguishable and correct

including the six non-similar pairs of IFSs, are adopted, and the data set is presented in Table 6.5. Table 6.5 shows the results of various measures and our proposed measure. In the table, several drawbacks of the existing measure can be seen which could be seen below.

Since the MD of the IFSs in Cases 1 and 2 are exactly in a similar pattern and the only differences can be found in ND, it stands to reason that the IFSs evaluation of Case 2 is less distanced from Case 1. In Case 2 NDs are the same, while in Case 2 they are different. From this vantage point, it makes sense and follows logic to claim that IFSs in Case 2 are closer to the IFSs in Case 1. Therefore, it makes sense that the distance value in Case 2 is less than that of Case 1. Many methods fell short of discovering this information. Table 6.5 shows that distance measures like D_{GR}, D_{GR1}, D_{GR2}, D_{GR3}, and D_{GR4} create a seemingly illogical choice. However, by getting the distance value in Case 2 less than that of Case 1, the suggested measures make a sensible choice.

Again, similar to the logic given for Cases 1 and 2, it makes sense to claim that the distance value in Case 3 is less than that of Case 4. Distance measures like D_{SK1}, D_{SK2}, and D_X could not distinguish between the IFSs in both cases. However, by getting the distance value in Case 3 less than that of Case 4, the suggested measures make a sensible choice.

It seems sense in general to assume that the IFSs in Case 5 are more similar than that of Case 6. It becomes apparent that the distance value of IFSs

Table 6.5 Comparison of distance measures

Distance	*Case 1*	*Case 2*	*Case 3*	*Case 4*	*Case 5*	*Case 6*
A	(0.4, 0.2)	(0.4, 0.2)	(0.4, 0.3)	(0.4, 0.3)	(0.4, 0.4)	(0.4, 0.4)
B	(0.5, 0.3)	(0.5, 0.2)	(0.5, 0.3)	(0.5, 0.2)	(0.3, 0.3)	(0.6, 0.4)
D_H	0.1	0.05	0.05	0.1	0.1	0.2
D_E	0.1	0.071	0.071	0.1	0.1	0.141
D_{SK1}	0.2	0.1	0.1	0.1	0.2	0.2
D_{SK2}	0.17	0.1	0.1	0.1	0.17	0.2
D_X	0.158	0.079	0.088	0.088	0.1554	N/A
D_{NC}	0.125	0.075	0.075	0.125	0.1253	0.1434
D_{GR}	0.06	0.10	0.09	0.121	0.05	0.16
D_{CD1}	0.09	0.0475	0.0475	0.1	0.09	0.09
D_{CD2}	0.0995	0.0499	0.0499	0.1	0.0995	0.0995
D_{CD3}	0.09	0.0672	0.0672	0.1	0.09	0.1273
D_{MP}	0.1429	0.0769	0.0667	0.1429	0.1429	0.111
D_{GC}	0.1038	0.0537	0.0501	0.1038	0.0446	0.0955
D_{GM}	0.1028	0.0520	0.0508	0.1028	0.0193	0.0988
D_{GR1}	0.01	0.0442	0.0458	0.1	0.01	0.0967
D_{GR2}	0.015	0.0412	0.0437	0.1	0.015	0.0950
D_{GR3}	0.0075	0.0456	0.0469	0.1	0.0075	0.0975
D_{GR4}	0.0115	0.0433	0.0452	0.1	0.0115	0.0962
D_A	0.073	0.04	0.043	0.09	0.0667	0.0933

in Case 5 is less than that in Case 6. In Table 6.5, distance measures such as D_H, D_{SK1}, D_{CD1}, and D_{CD2}, the distance values could not distinguish the pairs, whereas in D_{NC} and D_{MP}, the distance values could distinguish but not logically correct. Further, D_X violated property 1 of distance measure. But, in our proposed measure, the distance values in Case 5 are less than distance values in Case 6, thus making a logical selection.

Apart from this, several other drawbacks could be seen. Distance measures such as D_X produce undefined results for Profile 6. Distance measures such as D_H, D_E give the same distance values for Cases 2, 3 and 1, 4, 5, respectively. D_{SK1} gives the same distance values for Cases 1, 5, 6 and 2, 3, 4, respectively. D_{SK2} gives the same distance values for Cases 1, 5 and 2, 3, 4s respectively. D_{GR1}, D_{GR2}, D_{GR2}, D_{GR4} gives the same distance values for Cases 1, 5. D_{NC} gives the same distance values for Case 2, 3. D_{MP} gives the same distance values for Case 1, 4, 5. D_{GM}, D_{GC} gives the same distance values for Case 1, 4. D_{CD1}, D_{CD2} gives the same distance values for Case 2, 3 and 1, 5, 6, respectively. D_{CD3} gives the same distance values for Case 2, 3 and 1, 5, respectively.

As a distance measure to compare two IFS, it can be seen from the numerical examples from the previous subsection. The comparison table discussed above outperforms other measures. It can serve as a better alternative to the existing measures. Further, our proposed measure is a modified form of Hamming distance D_H proposed by Burillo and Bustince [15] based on which many other measures have been developed each with different limitations that is not so in our proposed measure.

Hence, our proposed measure can serve as a better alternative to the existing measures, which is evident from the above explanation.

6.4 APPLICATION IN PSYCHOPATHIC DIAGNOSIS

Psychopathy, characterized by traits such as manipulation, glibness, callousness, impulsivity, lack of empathy, irresponsibility, and violence, is assessed using Dr Hare's Psychopathic Checklist-Revised (PCL-R). This comprehensive 20-item scale assigns scores ranging from 0 to 3 for each trait, with a total score over 30 indicative of psychopathy. On this scale, 0 signifies the absence of any behavioural evidence, 1 indicates the presence of at least one example but inconsistent conduct, and 2 denotes the presence of a pattern of behaviour. However, the scale's limitations, including reliance on vague information and a rigid 3-point scale, may lead to oversights in nuanced behaviours. To address these drawbacks and enhance applicability and accuracy, a recommendation is made for adopting a more flexible rating scale. This adjustment allows for a nuanced evaluation, accommodating subtleties within the spectrum of observed behaviours in psychopathy assessments. The PCL-R personality traits and features are given below along with a brief explanation of each.

1. Superficial Charm and Glib (α_1): Psychopaths tend to have the ability to enchant others with their vivacious charm and charisma and be well-spoken.
2. Grandiose Self-Worth (α_2): Psychopaths exhibit extreme narcissism and egocentrism; they are unconcerned with social norms and regulations and frequently discuss their ambitious life goals.
3. Need for Stimulation (α_3): When their relationships become routine or banal, psychopaths become quickly bored and ceaselessly search for excitement. When this happens, they withdraw from their relationships.
4. Pathological Lying (α_4): Psychopaths lie regarding numerous elements of their life, no matter how insignificant, for no apparent reason at all.
5. Manipulative and cunning (α_5): As they can lie their way out of any situation, psychopaths will cover up lies after lies.
6. Lack of Guilt or Remorse (α_6): Since psychopaths frequently do not accept responsibility for their crimes and do not feel remorse for their conduct, a variety of justifications are always ready, regardless of the available evidence.

7. Shallow Affect (α_7): Psychopaths have very shallow emotional levels, and they may come across as cold, aloof, or emotionless.
8. Callousness and Lack of Empathy (α_8): Psychopaths are unable to comprehend other people's viewpoints because they are emotionally detached and devoid of emotions.
9. Parasitic Lifestyle (α_9): Individuals that lead parasitic lives are deceitful, self-centred, and reliant on others.
10. Poor Behaviour Controls (α_{10}): Psychopaths struggle with self-control, and they are prone to being easily provoked or having strong emotional reactions.
11. Promiscuous Sexual Behaviour (α_{11}): Psychopaths indulge in sexual promiscuity because it feeds their ego and gives them a sense of dominance.
12. Early Behaviour Problems (α_{12}): The majority of psychopaths start to exhibit their peculiar behaviour at a young age, and their difficulties often last throughout their whole lives.
13. Lack of Realistic or Long-Term Goals (α_{13}): Psychopaths live carelessly and don't try to make plans for their future since they are unable to adhere to a life plan.
14. Impulsivity (α_{14}): Psychopaths are self-centred and impulsive; they do not consider their options before acting.
15. Irresponsibility (α_{15}): Because they constantly find ways to get out of doing their actual task, psychopaths are unable to cling onto responsibility and lack pride in it.
16. Failure to Accept Responsibility for Own Actions (α_{16}): Psychopaths frequently commit crime and do not accept any responsibility for their actions.
17. Many Short-Term Marital Relationships (α_{17}): Relationships with other people are a struggle for psychopaths to forge, and this includes sexual relationships. Potential marriage failures result from this struggle to bond.
18. Juvenile Delinquency (α_{18}): Similar to many psychopaths who have early childhood habits, many psychopaths also engage in antisocial and criminal actions as youngsters that might result in a criminal record.
19. Revocation of Conditional Release (α_{19}): Psychopaths may violate the conditions of their probation or conditional releases as a result of a variety of offences or carelessness.
20. Criminal Versatility (α_{20}): Psychopaths frequently commit both violent and nonviolent crimes because of their impulsive tendencies, capacity for rapid boredom, and antisocial tendencies.

The PCL-SV, proposed as a swift and cost-effective method for assessing criminal psychopathy, consists of 12 items with scores ranging from 0 to 24. Table 6.6 outlines features, comparing them with equivalent PCL-R

Table 6.6 Characteristics examined in PCL-SV and the corresponding PCL-R characteristics

Sl No	PCL-SV characteristics	Corresponding PCL-R characteristics
1	*Superficial* (β_1)	(α_1)
2	*Grandiose* (β_2)	(α_2)
3	*Deceitful* (β_3)	(α_4), (α_5)
4	*Lacks remorse* (β_4)	(α_6)
5	*Lacks empathy* (β_5)	(α_7), (α_8)
6	*Doesn't accept responsibility* (β_6)	(α_{16})
7	*Impulsive* (β_7)	(α_3), (α_{14})
8	*Poor behaviour controls* (β_8)	(α_{10})
9	*Lacks goal* (β_9)	(α_9) (α_{13})
10	*Irresponsible* (β_{10})	(α_{15})
11	*Adolescent antisocial behaviour* (β_{11})	(α_{12}), (α_{18})
12	*Adult antisocial behaviour* (β_{12})	(α_{19}), (α_{20})

Table 6.7 Linguistic variables and their corresponding IF values

Linguistic labels (LV)	Corresponding IF values
VH	(1,0)
H	(0.7,0.3)
M	(0.5,0.5)
L	(0.3,0.7)
VL	(0,1)

characteristics. Notably, the PCL-SV reduces two PCL-R categories: Promiscuous Sexual Behaviour and Many Short-Term Marital Relationships. It is not intended as the sole diagnostic tool; recommended total cut scores are 12 or below (non-psychopathic), 13–17 (suggestive of psychopathy), and 18 and higher (a strong indication of psychopathy, requiring further evaluation with the complete PCL-R for confirmation). However, like the PCL-R, its application is constrained by ambiguous information and the use of only a 3-point scale.

As a result, the activities stated in IFSs background of the ideal scenario for any particular sort of crime may be utilized to establish which kind of behaviour the criminal possesses. The personality of the offender can therefore be expressed in the linguistic terms of his conduct under examination. For the majority of the time, the information gleaned from the perpetrator is nebulous in nature and might be expressed in linguistic terms. This is the primary justification for the use of distance measurements based on IFSs in the psychopathic diagnosis of offenders.6.7

From the information, an attempt is made to find whether the criminal had psychopathic behaviour. For this, the 12 characteristics of PCL-SV are taken as criteria for the determination of behaviour of the offender. The weighted vector of the characteristics is taken to be w = $(w_1, w_2, \ldots, w_{12})^T$, where $0 \leq w_i \leq 1$ and $\sum_{i=1}^{12} w_i = 1$. The importance of each activity is viewed generally as being equal, i.e., $w_i = (1/12)$; i = 1, 2, …, 12.

The characteristics are represented by IFS as follows by linguistic labels (Table 6.8):

$$P = \{(\beta_1, LV), (\beta_2, LV), (\beta_3, LV), (\beta_4, LV), (\beta_5, LV), (\beta_6, LV), (\beta_7, LV), (\beta_8, LV), (\beta_9, LV), (\beta_{10}, LV), (\beta_{11}, LV), (\beta_{12}, LV)\}$$

Further, the ideal psychopath could be represented by

$$I = \{(\beta_1, VH), (\beta_2, VH), (\beta_3, VH), (\beta_4, VH), (\beta_5, VH), (\beta_6, VH), (\beta_7, VH), (\beta_8, VH), (\beta_9, VH), (\beta_{10}, VH), (\beta_{11}, VH), (\beta_{12}, VH)\}$$

Now, Similarity is given by $S_A(P_1, I_1) = 1 - D_A(P_1, I_1)$

According to the PCL-SV checklist, getting a score below 12 out of 24 will qualify an offender as a non-psychopath. So, if $S_A(P_1, I_1) < (12/24) = 0.5$, then the offender is not a psychopath.

If $S_A(P_1, I_1) > 0.5$, then we use the PCL-R guide. For this, the 20 characteristics of PCL-R are taken as measure for defining the behaviour of the offender. The weighted vector of the characteristics is taken to be w = $(w_1, w_2, \ldots, w_{12})^T$, where $0 \leq w_i \leq 1$ and $\sum_{i=1}^{20} w_i = 1$. The importance of each activity is viewed generally as being equal, $w_i = 1/20$; I = 1, 2, …, 20.

The characteristics are represented by IFS as follows by linguistic labels (Table 6.8):

$$P = \{(\alpha_1, LV), (\alpha_2, LV), (\alpha_3, LV), (\alpha_4, LV), (\alpha_5, LV), (\alpha_6, LV), (\alpha_7, LV), (\alpha_8, LV), (\alpha_9, LV), (\alpha_{10}, LV), (\alpha_{11}, LV), (\alpha_{12}, LV), (\alpha_{13}, LV), (\alpha_{14}, LV), (\alpha_{15}, LV), (\alpha_{16}, LV), (\alpha_{17}, LV), (\alpha_{18}, LV), (\alpha_{19}, LV), (\alpha_{20}, LV)\}$$

Further, the ideal psychopath could be represented by

$$I = \{(\alpha_1, VH), (\alpha_2, VH), (\alpha_3, VH), (\alpha_4, VH), (\alpha_5, VH), (\alpha_6, VH), (\alpha_7, VH), (\alpha_8, VH), (\alpha_9, VH), (\alpha_{10}, VH), (\alpha_{11}, VH), (\alpha_{12}, VH), (\alpha_{13}, VH), (\alpha_{14}, VH), (\alpha_{15}, VH), (\alpha_{16}, VH), (\alpha_{17}, VH), (\alpha_{18}, VH), (\alpha_{19}, VH), (\alpha_{20}, VH)\}$$

Now, similarity is given by $S_A(P, I) = 1 - D_A(P, I)$.

According to the PCL-R checklist, getting a score of 30 out of 40 will qualify an offender as a psychopath. So, if $S_A(P,I) > (30/40) = 0.75$, then the offender is a psychopath.

6.4.1 Algorithm

The algorithm of the psychopathic diagnosis is described as follows:

Step 1: Express the 12 characteristics of PCL-SV in linguistic terms for the determination of behaviour of the offender.
Step 2: Plot the values of the linguistic terms to their corresponding IFS values of the offender set and ideal psychopath set based on PCL-SV.
Step 3: Use the proposed measure to find the similarity between the offender set and ideal psychopath set based on PCL-SV.
Step 4: If the value is less than 0.5, the offender is not psychopath and there is no need to continue further. If it is more than 0.5, express the 20 characteristics of PCL-R in linguistic terms for the determination of behaviour of the offender.
Step 5: Plot the values of the linguistic terms to their corresponding IFS values of the offender set and ideal psychopath set based on PCL-R.
Step 6: Use the proposed measure to find the similarity between the offender set and ideal psychopath set based on PCL-R. If the similarity values are less than 0.75, the offender is not psychopath. If it is more than 0.75, the offender is a psychopath.

6.4.2 An analysis of a psychopathic diagnostic case

For psychopathic analysis, we take into account the serial murderer John Gacy's case (Hermann, Morrison, Sor, & Norman, 1983).

This in-depth exploration delves into the tumultuous life of John Wayne Gacy, unravelling his challenging upbringing, strained relationships, and the psychological factors that led to a series of heinous crimes. Gacy's journey from a troubled youth to a notorious figure in criminal history provides profound insights into the complex web of human behaviour and the devastating consequences of unaddressed emotional turmoil.

Gacy's challenging childhood, marked by a strained relationship with his father, triggered emotional repression and efforts to conceal his homosexual feelings. Leaving home at 19, he took on various jobs and joined the Jaycees organization, where his peculiar behaviour hinted at underlying issues. Marriage to Marlynn Myers and their move to Waterloo, Iowa, seemingly portrayed a typical life. However, legal troubles emerged when Gacy faced accusations of coercing teens into sex, resulting in a guilty plea, a 10-year sentence, and eventual divorce due to brutal treatment related to his homosexual pursuits. In prison, Gacy showcased exceptional adaptability as a chef, skilfully navigating conflicts and earning favour with fellow inmates. Despite severe depression following his father's death, he endured prison life, completing 18 months of a 10-year term before being released on parole.

Post-prison, Gacy's life in Chicago unfolded with cohabitation with his mother, the establishment of a construction business, and marriage to Carole Huff. However, his disclosure of bisexuality led to divorce after four years. Gacy's double life expanded to hosting extravagant parties, entertaining children as Pogo the Clown, and engaging in various sexual activities. House renovations served as a deceptive cover for gruesome activities, targeting young boys for heinous acts. Confessing to murdering his first victim using a disturbing rope trick, he callously buried 29 victims beneath his house, concealing his violent tendencies behind lies about his sexuality.

Confessions unveiled a chilling pattern, exposing that Gacy's killings were not solely driven by sex-related motives but also occurred when victims threatened exposure. Despite pleas of innocence, Gacy showed no remorse, questioning victims' characters. Violating parole terms, he continued his heinous crimes, leaving authorities oblivious to the ongoing atrocities. Gacy intricately wove a facade of charm, grandiosity, and deceit, masking his sinister psychopathic tendencies. Projecting superficial charm, Gacy actively participated in charitable endeavours, strategically leveraging his public image for personal gain, revealing a chilling lack of personal connection to his altruistic actions. His grandiosity, evident in videotaped interviews during detention, reflected an inflated ego commensurate with his ambitions for power.

Deceit became a defining feature as Gacy engaged in pathological lying to conceal his murderous intentions. From hiding his true self from his wife to luring victims with false promises of employment, Gacy navigated a web of lies that extended even to misleading investigators and associates, maintaining a double life with alarming dexterity. His manipulative and cunning tactics, epitomized by the persona of Pogo the Clown, allowed him to establish rapport with victims' families and facilitate his gruesome crimes. Gacy's psychopathic profile revealed a profound absence of remorse and empathy, as he callously disregarded the pleas of victims and deflected blame, even questioning the innocence of a child. Shallow affect permeated his emotional responses, evident in his childhood and resilient despite personal losses. Transitioning from impulsive acts to calculated violence, Gacy's psychopathy manifested in poor behaviour controls, marked by violent outbursts and a lack of restraint following personal setbacks. A complex blend of adolescent antisocial behaviours, a lack of realistic goals, and recurring encounters with the law characterized Gacy's criminal versatility. Despite achieving success in political and business spheres, his irresponsibility in personal relationships underscored the enduring influence of psychopathy on his life. Gacy's psychopathic profile paints a disturbing picture of a man whose charming exterior concealed a malevolent and calculated psyche.

In conclusion, this study provides comprehensive insights into Gacy's life, emphasizing the intricate dynamics of human behaviour, the impact of emotional trauma, and the challenges in identifying individuals with criminal propensities. The lessons drawn contribute to a broader understanding of criminal psychology and underscore the importance of mental health interventions in society.

Based on the proposed algorithm, we analyse the psychopathic analysis as follows:

For Step 1: The characteristics are represented by IFS as follows by linguistic labels based on the list of personality traits and features evaluated in the PCL-SV (Table 6.6) as well as from the case study above. So,

$P_1 = \{(\beta_1, VH), (\beta_2, VH), (\beta_3, VH), (\beta_4, VH), (\beta_5, VH), (\beta_6, H), (\beta_7, H), (\beta_8, VH), (\beta_9, VL), (\beta_{10}, M), (\beta_{11}, H), (\beta_{12}, H)\}$

For Step 2, we plot the corresponding linguistic terms.

$P_1 = \{(1,0),(0.7,0.3),(1,0),(1,0),(1,0),(0.7,0.3),(0.7,0.3),(1,0),(0,1),(0.5,0.5), (0.7,0.3),(0.7,03)\}$

Similarly, for an ideal psychopath,

$I_1 = \{(1,0), (1,0), (1,0), (1,0), (1,0), (1,0), (1,0), (1,0), (1,0), (1,0), (1,0), (1,0)\}$

For Step 3, using the proposed similarity measure we get, $S_A(P_1, I_1) = 0.75 > 0.5$.

As per Step 4, it can be concluded that the criminal has potential psychopathic behaviour which needs further investigation by the PCL-R list. So, we consider the characteristics of the PCL-R list.

For Step 5, based on the list of personality traits and characteristics examined in the PCL-R, as well as a clear outline for each, the qualities are represented by IFS as follows by linguistic labels. So,

$P = \{(\alpha_1, VH), (\alpha_2, H), (\alpha_3, VH), (\alpha_4, VH), (\alpha_5, VH), (\alpha_6, VH), (\alpha_7, VH), (\alpha_8, VH), (\alpha_9, VL), (\alpha_{10}, VH), (\alpha_{11}, VH), (\alpha_{12}, VH), (\alpha_{13}, VL), (\alpha_{14}, M), (\alpha_{15}, M), (\alpha_{16}, H), (\alpha_{17}, VH), (\alpha_{18}, M), (\alpha_{19}, VH), (\alpha_{20}, H)\}$

i.e.,

$P = \{(1,0),(0.7,0.3),(1,0),(1,0),(1,0),(1,0),(1,0),(1,0),(0,1),(1,0),(1,0),(1,0),(0,1), (0.5,0.5),$

Also,

I = {(1,0), (1,0), (1,0), (1,0), (1,0), (1,0), (1,0), (1,0), (1,0), (1,0), (1,0), (1,0), (1,0), (1,0), (1,0), (1,0), (1,0), (1,0), (1,0), (1,0)}

For Step 6, Similarity, $S_A(P, I) = 0.78 > 0.75$.

Hence, it can be concluded that the criminal had psychopathic behaviour which is validated by the investigators investigating the case (Hermann, Morrison, Sor, & Norman, 1983).

6.4.3 Discussion on the proposed approach with the existing one

In the previous segment using the proposed approach, we get the PCL-SV value as 0.75 and PCL-R value as 0.78 which gives a quite precise evaluation

Now the existing PCL-SV based on the scores 0, 1, and 2 is given in Table 6.8.

In Table 6.8, we get the PCL-SV scores in between 14 and 19 out of 24, i.e., in between 14/24=0.58 and 19/24=0.79. Thus, we do not get a precise value and get a value in the range of 0.58–0.79.

Again, the existing PCL-R based on the scores 0, 1, and 2 is given in Table 6.9.

Similar to the PCL-SV, in Table 6.9, we get the PCL-R scores in between 28 and 31 out of 40, i.e., in between 28/40=0.70 and 31/40=0.78. Thus, we do not get a precise value and get a value in the range of 0.70–0.78.

In the findings, we can see that using the PCL-SV and PCL-R based on the scores 0, 1, and 2, we do not get a precise value. This is due to the fact

Table 6.8 Characteristics analyzed in PCL-SV as their probable scores (0, 1, and 2)

Sl No.	*PCL-SV characteristics*	*Probable scores (0, 1, 2)*
1	β_1	2
2	β_2	1, 2
3	β_3	2
4	β_4	2
5	β_5	2
6	β_6	1, 2
7	β_7	1, 2
8	β_8	2
9	β_9	0
10	β_{10}	1
11	β_{11}	0, 1
12	β_{12}	0, 1

Table 6.9 Characteristics analyzed in PCL-R as their probable scores (0, 1, and 2)

	PCL-R characteristics	*Probable scores (0, 1, 2)*
1	α_1	2
2	α_2	1,2
3	α_3	2
4	α_4	2
5	α_5	2
6	α_6	2
7	α_7	2
8	α_8	2
9	α_9	0
10	α_{10}	2
11	α_{11}	2
12	α_{12}	2
13	α_{13}	0
14	α_{14}	1
15	α_{15}	1
16	α_{16}	0,1
17	α_{17}	2
18	α_{18}	1
19	α_{19}	2
20	α_{20}	0, 1

that just confining to 3-point scale limits its applicability due to the presence of uncertainty in the form of vague information. But, using our proposed approach, we get a precise value as our approach can better handle the vague information. Thus, our proposed approach is more technically sound and logical.

6.5 CONCLUSION

Crime is a foremost concern for law enforcement due to the surge in crimes. Distance and similarity measures have been an imperative facet of decision-making. Countless distance/similarity approaches have been developed using IFS. Here, a new approach is used to describe a novel distance measure. A comparative study of our proposed measure and the prevailing ones is done. It is seen that the prevailing measures have some limits. So our proposed measure can serve as a better substitute to prevailing forms. A new technique is proposed to evaluate the psychopathic tendencies of criminals using linguistic variables that are further elaborated by a case study of a serial killer. The suggested approach can be utilized to address challenges in psychopathic analysis related to MADM. Using the method, data on the psychopathic tendencies of the serious offender could be maintained which will help in the criminal investigation process and some decisions such as bail could be taken considering the psychopathic tendencies of the offender to prevent serial crimes.

6.6 FUTURE WORK/TARGET

The field of criminal investigation presents complex decision-making challenges, often characterized by uncertainty and vague information. Fuzzy methods and their derivatives have shown promise in addressing these challenges, but further advancements are needed in this domain. The role of distance and similarity measures is particularly significant in decision-making processes, offering considerable potential for improving decision outcomes. By leveraging Intuitionistic Fuzzy Sets (IFS), there is an opportunity to develop new similarity and distance measures that can enhance decision-making capabilities. Additionally, the application of linguistic variables holds promise in explaining various mental disorders within the context of decision-making. This suggests a future scope for developing new measures and methodologies that utilize linguistic variables to effectively address decision-making problems in the field of criminal investigation.

In summary, the future direction of this research lies in the development of novel measures and methodologies that leverage fuzzy methods, IFS, and linguistic variables to tackle the complexities of decision-making in criminal investigation. These advancements have the potential to improve the accuracy and effectiveness of decision-making processes in the field, ultimately contributing to more informed and successful outcomes.

REFERENCES

Atanassov, K. T., & Stoeva, S. (1986). Intuitionistic fuzzy sets. *Fuzzy Sets and Systems*, 20(1), 87–96.

Atanassov, K. T. (1999). *Intuitionistic fuzzy sets* (pp. 1–137). Physica-Verlag HD.

Atanassov, K. T. (2012). *On intuitionistic fuzzy sets theory* (Vol. 283). Springer.

Burillo, P., & Bustince, H. (1996). Entropy on intuitionistic fuzzy sets and on interval-valud fuzzy sets. *Fuzzy Sets and Systems*, 78(3), 305–316.

Chen, C., & Deng, X. (2020). Several new results based on the study of distance measures of intuitionistic fuzzy sets. *Iranian Journal of Fuzzy Systems*, 17(2), 147–163.

Cleckley, H. (1941). *The mask of sanity*. The CV Mosby Company.

Fox, J. A., & Levin, J. (2007). Extreme killing: Understanding serial and mass murder. *Sociology*, 41(1), 174–176. http://www.jstor.org/stable/42856971

Gao, Y., & Raine, A. (2010). Successful and unsuccessful psychopaths: A neurobiological model. *Behavioral Sciences & the Law*, 28(2), 194–210.

Gohain, B., Dutta, P., Gogoi, S., & Chutia, R. (2021). Construction and generation of distance and similarity measures for intuitionistic fuzzy sets and various applications. *International Journal of Intelligent Systems*, 36(12), 7805–7838.

Gohain, B., Chutia, R., & Dutta, P. (2022). Distance measure on intuitionistic fuzzy sets and its application in decision-making, pattern recognition, and clustering problems. *International Journal of Intelligent Systems*, 37(3), 2458–2501.

Garg, H., & Rani, D. (2021). Novel similarity measure based on the transformed right-angled triangles between intuitionistic fuzzy sets and its applications. *Cognitive Computation*, 13(2), 447–465.

Garg, H., & Rani, D. (2022). Novel distance measures for intuitionistic fuzzy sets based on various triangle centers of isosceles triangular fuzzy numbers and their applications. *Expert Systems with Applications*, 191, 116228.

Hare, R. D. (1996). Psychopathy: A clinical construct whose time has come. *Criminal Justice and Behavior*, 23(1), 25–54.

Hare, R. D. (2003a). *Psychopathy checklist—Revised.* Psychological Assessment.

Hare, R. D. (2003b). *Hare PCL-R: Rating booklet.* Multi-Health Systems.

Hart, S. D., Cox, D. N., & Hare, R. D. (1995). *Hare psychopathy checklist: Screening version (PCL: SV).* Multi-Heath Systems.

Hermann, D. H., Morrison, H. L., Sor, Y., & Norman, J. A. (1983). People of the state of Illinois vs. John Gacy: The functioning of the insanity defense at the limits of the criminal law. *West Virginia Law Review*, 86, 1169.

Mahanta, J., & Panda, S. (2021). A novel distance measure for intuitionistic fuzzy sets with diverse applications. *International Journal of Intelligent Systems*, 36(2), 615–627.

Norris, C. S. (2011). *Psychopathy and gender of serial killers: A comparison using the PCL-R* (Doctoral dissertation, East Tennessee State University).

Nguyen, X. T., & Chou, S. Y. (2021). Novel similarity measures, entropy of intuitionistic fuzzy sets and their application in software quality evaluation. *Soft Computing*, 1–12.

Szmidt, E., & Kacprzyk, J. (2000). Distances between intuitionistic fuzzy sets. *Fuzzy Sets and Systems*, 114(3), 505–518.

Xiao, F. (2019). A distance measure for intuitionistic fuzzy sets and its application to pattern classification problems. *IEEE Transactions on Systems, Man, and Cybernetics: Systems*, 51(6), 3980–3992.

Zadeh, L. A. (1965). Fuzzy sets. *Information and Control*, 8(3), 338–353.

Zhang, H. (2014). Linguistic intuitionistic fuzzy sets and application in MAGDM. *Journal of Applied Mathematics*, 2014, 1–11.

Chapter 7

A novus distance measure of intuitionistic fuzzy sets

An application of seaport-dry port selection via fuzzy SWOT analysis

Arijit Mondal and Sankar Kumar Roy

7.1 INTRODUCTION

For the purposes of human cognition, physical entities in the real world are typically described in broad terms, which causes information to appear hazy and ambiguous. The cornerstone of the fuzzy set (FS) theory was developed by Zadeh (1965) to deal with fuzzy and uncertain properties in real-world applications. Following that, FS theory attracted a lot of attention from scientists (Gogoi and Chutia, 2022; Qi, 2021), yet fuzzy sets theory still has several flaws. Zadeh (1975) created Type-2 fuzzy set based on FS in 1975 to get over the difficulties of various fuzziness and uncertainty in real-world applications. However, the complexity of Type-2 fuzzy sets makes them challenging to use, greatly limiting their usefulness. Later, Atanassov (1986) proposed the intuitionistic fuzzy set (IFS) theory in 1986, which can be seen as another separately introduced expression to address some FS theory shortcomings. By including non-membership and hesitation degree functions, IFS is better able to describe phenomena than ordinary FS theory. After IFS was created, its benefits made it a useful and potent tool for handling ambiguous and inaccurate information in a variety of real-world situations (Jiang et al., 2019; Mondal et al., 2021; Ngan et al., 2018).

Distance measures, divergence measures, and similarity measures of IFSs have drawn numerous scholars (Jiang et al., 2019; Maheshwari and Srivastava, 2016; Ngan et al., 2018; Wang and Xin, 2005) from a variety of fields due to their significance as fuzzy mathematics topics. Based on the geometric interpretation of IFSs, Szmidt and Kacprzyk (2000) presented four IFS distance measurements, each of which has certain advantageous geometric characteristics. Later, these distance measurements were revised by Grzegorzewski (2004) and Szmidt and Kacprzyk (2004). In addition to this, Wang and Xin (2005), Park et al. (2009), Jiang et al. (2019), and Ngan et al. (2018) provided a number of distance and similarity measurements for IFSs. IFS divergence measurements were explored by Maheshwari and Srivastava (2016). The divergence measure can be envisaged as a special case of distance measure. A new spherical distance was developed by Yang and Chiclana (2009) and was applied in decision analysis. Hatzimichailidis

 DOI: 10.1201/9781003497219-7

et al.'s (2012) study was expanded by Luo and Zhao (2018) and was used for medical diagnostics.

After reviewing the available distance measures between IFSs, we discovered that most distances are fundamentally linear, while some of the methods cannot completely fulfil the axioms of a distance measure. One of the major drawbacks of the previous distance measures is that they were not able to encounter the interrelationship of the criteria. However, the interaction among criteria is an evident matter in practical problems. For example, in a medical diagnosis issue, the symptoms "fever" and "cough" are not independent, but rather highly dependent on one another. As a result, the existing distance measures are unable to adequately explain the judgement of decision-maker (DM) or might even produce paradoxical results. As a result, the question of how to assess the distance or resemblance between IFSs remains an intriguing one.

Motivated by the aforementioned shortcoming of previous distance measures and Chao et al. (2021), Mondal and Roy (2023), and Mondal et al. (2023d), we suggest a new distance measure termed intuitionistic fuzzy Mahalanobis distance (IFMD), where the correlation of criteria is taken into account. In this context, some statistical concepts including intuitionistic fuzzy (IF)-deviation, IF-variance, IF-covariance, and IF-correlation coefficient are first introduced. After that, in light of the CRTIC method Diakoulaki et al. (1995), we propose a criteria weight determination method to reduce decision-making risks of subjective weights. After that, the IFMD is put up in consideration of the conventional distances (Hamming, Euclidean) and the conventional Mahalanobis distance (MD) Mahalanobis (2018). Besides satisfying the axioms of a distance measure, IFMD takes the interdependency of criteria and their adaptive weights into account. These features enable IFMD to provide more accurate distance measurements compared to other distance measures.

In a true multi-criteria decision-making (MCDM) framework, the psychological behaviours of people are included in addition to the mathematical formulation (Mondal et al., 2023b,c). However, current researches rarely take these psychological elements into account in the framework for making decisions, which might lead to unsatisfactory outcomes. Recognizing that regret theory (RT) is a useful tool for expressing the psychological behaviour of DM with regard to regret and joy, we present a unique MCDM method based on RT. Additionally, in order to choose the best option, we consider the gaining and losing qualities of the entities in order to prevent the compensating of poor performance of certain criteria by good performance of other criteria. By recognizing the gaining and losing tendencies in the suggested sorting process, DM can ultimately choose multiple possibilities. In a nutshell, the major contributions of the study are as follows:

- A new distance measure of IFSs is constructed that takes attributes' adaptive weights and envisages the influence of criteria correlation.

- A novel attribute weighting method is provided that takes the interplay of criteria into account and is based on correlation and variance.
- A new MCDM method is introduced based on RT. After that, gain and loss scores are used to calculate the relative overall score of each entity, and the entity with the greatest relative overall score is selected as the optimal option for the MCDM problem.
- Two practical examples including pattern recognition and cluster analysis problems are employed to show the applicability of IFMD. Thereafter, a case study of the seaport-dry port selection problem in West Bengal is also explored. The criteria of the concerned problem are determined by fuzzy SWOT analysis. Lastly, two large-scale data sets are studied to show the diversity of the proposed method.

The rest of the study is formatted as follows: Section 7.2 reveals some preliminary concepts regarding this study. Section 7.3 presents the weight determination method and the procedure of IFMD calculation. Section 7.4 shows some applications of the proposed distance measure. Section 7.5 develops the proposed MCDM method. Section 7.6 shows the application of the MCDM method in a seaport-dry port selection problem and two large-scale data sets. Lastly, Section 7.7 demonstrates the conclusions and some future research scopes.

7.2 PREREQUISITES

Fundamental concepts regarding IFS, MD, and regret theory are described in this section.

Definition 7.1 (Atanassov, 1986) Let X be a set of points and let $x \in X$. A membership function $\mu_A(x)$ and a non-membership function $\nu_A(x)$ with domain A and range [0, 1] define an IFS A in X. Thus, an IFS can be mathematically illustrated as a triplet as follows:

$$A = \{\langle x, \mu_A(x), \nu_A(x)\rangle | x \in X\}.$$

What's more, $\pi_A(x) = 1 - \mu_A(x) - \nu_A(x)$ is said to be the hesitancy degree of x to A.

Definition 7.2 (Atanassov, 1986) Let A and B be two IFSs in the universal set X with the membership functions μ_A, μ_B, and non-membership functions ν_A, ν_B, respectively. Then, some set operations between them are defined in the following:

- Complement of A: $A^C = \{\langle x, \nu_A(x), \mu_A(x)\rangle | x \in X\}$.

- Intersection of A and B: $A \cap B = \{\langle x, \min\{\mu_A(x), \mu_B(x)\}, \max\{\nu_A(x), \nu_B(x)\}\rangle | x \in X\}$
- Union of A and B: $A \cup B = \{\langle x, \max\{\mu_A(x), \mu_B(x)\}, \min\{\nu_A(x), \nu_B(x)\}\rangle | x \in X\}$
- Equality relation: $A = B \Leftrightarrow \mu_A(x) = \mu_B(x), \nu_A(x) = \nu_B(x), \forall x \in X.$
- Inclusion relation: $A \subseteq B \Leftrightarrow \mu_A(x) \leq \mu_B(x), \nu_A(x) \geq \nu_B(x), \forall x \in X.$

Definition 7.3 (Wang and Xin (2005) Let A, B, and C be three IFSs in the universal set X. Let d be a mapping such that $d : X^I \times X^I \to \mathbb{R}$, where X^I is the set of all IFSs in X and $\mathbb{R}$ is the set of real numbers. d is said to be a distance measure of IFSs if it accomplishes the following properties:

(1) $d(A,B) \geq 0$.

(2) $d(A,B) = 0$ if and only if $A = B$.

(3) $d(A,B) = d(B,A)$.

(4) $A \subseteq B \subseteq C$, then $d(A,C) \geq d(A,B)$ and $d(A,C) \geq d(B,C)$.

Definition 7.4 Let two objects s and t be represented by two vectors $\mathbf{x}_s$ and $\mathbf{x}_t$ with n variables, and let Σ illustrate the $n \times n$ covariance matrix of n variables, then the MD between s and t is defined as:

$$d_{MD}(s,t) = \sqrt{(\mathbf{x}_s - \mathbf{x}_t)^T \Sigma^{-1} (\mathbf{x}_s - \mathbf{x}_t)},$$

where $(\mathbf{x}_s - \mathbf{x}_t)^T$ interprets the transpose of the vector $(\mathbf{x}_s - \mathbf{x}_t)$ and Σ^{-1} is the inverse of the covariance matrix Σ.

7.2.1 Regret theory

Bell (1982) and Loomes and Sugden (1982) proposed RT as an alternative decision behavioural theory. RT describes how DM feels both regret and joy while deciding between two options.

Definition 7.5 (Loomes and Sugden, 1982) The regret–rejoice function $R(\Delta u_x)$ for choosing the object x over y is defined as follows:

$$R(\Delta u_x) = 1 - e^{-\delta \Delta u_x}, 0 \leq \delta < \infty. \tag{7.1}$$

Here, u_x is the utility value of x, $\Delta u_x = u(x) - u(y)$ is the utility difference of two objects x and y, and δ is the regret aversion coefficient of DM.

Definition 7.6 (Loomes and Sugden, 1982) Let x and y be two objects and $\Delta u_x = u(x) - u(y)$, then the perceived utility value of x with respect to y is obtained as follows:

$$Pu_y(x) = u(x) + R(\Delta u_x). \tag{7.2}$$

7.2.2 Existing distance measures for IFSs

This subsection provides a quick review of six existing distance measurements for IFSs. Let A and B be two IFSs in the universal set $X = \{x_1, x_2, \ldots, x_m\}$ with the membership functions μ_A, μ_B and non-membership functions ν_A, ν_B, respectively. In what follows, the existing distance measures are mentioned.

1. The Hamming distance (Szmidt and Kacprzyk, 2000):

$$d_H(A,B) = \frac{1}{2m}\sum_{i=1}^{m}\left(\left|\mu_A(x_i) - \mu_B(x_i)\right| + \left|\nu_A(x_i) - \nu_B(x_i)\right|\right).$$

2. The Euclidean distance (Szmidt and Kacprzyk, 2000):

$$d_E(A,B) = \left[\frac{1}{2m}\sum_{i=1}^{m}\left(\left(\mu_A(x_i) - \mu_B(x_i)\right)^2 + \left(\nu_A(x_i) - \nu_B(x_i)\right)^2\right)\right]^{(1/2)}.$$

3. Szmidt and Kacprzyk's (2004) distance measure:

$$d_{SK}(A,B) = \frac{1}{m}\sum_{i=1}^{m}\left(\frac{\left|\mu_A(x_i) - \mu_B(x_i)\right| + \left|\nu_A(x_i) - \nu_B(x_i)\right| + \left|\pi_A(x_i) - \pi_B(x_i)\right|}{\left|\mu_A(x_i) - \nu_B(x_i)\right| + \left|\nu_A(x_i) - \mu_B(x_i)\right| + \left|\pi_A(x_i) - \pi_B(x_i)\right|}\right).$$

4. The Hausdorff distance measure (Grzegorzewski, 2004):

$$d_{HD}(A,B) = \frac{1}{m}\sum_{i=1}^{m}\max\left\{\left|\mu_A(x_i) - \mu_B(x_i)\right|, \left|\nu_A(x_i) - \nu_B(x_i)\right|\right\}.$$

5. Wang and Xin's (2005) distance measure:

$$d_{WX}(A,B) = \frac{1}{4m}\sum_{i=1}^{m}\begin{pmatrix}\left|\mu_A(x_i) - \mu_B(x_i)\right| + \left|\nu_A(x_i) - \nu_B(x_i)\right| \\ +2\max\left\{\left|\mu_A(x_i) - \mu_B(x_i)\right|, \left|\nu_A(x_i) - \nu_B(x_i)\right|\right\}\end{pmatrix}.$$

6. Shen et al.'s (2018) distance measure:

$$d_S(A,B)=\frac{1}{m}\sum_{i=1}^{m}\left(\frac{\left(\mu_A(x_i)-\mu_B(x_i)\right)^2+\left(\nu_A(x_i)-\nu_B(x_i)\right)^2}{2}\right)^{(1/2)}.$$

7. Maheshwari and Srivastava's (2016) divergence measure:

$$d_{MS}(A,B)=-\log_2\left(\frac{1}{2}+\frac{1}{2m}\sum_{i=1}^{m}\left(\sqrt{\mu_A(x_i)\mu_B(x_i)}\right)+\sqrt{\nu_A(x_i)\nu_B(x_i)}+\sqrt{\pi_A(x_i)\pi_B(x_i)}\right).$$

8. Park et al.'s (2009) distance measure:

$$d_P(A,B)=\frac{1}{4m}\sum_{i=1}^{m}\left(\begin{array}{l}\left|\mu_A(x_i)-\mu_B(x_i)\right|+\left|\nu_A(x_i)-\nu_B(x_i)\right|+\left|\pi_A(x_i)-\pi_B(x_i)\right|\\ +2\max\left\{\left|\mu_A(x_i)-\mu_B(x_i)\right|,\left|\nu_A(x_i)-\nu_B(x_i)\right|,\left|\pi_A(x_i)-\pi_B(x_i)\right|\right\}\end{array}\right).$$

7.3 IFDM

The relevant concepts regarding the development of MD of IFSs are discussed at first in this section.

Let us consider an MCDM problem with a set of m entities $E=\{e_1,e_2,\dots,e_m\}$ and a set of n criteria $C=\{c_1,c_2,\dots,c_n\}$. Moreover, $\tilde{z}_{st}=\langle\mu_{st},\nu_{st}\rangle$ depicts the IFN assessment value of e_s with reference to c_t. Therefore, $(\tilde{z}_{st})_{m\times n}$ symbolizes the IF decision matrix (IFDM) for the mentioned MCDM problem. Thus, the entities of the IFDM indicate IFSs defined on the criteria set C. This implies that $e_s=\{\langle c_t,\mu_{st},\nu_{st}\rangle | c_t\in C\}$ is an IFS for all s.

Example 7.1 Let us consider an example of pattern recognition problem under IF uncertainty from Jiang et al. (2019). The IFDM of three patterns under three elements is illustrated in Table 7.1. The IFS concerning the first entity is $A_1=\{\langle x_1,1.0,0.0\rangle,\langle x_2,0.8,0.0\rangle,\langle x_3,0.7,0.1\rangle\}$.

Table 7.1 The IFDM for pattern recognition (Jiang et al., 2019)

	x_1	x_2	x_3
A_1	⟨1.0,0.0⟩	⟨0.8,0.0⟩	⟨0.7,0.1⟩
A_2	⟨0.8,0.1⟩	⟨1.0,0.0⟩	⟨0.9,0.0⟩
A_3	⟨0.6,0.2⟩	⟨0.8,0.0⟩	⟨1.0,0.0⟩

7.3.1 Relevant concepts

Definition 7.7 Suppose $\{e_1, e_2, \ldots, e_m\}$ is a collection of IFSs defined on a collection of criteria $C = \{c_1, c_2, \ldots, c_n\}$ in an IFDM, then IF-mean of the IFDM is illustrated as:

$$\bar{\mathbb{M}} = \{\langle c_t, \bar{\mu}(c_t), \bar{v}(c_t)\rangle | c_t \in C\},$$

where $\bar{\mu}(c_t) = (1/m)\sum_{s=1}^{m} \mu_{st}$ and $\bar{v}(c_t) = (1/m)\sum_{s=1}^{m} v_{st}$.

Definition 7.8 Suppose $\{e_1, e_2, \ldots, e_m\}$ is a collection of IFSs defined on a collection of criteria $C = \{c_1, c_2, \ldots, c_n\}$ in an IFDM, then the IF-deviation of criterion c_t with reference to entity e_s is interpreted as:

$$Dv_{e_s}(c_t) = |\mu_{st} - \bar{\mu}(c_t)| - |v_{st} - \bar{v}(c_t)|, (s = 1, 2, \ldots, m; t = 1, 2, \ldots, n).$$

Example 7.2 In Example 7.1, the IF-mean of the IFDM is derived based on Definition 7.7 as:

$\{\langle c_1, 0.80, 0.10\rangle, \langle c_2, 0.87, 0.0\rangle, \langle c_3, 0.87, 0.03\rangle\}$. According to Definition 7.8, the IF-deviations of the first criterion with reference to three entities are computed as: $Dv_{e_1}(c_1) = 0.3, Dv_{e_2}(c_1) = 1.11 \times 10^{-16}, Dv_{e_3}(c_1) = -0.03$.

Definition 7.9 Suppose $\{e_1, e_2, \ldots, e_m\}$ is a collection of IFSs defined on a collection of criteria $C = \{c_1, c_2, \ldots, c_n\}$ in an IFDM, then the IF-variance of criteria c_t is given as:

$$Var_{IF}(c_t) = \frac{1}{m-1}\sum_{s=1}^{m}\left(Dv_{e_s}(c_t)\right)^2.$$

Example 7.3 In Example 7.1, the IF-variance of the first criterion is calculated with the help of Example 7.2 as: $(1/2)\ (0.3^2 + (1.11 \times 10^{-16})^2 + (-0.03)^2) = 0.09$.

Definition 7.10 Suppose $\{e_1, e_2, \ldots, e_m\}$ is a collection of IFSs defined on a collection of criteria $C = \{c_1, c_2, \ldots, c_n\}$ in an IFDM, then the IF-covariance of criteria c_t and c_u is defined as:

$$Cov_{IF}\left(c_t,c_u\right)=\frac{1}{m-1}\sum_{s=1}^{m}Dv_{e_s}\left(c_t\right)Dv_{e_s}\left(c_u\right).$$

Proposition 7.3.1 Suppose $\{e_1,e_2,\ldots,e_m\}$ is a collection of IFSs defined on a collection of criteria $C=\{c_1,c_2,\ldots,c_n\}$ in an IFDM, then

1. $Var_{IF}\left(c_t\right)>0,\forall c_t\in C$.
2. $Cov_{IF}\left(c_t,c_t\right)=Var_{IF}\left(c_t\right)$.
3. $Cov_{IF}\left(c_t,c_u\right)=Cov_{IF}\left(c_u,c_t\right)$.

Definition 7.11 Suppose $\{e_1,e_2,\ldots,e_m\}$ is a collection of IFSs defined on a collection of criteria $\{c_1,c_2,\ldots,c_n\}$ in an IFDM, then the IF-covariance matrix for the IFDM is provided as:

$$\Sigma_{IF}=\begin{bmatrix} Var_{IF}\left(c_1\right) & Cov_{IF}\left(c_1,c_2\right) & \cdots & Cov_{IF}\left(c_1,c_n\right)\\ Cov_{IF}\left(c_2,c_1\right) & Var_{IF}\left(c_2\right) & \cdots & Cov_{IF}\left(c_2,c_n\right)\\ \vdots & \vdots & \ddots & \vdots\\ Cov_{IF}\left(c_n,c_1\right) & Cov_{IF}\left(c_n,c_1\right) & \cdots & Var_{IF}\left(c_n\right)\end{bmatrix}.$$

Example 7.4 In Example 7.1, the IF-covariance matrix or the IFDM is computed in accordance with Definitions 7.10 and 7.11 as:

$$\Sigma_{IF}^{1}=\begin{bmatrix} 0.09 & 3.46\times10^{-18} & -0.06\\ 3.46\times10^{-18} & 0.013 & 0.01\\ -0.06 & 0.01 & 0.04\end{bmatrix}.$$

7.3.2 Attribute weight determination

In the following part, we present a novel, CRITIC (Diakoulaki et al., 1995)-based method for assessing the attribute weight. For this, the following definition is provided to illustrate the correlation coefficient between two criteria.

Definition 7.12 Suppose $\{e_1,e_2,\ldots,e_m\}$ is a collection of IFSs defined on a collection of criteria $C=\{c_1,c_2,\ldots,c_n\}$ in an IFDM, then the IF-correlation coefficient of the criteria c_t and c_u is provided as:

$$Cor_{IF}(c_t,c_u) = \frac{Cov_{IF}(c_t,c_u)}{\sqrt{Var_{IF}(c_t)}\sqrt{Var_{IF}(c_u)}}.$$

Since a larger IF-variance value reflects a stronger contrast strength of a characteristic. The IF-variance therefore illustrates the value that a criterion adds to the decision-making process. On the other side, if a criterion's IF-covariance value is lower than that of other criteria, it indicates that the criterion has a higher discordance. In order to calculate how much information criterion c_t generates, the following formula is used:

$$I(c_t) = Var_{IF}(c_t)\sum_{u=1}^{n}\left|Cor_{IF}(c_t,c_u)\right|.$$

The larger the value of $I(c_t)$, the greater is the importance of the attribute c_t to the decision-making framework. The more $I(c_t)$ is bigger, the more significant the criterion c_t is to the framework for making decisions. Finally, the following equation is used to obtain the normalized weights of the criteria:

$$w_t = \frac{I(c_t)}{\sum_{t=1}^{n} I(c_t)}. \tag{7.3}$$

Example 7.5 In Example 7.1, the values of $I(c_t)$ for the criteria are derived as $I(c_1) = 0.15, I(c_2) = 0.02, I(c_3) = 0.11$. Consequently, the weights of the criteria are calculated as: $w_1 = 0.5357$, $w_2 = 0.0714$, and $w_3 = 0.3929$.

7.3.3 IFMD calculation

With help of the prerequisites which are defined in Section 7.3.1, the IFMD between two IFSs is defined in what follows.

Definition 7.13 Suppose $\{e_1, e_2, \ldots, e_n\}$ is a collection of IFSs defined on a collection of criteria $C = \{c_1, c_2, \ldots, c_n\}$ in an IFDM. What's more, suppose $W = \{w_1, w_2, \ldots, w_n\}$ is the weight vector with respect to C. Then, the IFMD between two IFSs e_s and e_v is provided as follows:

$$\hat{d}_{MD}(e_s,e_v)=\left(\begin{array}{c}\left[\left\{\mathbf{W}\left(\left|\mu_{e_s}(\mathbf{c})-\mu_{e_v}(\mathbf{c})\right|^{\rho}+\left|\nu_{e_s}(\mathbf{c})-\nu_{e_v}(\mathbf{c})\right|^{\rho}\right)\right\}^{\left(\frac{1}{\rho}\right)}\right]^{T}\\ \Sigma_{IF}^{-1mp}\left[\left\{\mathbf{W}\left(\left|\mu_{e_s}(\mathbf{c})-\mu_{e_v}(\mathbf{c})\right|^{\rho}+\left|\nu_{e_s}(\mathbf{c})-\nu_{e_v}(\mathbf{c})\right|^{\rho}\right)\right\}^{\left(\frac{1}{\rho}\right)}\right]\end{array}\right)^{(1/2)}.$$

Here, $\left[\left\{\mathbf{W}\left(\left|\mu_{e_s}(\mathbf{c})-\mu_{e_v}(\mathbf{c})\right|^{\rho}+\left|\nu_{e_s}(\mathbf{c})-\nu_{e_v}(\mathbf{c})\right|^{\rho}\right)\right\}^{\left(\frac{1}{\rho}\right)}\right]$ represents the row vector:

$$\Big[\{w_1(\left|\mu_{e_s}(c_1)-\mu_{e_v}(c_1)\right|^{\rho}+\left|\nu_{e_s}(c_1)-\nu_{e_v}(c_1)\right|^{\rho})\}^{(1/\rho)},\{w_2(\left|\mu_{e_s}(c_2)-\mu_{e_v}(c_2)\right|^{\rho}+ \left|\nu_{e_s}(c_2)-\nu_{e_v}(c_2)\right|^{\rho})\}^{(1/\rho)},\ldots,\{w_n\left|\mu_{e_s}(c_n)-\mu_{e_v}(c_n)\right|^{\rho}+\left|\nu_{e_s}(c_n)-\nu_{e_v}(c_n)\right|^{\rho}\}^{(1/\rho)}\Big],$$

and Σ_{IF}^{-1mp} is the Moore-Penrose pseudo-inverse of the IF-covariance matrix.

Remark 7.1 One thing to keep in mind is that the inverse of the IF-covariance matrix could not always exist. Therefore, instead of employing its fundamental inverse as suggested by the original MD measure Mahalanobis (2018), the Moore-Penrose pseudo-inverse of the IF- covariance matrix is used in the calculation of IFMD.

The IFMD is further normalized in the following to eradicate the biases.

$$\hat{d}^{*}_{MD}(e_s,e_v)=\frac{\hat{d}_{MD}(e_s,e_v)}{\sqrt{\sum_{s,v}\left(\hat{d}_{MD}(e_s,e_v)\right)^2}}.$$

Remark 7.2 For $\rho=1$ and $\rho=2$, the weighted normalized IFMD is called the weighted normalized Hamming distance-based IFMD $\left(\hat{d}^{*}_{HMD}\right)$ and the weighted normalized Euclidean distance-based IFMD $\left(\hat{d}^{*}_{EMD}\right)$, respectively.

Theorem 7.1 The normalized IFMD $\hat{d}^{*}_{MD}$ satisfies the following characteristics.

(1) $0\le\hat{d}^{*}_{MD}(e_s,e_v)\le 1.$

(2) $\hat{d}^{*}_{MD}(e_s,e_v)=0$ *if and only if* $e_s=e_v$.

(3) $\hat{d}^*_{MD}(e_s, e_v) = \hat{d}^*_{MD}(e_v, e_s)$.

(4) *If three IFSs* e_s, e_t, e_v *satisfy* $e_s \subseteq e_t \subseteq e_v$, *then* $\hat{d}^*_{MD}(e_s, e_v) \geq \hat{d}^*_{MD}(e_s, e_t)$ *and* $\hat{d}^*_{MD}(e_s, e_v) \geq \hat{d}^*_{MD}(e_t, e_v)$.

Proof: The proofs of (1), (2), and (3) of Theorem 7.1 are obvious. To prove (4), we assume three IFSs e_s, e_t, e_v with $e_s \subseteq e_t \subseteq e_v$. Therefore, we get

$$\left|\mu_{e_s}(c) - \mu_{e_v}(c)\right|^p \geq \left|\mu_{e_s}(c) - \mu_{e_t}(c)\right|^p \text{ and } \left|v_{e_s}(c) - v_{e_v}(c)\right|^p \geq \left|v_{e_s}(c) - v_{e_t}(c)\right|^p.$$

Therefore, we obtain $\hat{d}_{MD}(e_s, e_v) \geq \hat{d}_{MD}(e_s, e_t)$. Moreover, $\sqrt{\sum_{s,v}\left(\hat{d}_{MD}(e_s, e_v)\right)^2}$ remains the same for all of the distance measures. Therefore, we derive $\hat{d}^*_{MD}(e_s, e_v) \geq \hat{d}^*_{MD}(e_s, e_t)$. Similarly, we can prove $\hat{d}^*_{MD}(e_s, e_v) \geq \hat{d}^*_{MD}(e_t, e_v)$. Thus, $\hat{d}^*_{MD}$ satisfies all of the properties of a distance measure.

> **Example 7.6** Consider the IFDM in Example 7.1 and another IFS $B = \{\langle x_1, 0.5, 0.3\rangle, \langle x_2, 0.6, 0.2\rangle, \langle x_3, 0.8, 0.1\rangle\}$. The sample B needs to be classified into one of the given patterns $A_1, A_2, or\ A_3$. Now, the Hamming distance-based IFMDs between the patterns and the sample are computed as: $\hat{d}^*_{HMD}(A_1, B) = 0.8386, \hat{d}^*_{HMD}(A_2, B) = 0.5080, \hat{d}^*_{HMD}(A_3, B) = 0.1967$. On the other hand, the Euclidean distance-based IFMDs between the patterns and the sample are computed as: $\hat{d}^*_{EMD}(A_1, B) = 0.6868, \hat{d}^*_{EMD}(A_2, B) = 0.5958, \hat{d}^*_{EMD}(A_3, B) = 0.4164$. The sample B should be classified into that patterns for which the distance is the smallest. From both cases, we see that B should be classified into pattern A_3.

7.4 APPLICATION OF THE PROPOSED DISTANCE MEASURE

The rationality and performance of the proposed distance measure are verified in this section by three kinds of experiments including numerical experiments, pattern recognition, and clustering problem.

> **Example 7.7 Application in pattern recognition:** Let us consider an example of pattern recognition problem under IF uncertainty from Jiang et al. (2019). There are three patterns A_1, A_2, and A_3 and a test sample B. Three patterns under three criteria along with the test sample

Table 7.2 The patterns and test sample (Jiang et al., 2019)

Pattern	Criteria		
	x_1	x_2	x_3
A_1	⟨0.34,0.34⟩	⟨0.19,0.48⟩	⟨0.02,0.12⟩
A_2	⟨0.35,0.33⟩	⟨0.20,0.47⟩	⟨0.00,0.14⟩
A_3	⟨0.33,0.35⟩	⟨0.21,0.46⟩	⟨0.01,0.13⟩
Test sample B	⟨0.37,0.31⟩	⟨0.23,0.44⟩	⟨0.04,0.10⟩

are illustrated in Table 7.2. The weights of three criteria are determined with the help of Section 7.3.2 as: w_1 = 0.3462, w_2 = 0.1923, and w_3 = 0.4615. Therefore, the classification results under different distance measures are displayed in Table 7.3.

According to the minimum value principle, the sample B should be classified into that pattern for which the distance is the smallest. In Table 7.3, we see that most of the distance measures cannot classify B into one of the patterns because the minimal distance value is equal for the two given patterns. In contrast, the proposed distance measures $\left(\hat{d}^*_{HMD}, \hat{d}^*_{EMD}\right)$ and d_{MS} classify B into pattern A_1 which also support the result of Jiang et al. (2019). On the other hand, the distance measure d_{SK} classifies B into pattern A_2, which contrasts with the outcomes of other measures. Thus, the proposed distance measures perform better than other existing distance measures for this experiment.

Example 7.8 Application in cluster analysis: The proposed distance measure can be utilized in clustering problems under IF environment. For this purpose, an example of a clustering problem from Zeshui

Table 7.3 The classification results under different distance measures

Measures	(A_1, B)	$(A2, B)$	$(A3, B)$	Classification result
d_H	0.2700	0.2700	0.2700	Cannot be obtained
d_E	0.1909	0.1909	0.2156	Cannot be obtained
d_{SK}	0.4700	0.3417	0.8311	A_2
d_{HD}	0.0300	0.0300	0.0367	Cannot be obtained
d_{WX}	0.2700	0.2700	0.3150	Cannot be obtained
d_s	0.0300	0.0300	0.0339	Cannot be obtained
d_P	0.2700	0.2700	0.3300	Cannot be obtained
d_{MS}	−2.3201	−2.3121	−2.3185	A_1
$\hat{d}^*_{HMD}$	0.2679	0.5628	0.7820	A_1
$\hat{d}^*_{EMD}$	0.2781	0.4543	0.8463	A_1

(2009) is considered here. The problem consists of five building materials: A_1 : sealant, A_2: floor varnish, A_3: wall paint, A_4: carpet, and A_5: polyvinyl chloride flooring. Moreover, each of the materials has eight criteria: $x_i, 1 \le i \le 8$. Therefore, the IFDM for the problem is given in Table 7.4.

Next, the IFSCA (Xu et al., 2008) is employed to cluster five building materials in the following steps:

Step 1: The proposed distance measure-based similarity measure, i.e., $\hat{S}^*_{HMD} = 1 - \hat{d}^*_{HMD}$ is applied to obtain a similarity matrix S as follows:

$$S = \begin{pmatrix} 1 & 0.646 & 0.430 & 0.353 & 0.638 \\ 0.646 & 1 & 0.798 & 0.569 & 0.675 \\ 0.430 & 0.798 & 1 & 0.778 & 0.599 \\ 0.353 & 0.569 & 0.778 & 1 & 0.820 \\ 0.638 & 0.675 & 0.599 & 0.820 & 1 \end{pmatrix}.$$

Step 2: After that, $S^2$is determined according to Xu et al. (2008) in the following:

$$S^2 = \begin{pmatrix} 1 & 0.646 & 0.646 & 0.638 & 0.646 \\ 0.646 & 1 & 0.798 & 0.778 & 0.675 \\ 0.646 & 0.798 & 1 & 0.778 & 0.778 \\ 0.638 & 0.778 & 0.778 & 1 & 0.820 \\ 0.646 & 0.675 & 0.778 & 0.820 & 1 \end{pmatrix}.$$

We see that $S^2 \subseteq S$ does not hold, i.e., the similarity matrix S^2 is not an equivalent similarity matrix. Therefore, we further calculate S^4 in the same way.

$$S^4 = \begin{pmatrix} 1 & 0.646 & 0.646 & 0.646 & 0.646 \\ 0.646 & 1 & 0.798 & 0.778 & 0.778 \\ 0.646 & 0.798 & 1 & 0.778 & 0.778 \\ 0.638 & 0.778 & 0.778 & 1 & 0.820 \\ 0.646 & 0.778 & 0.778 & 0.820 & 1 \end{pmatrix}.$$

Now, we find that S^4 is an equivalent similarity matrix.

Step 3: Based on the study of Xu et al. (2008), the possible clustering results according to different values of λ are depicted in Table 7.5.

Table 7.4 The IFDM for the clustering problem

	x_1	x_2	x_3	x_4	x_5	x_6	x_7	x_8
A_1	⟨0.20,0.50⟩	⟨0.10,0.80⟩	⟨0.50,0.30⟩	⟨0.90,0.00⟩	⟨0.40,0.35⟩	⟨0.10,0.90⟩	⟨0.30,0.50⟩	⟨1.00,0.00⟩
A_2	⟨0.50,0.40⟩	⟨0.60,0.15⟩	⟨1.00,0.00⟩	⟨0.15,0.65⟩	⟨0.00,0.80⟩	⟨0.70,0.15⟩	⟨0.50,0.30⟩	⟨0.65,0.20⟩
A_3	⟨0.45,0.35⟩	⟨0.60,0.30⟩	⟨0.90,0.00⟩	⟨0.10,0.80⟩	⟨0.20,0.70⟩	⟨0.60,0.20⟩	⟨0.15,0.80⟩	⟨0.20,0.65⟩
A_4	⟨1.00,0.00⟩	⟨1.00,0.00⟩	⟨0.85,0.10⟩	⟨0.75,0.15⟩	⟨0.20,0.80⟩	⟨0.15,0.85⟩	⟨0.10,0.70⟩	⟨0.30,0.70⟩
A_5	⟨0.90,0.00⟩	⟨0.90,0.10⟩	⟨0.80,0.10⟩	⟨0.70,0.20⟩	⟨0.50,0.15⟩	⟨0.30,0.65⟩	⟨0.15,0.75⟩	⟨0.70,0.30⟩

Table 7.5 The possible clustering results of five building materials

No.	*Confidence level*	*Clustering result*
1	$0 < \lambda \leq 0.646$	$\{A_1, A_2, A_3, A_4, A_5\}$
2	$0.646 < \lambda \leq 0.778$	$\{A_1\}, \{A_2, A_3, A_4, A_5\}$
3	$0.778 < \lambda \leq 0.798$	$\{A_1\}, \{A_2, A_3\}, \{A_4, A_5\}$
4	$0.798 < \lambda \leq 0.820$	$\{A_1\}, \{A_2, A_3\}, \{A_4, A_5\}$
5	$0.820 < \lambda \leq 1$	$\{A_1\}, \{A_2\}, \{A_3\}, \{A_4\}, \{A_5\}$

The number of clusters in the clustering process can be established in accordance with the real applications. Other studies (Jiang et al., 2019; Xu et al., 2008) also support the clustering outcomes of the suggested measure, demonstrating its viability.

7.5 NEW MCDM METHOD BASED ON PROPOSED DISTANCE

In this section, a novel understanding of the relative benefit scores of the entities is produced using RT-based gain and loss scores. Since an entity's worth increases as it gets further away from the worst option. As a result, the MD of an entity and the worst entity are used to define its utility. On the basis of Peng et al. (2019), the utility function $u(e_s)$ of an entity e_s is defined as:

$$u(e_s) = \frac{1 - \exp\left(-\theta \hat{d}^{*}_{MD}\left(e_s, e^{-}\right)\right)}{\theta}, \tag{7.4}$$

where θ is the risk resistance coefficient of DM satisfying $0 < \theta < 1$.

Definition 7.14 The regret–rejoice function of the entity e_s over e_t is provided as:

$$R_{e_t}(e_s) = 1 - \exp\left(-\delta \Delta u_s^t\right), \tag{7.5}$$

where $\Delta u_s^t = u(e_s) - u(e_t)$ and $0 \leq \delta < \infty$ is the regret resistance coefficient.

Definition 7.15 The perceived utility function of the entity e_s with respect to the entity e_t is illustrated as:

$$Pu_{e_t}(e_s) = u(e_s) + R_{e_t}(e_s). \tag{7.6}$$

A $m \times m$ matrix can be deduced by taking use of the perceived utility functions of the entities.

Definition 7.16 The utility matrix for the MCDM is deduced as follows:

$$\mathbb{U}_{m\times m}=\begin{bmatrix} Pu_{e1}(e_1) & Pu_{e2}(e_1) & \cdots & Pu_{em}(e_1)\\ Pu_{e1}(e_2) & Pu_{e2}(e_2) & \cdots & Pu_{em}(e_2)\\ \cdots & \cdots & \ddots & \cdots\\ Pu_{e1}(e_m) & Pu_{e2}(e_m) & \cdots & Pu_{em}(e_m)\end{bmatrix}$$

In the utility matrix $\mathbb{U}_{m\times m}$, the rows reflect the gains of the entities, while the columns represent the losses. Thus, the net gain and loss scores of the entity e_s are defined as:

$$\mathbb{G}(e_s)=\frac{1}{m}\sum\nolimits_{t=1}^{m}Pu_{e_t}(e_s),\ \mathbb{L}(e_s)=\frac{1}{m}\sum\nolimits_{t=1}^{m}Pu_{e_s}(e_t), \tag{7.7}$$

Definition 7.17 With the assistance of net gain and loss scores, the relative benefit score of entity e_s is defined in Eq. (7.8).

$$\mathbb{BS}(e_s)=\frac{\mathbb{G}(e_s)}{\mathbb{G}(e_s)+\mathbb{L}(e_s)}. \tag{7.8}$$

The relative benefit score of an entity that is defined in Eq. (7.8) represents the achievement of an entity. Therefore, the larger the value of $\mathbb{BS}(e_s)$ for an entity, the higher the priority that entity should receive. Therefore, according to the values of $\mathbb{BS}(e_s)$, the entities are ranked.

7.6 APPLICATION OF PROPOSED MCDM METHOD

In this section, the diversity and superiority of the proposed distance-based MCDM method are shown. For this purpose, firstly, a case study of seaport-dry port selection problem by SWOT analysis is investigated, then two large-scale data sets are explored. Therefore, the comparative analysis is conducted and some advantages of the proposed method are revealed.

7.6.1 Case study: seaport-dry port selection

In India, containerization is growing at an annual growth rate of 7% from 2010 to 2019, compared to the world's 4%, according to data from the World Bank. India is a peninsular nation whose coastline is home to 12 large ports and 187 intermediate and small ports. India's dry ports are referred to as inland container terminals and container goods stations. The development of the current dry ports in India has been undertaken by inland parties, shipping lines, transportation firms, or governmental organizations, mostly in order to serve the local market. However, according to the audit assessment, dry ports in India are not being used to their full potential and are at risk of closing down because of poor operations. When choosing a location for a dry port, numerous factors must be taken into account, but in India, just land availability and a market analysis are taken into account. In order to relieve congestion and lower overall transportation costs, dry ports must be properly located. An organized process comprising pertinent location criteria and their prioritization must be used to locate a dry port. Fuzzy SWOT analysis is a crucial decision-support tool that is frequently employed to carefully examine the internal and external factors of an organization. Fuzzy SWOT seeks to minimize threats and weaknesses while maximizing opportunities and strengths. In this study, we conduct a fuzzy SWOT analysis to determine the criteria for selecting optimal dry port locations out of eight possible dry port locations in West Bengal. Along the banks of the Bhagirathi River, eight potential locations are chosen: Hooghly (DP1), Khidderpore (DP2), Sonarpur (DP3), Digha (DP4), Haldia (DP5), Contai (DP6), Mandarmoni (DP7), and Kalinagar (DP8). Fuzzy SWOT analysis for these locations is shown in Figure 7.1. It should be noted that the criteria are selected according to the studies of (Nguyen and Notteboom, 2016; Ng and Gujar, 2009; Roso, 2007; Tadi´c et al., 2020). What is more, linguistic terms are used to assess the entities with respect to 15 criteria. The IFNs corresponding to the linguistic terms are adopted from Memari et al. (2019). Thereafter, the IFDM for 8 seaport-dry ports with respect to 15 criteria is published in Table 7.6. Therefore, the following steps are gone through to find the optimal dry port by the proposed MCDM method.

Step 1: The criteria W1, W2, W3, T1, T2, and T3 are of cost type; therefore, the IFDM is normalized according to Eq. (7.9).

$$\tilde{\tilde{z}} = \begin{cases} \langle \mu_{st}, \nu_{st} \rangle, & \text{for benefit criterion,} \\ \langle \nu_{st}, \mu_{st} \rangle, & \text{for cost criterion,} \end{cases} \tag{7.9}$$

Step 2: The worst entity is derived according to Eq. (7.10).

$$e^- = \left\{ \langle \mu_t^-, \nu_t^- \rangle \mid t = 1, 2, \ldots, n \right\}, \tag{7.10}$$

where $\mu_t^- = \min_s \{\mu_{st}\}, \nu_t^- = \max_s \{\nu_{st}\}$.

Step 3: According to Section 7.3.2, the weights of the criteria are obtained as:

$$\begin{aligned} &w_1 = 0.0387, w_2 = 0.0605, w_3 = 0.0692, w_4 = 0.0274, \\ &w_5 = 0.1063, w_6 = 0.0484, w_7 = 0.0493, w_8 = 0.0875, \\ &w_9 = 0.0305, w_{10} = 0.1576, w_{11} = 0.0425, w_{12} = 0.0827, \\ &w_{13} = 0.0637, w_{14} = 0.0499, w_{15} = 0.0856. \end{aligned}$$

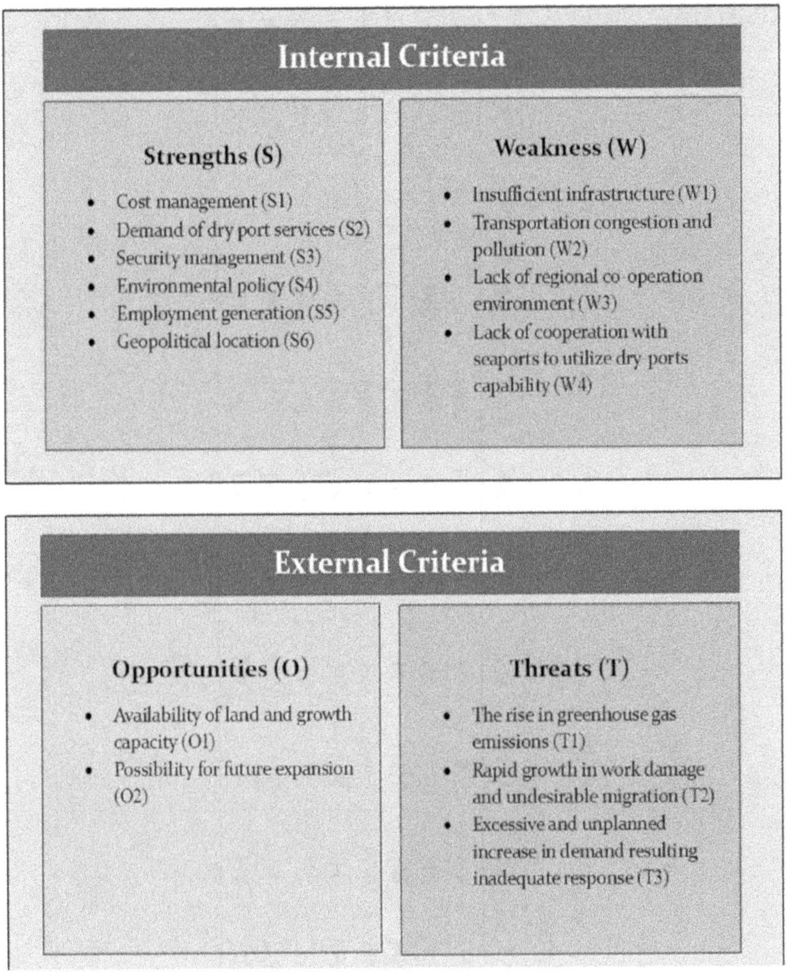

Internal Criteria

Strengths (S)

- Cost management (S1)
- Demand of dry port services (S2)
- Security management (S3)
- Environmental policy (S4)
- Employment generation (S5)
- Geopolitical location (S6)

Weakness (W)

- Insufficient infrastructure (W1)
- Transportation congestion and pollution (W2)
- Lack of regional co-operation environment (W3)
- Lack of cooperation with seaports to utilize dry ports capability (W4)

External Criteria

Opportunities (O)

- Availability of land and growth capacity (O1)
- Possibility for future expansion (O2)

Threats (T)

- The rise in greenhouse gas emissions (T1)
- Rapid growth in work damage and undesirable migration (T2)
- Excessive and unplanned increase in demand resulting inadequate response (T3)

Figure 7.1 Fuzzy SWOT analysis for dry port locations.

Step 4: By following Section 7.5, the utility matrix for the IFDM is determined as:

$$\mathbb{U}_{8\times 8} = \begin{bmatrix} 0.2952 & .02459 & 0.2386 & 0.2822 & 0.2600 & 0.3270 & 0.3425 & 0.3395 \\ 0.5026 & 0.4556 & 0.4486 & 0.4903 & 0.4691 & 0.5329 & 0.5477 & 0.5448 \\ 0.5326 & 0.4859 & 0.4790 & 0.5203 & 0.4993 & 0.5626 & 0.5773 & 0.5745 \\ 0.3511 & 0.3024 & 0.2951 & 0.3383 & 0.3163 & 0.3823 & 0.3978 & 0.3948 \\ 0.4447 & 0.3971 & 0.3899 & 0.4321 & 0.4106 & 0.4753 & 0.4904 & 0.4875 \\ 0.1551 & 0.1041 & 0.0965 & 0.1416 & 0.1187 & 0.1878 & 0.2039 & 0.2008 \\ 0.0841 & 0.0324 & 0.0247 & 0.0705 & 0.0472 & 0.1174 & 0.1338 & 0.1305 \\ 0.0979 & 0.0463 & 0.0386 & 0.0843 & 0.0610 & 0.1311 & 0.1474 & 0.1442 \end{bmatrix}$$

Table 7.6 The IFDM for seaport-dry port selection

	S1	*S2*	*S3*	*S4*	*S5*	*S6*	*W1*	*W2*	*W3*	*W4*	*O1*	*O2*	*T1*	*T2*	*T3*
DP1	H	H	MH	M	M	VH	M	H	M	MH	MH	H	M	ML	MH
DP2	VVH	VH	H	MH	H	EH	M	ML	H	VL	VH	VVH	ML	L	VVL
DP3	VH	VVH	MH	M	MH	EH	MH	M	H	L	H	VH	L	ML	VL
DP4	H	MH	H	M	MH	VVH	MH	H	VH	M	MH	H	ML	H	MH
DP5	EH	H	MH	MH	H	H	MH	H	VH	MH	H	MH	VL	L	MH
DP6	H	H	M	ML	H	VH	H	VH	MH	VH	MH	H	H	ML	MH
DP7	H	M	L	ML	VL	H	VH	VVH	MH	EH	M	ML	VVL	M	VL
DP8	H	MH	M	MH	L	VH	H	VH	MH	VH	H	M	L	MH	H

Therefore, the net gain scores of the entities of the entities are calculated as:

$$\mathbb{G}(e_1)=0.2914,\mathbb{G}(e_2)=0.4990,\mathbb{G}(e_3)=0.5289,\mathbb{G}(e_4)=0.3473,\mathbb{G}(e_5)$$
$$=0.4409,\mathbb{G}(e_6)=0.1511,\mathbb{G}(e_7)=0.0801,\mathbb{G}(e_8)=0.0938$$

while the loss scores of them are computed as:

$$\mathbb{L}(e_1)=0.3079,\mathbb{L}(e_2)=0.2587,\mathbb{L}(e_3)=0.2514,\mathbb{L}(e_4)=0.2950,$$
$$\mathbb{L}(e_5)=0.2728,\mathbb{L}(e_6)=0.3396,\mathbb{L}(e_7)=0.3551,\mathbb{L}(e_8)=0.3512.$$

Step 5: Finally, the relative benefit scores of the entities are calculated according to Eq. (7.8) as:

$$\mathbb{BS}(e_1)=0.4862,\mathbb{BS}(e_2)=0.6585,\mathbb{BS}(e_3)=0.6779,\mathbb{BS}(e_4)=0.5407,$$
$$\mathbb{BS}(e_5)=0.6178,\mathbb{BS}(e_6)=0.3079,\mathbb{BS}(e_7)=0.1840,\mathbb{BS}(e_8)=0.2104.$$

Hence, the ranking order of the dry ports is DP3 $\succ$ DP2 $\succ$ DP5 $\succ$ DP4 $\succ$ DP1 $\succ$ DP6 $\succ$ DP8 $\succ$ DP7. Thus, DP3 is the best entity, i.e., Sonarpur is the best location for establishing dry port, followed by Khidderpore.

7.6.1.1 Comparative analysis

The performance of the proposed method is verified by conducting a comparative analysis with four existing methods, namely TOPSIS (Memari et al., 2019; VIKOR Krishankumar et al., 2020; TODIM Zhang et al., 2022; PROMETHEE Krishankumar et al., 2017). The results of different methods are revealed in Table 7.7.

From the results, we see that the optimal entity is the same for all methods except for TODIM. However, there are distinctions in the ranking of other entities. Since the proposed method's computing process is entirely distinct from that of previous methods. For instance, in TOPSIS, distances from the best and the worst entities are calculated first, and then based on the closeness coefficient, the entities are ranked. On the other hand, in VIKOR, the entities are ranked according to the values of individual regret, group utility, and compromise degree. However, in TODIM, first, the dominance degrees of the entities are calculated and then according to the

Table 7.7 Comparative analysis among some existing methods

Method	*Ranking order*
Proposed method	*DP3>DP2>DP5>DP4>DP1>DP6>DP8>DP7*
TOPSIS (Memari et al., 2019),	*DP3>DP2>DP5>DP4>DP1>DP8>DP6>DP7*
VIKOR (Krishankumaret al., 2020),	*DP3>DP2>DP5>DP1>DP4>DP6>DP8>DP7*
TODIM (Zhang et al., 2022)	*DP2>DP3>DP5>DP1>DP6>DP4>DP8>DP7*

overall values of the entities, they are ranked. Lastly, in PROMETHEE, the entities are ranked according to their preference degrees and the net flows.

7.6.2 Exploration of two large data sets

This section discusses the findings after augmenting two more data sets from the KEEL and UCI databases. The *South African Heart* data set (https://sci2s.ugr.es/keel/datasets.php) consists of 462 entities and 8 criteria including systolic blood pressure, cumulative tobacco (kg), low-density lipoprotein cholesterol, adiposity, type-A behaviour, obesity, current alcohol consumption, and age at onset. On the other side, the *Blood Transfusion* data set (https://archive.ics.uci.edu/ml/datasets/Blood+Transfusion +Servic e+Center) has 748 entities and four criteria including R, F, M, and T. The ranking of five different methods under two data sets is shown in Figure 7.2.

7.6.2.1 *Comparative analysis*

From Figures 7.2 and 7.3, the following conclusions are made:

(i) Form Figure 7.2, we see that although different methods provide different optimal entities for the *South African Heart* case, the overall fitting degree between the proposed method and other methods is high. Particularly, the optimal entity of the proposed method (e_{92}) is 2nd, 4th, 18th, and 6th in TOPSIS, VIKOR, TODIM, and PROMETHEE, respectively. This indicates that the ranking result of the proposed method has not been turned around. Furthermore, Figure 7.3 shows that the Spearman correlation coefficients (SCCs)

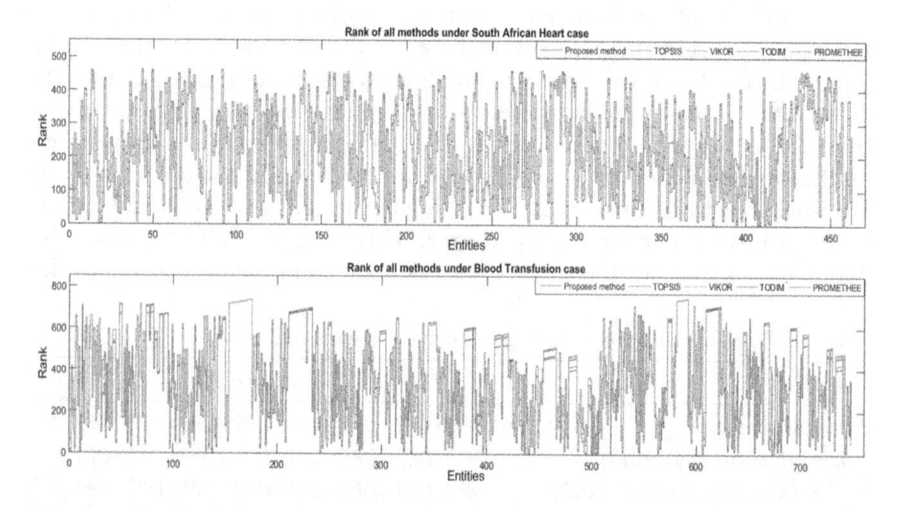

Figure 7.2 Ranking results of different methods under two data sets.

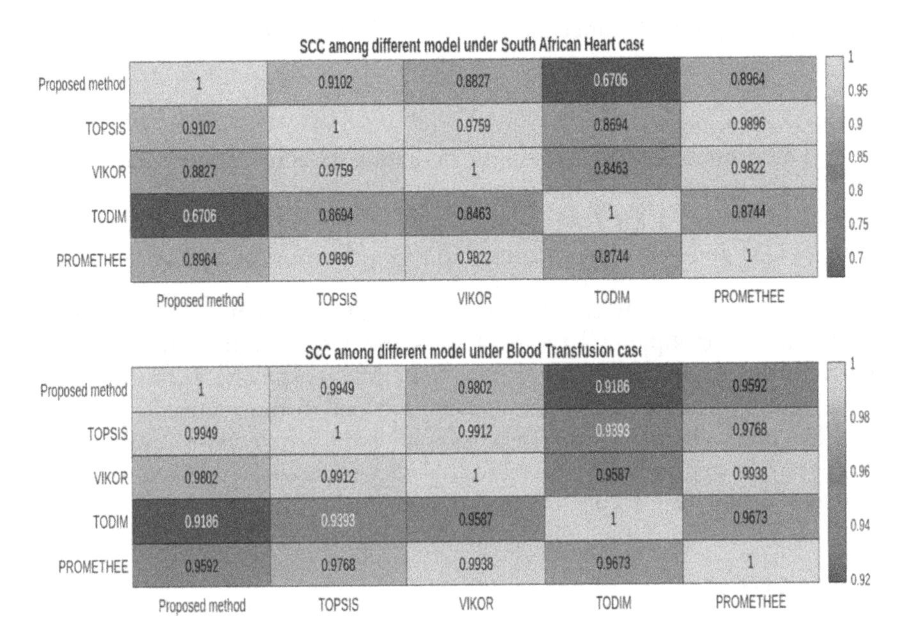

Figure 7.3 SCCs among different methods under two data sets.

between the proposed method and other methods are greater than 0.67. This proves that the proposed method and other methods have a strong correlation.

(ii) Figure 7.2 also demonstrates that the optimal entity for the *Blood Transfusion* in the proposed method (e_{138}) coincides with TOPSIS and is 4th, 9th, and 2nd in VIKOR, TODIM, and PROMETHEE. Moreover, according to Figure 7.3, the SCCs between the proposed method and other methods are greater than 0.92. This implies a strong relationship between the proposed method and other methods.

7.7 CONCLUSION

This study has aimed to encounter interaction among criteria while measuring distance between IFSs. For this purpose, a new distance measure of IFS, namely, IFMD has been put forward. In doing so, some statistical features of IFS have been introduced. After that, a novel method for determining criteria weights has been developed based on the CRITIC method. It has been seen that several existing distance measures have been unable to distinguish different IFSs, while the proposed distance measure has been able to do so. Thereafter, a new MCDM method with RT has been developed based on the proposed distance measure. Hence, the key advantages of the proposed study are as follows:

- The proposed distance measure takes the interrelationship among criteria into account.
- The proposed distance measure satisfies all of the axioms for a distance measure.
- The proposed criteria weight calculation method reduces the subjective risk in decision -making.
- The proposed MCDM method involves the psychological behaviour of DM and considers gaining and losing features of entities.

In the future, the study can be extended to some other uncertain environments like q-orthopair fuzzy set, Type-2 fuzzy set, picture fuzzy set, etc. Furthermore, the psychological behaviour of DM can be encountered in the distance measure in the future. The proposed distance measure and the MCDM method can be applied to supply chain problems (Mondal et al., 2023a; Mondal and Roy, 2021, 2022), decision-making problems (Mondal and Roy, 2023), and so on.

REFERENCES

Atanassov, K. (1986). Intuitionistic fuzzy sets. *Fuzzy Sets and Systems*, 20(1):87–96.

Bell, D. E. (1982). Regret in decision making under uncertainty. *Operations Research*, 30(5):961–981.

Chao, K., Zhao, H., and Xu, Z. (2021). Hesitant Mahalanobis distance with applications to estimating the optimal number of clusters. *International Journal of Intelligent Systems*, 36(9):5264–5306.

Diakoulaki, D., Mavrotas, G., and Papayannakis, L. (1995). Determining objective weights in multiple criteria problems: The critic method. *Computers & Operations Research*, 22(7):763–770.

Gogoi, M. K. and Chutia, R. (2022). Fuzzy risk analysis based on a similarity measure of fuzzy numbers and its application in crop selection. *Engineering Applications of Artificial Intelligence*, 107:104517.

Grzegorzewski, P. (2004). Distances between intuitionistic fuzzy sets and/or interval-valued fuzzy sets based on the Hausdorff metric. *Fuzzy Sets and Systems*, 148(2):319–328.

Hatzimichailidis, A. G., Papakostas, G. A., and Kaburlasos, V. G. (2012). A novel distance measure of intuitionistic fuzzy sets and its application to pattern recognition problems. *International Journal of Intelligent Systems*, 27(4):396–409.

Jiang, Q., Jin, X., Lee, S.-J., and Yao, S. (2019). A new similarity/distance measure between intuitionistic fuzzy sets based on the transformed isosceles triangles and its applications to pattern recognition. *Expert Systems with Applications*, 116:439–453.

Krishankumar, R., Premaladha, J., Ravichandran, K., Sekar, K., Manikandan, R., and Gao, X. (2020). A novel extension to VIKOR method under intuitionistic fuzzy context for solving personnel selection problem. *Soft Computing*, 24:1063–1081.

Krishankumar, R., Ravichandran, K., and Saeid, A. B. (2017). A new extension to PROMETHEE under intuitionistic fuzzy environment for solving supplier selection problem with linguistic preferences. *Applied Soft Computing*, 60:564–576.

Loomes, G. and Sugden, R. (1982). Regret theory: An alternative theory of rational choice under uncertainty. *The Economic Journal*, 92(368):805–824.

Luo, M. and Zhao, R. (2018). A distance measure between intuitionistic fuzzy sets and its application in medical diagnosis. *Artificial Intelligence in Medicine*, 89:34–39.

Mahalanobis, P. C. (2018). On the generalized distance in statistics. *Sankhya: The Indian Journal of Statistics, Series A (2008-)*, 80:S1–S7.

Maheshwari, S. and Srivastava, A. (2016). Study on divergence measures for intuitionistic fuzzy sets and its application in medical diagnosis. *Journal of Applied Analysis and Computation*, 6(3):772–789.

Memari, A., Dargi, A., Jokar, M. R. A., Ahmad, R., and Rahim, A. R. A. (2019). Sustainable supplier selection: A multi-criteria intuitionistic fuzzy TOPSIS method. *Journal of Manufacturing Systems*, 50:9–24.

Mondal, A., Giri, B. K., and Roy, S. K. (2023a). An integrated sustainable bio-fuel and bio-energy supply chain: A novel approach based on DEMATEL and fuzzy-random robust flexible programming with me measure. *Applied Energy*, 343:121225.

Mondal, A. and Roy, S. K. (2021). Multi-objective sustainable opened-and closed-loop supply chain under mixed uncertainty during covid-19 pandemic situation. *Computers & Industrial Engineering*, 159:107453.

Mondal, A. and Roy, S. K. (2022). Application of Choquet integral in interval type-2 Pythagorean fuzzy sustainable supply chain management under risk. *International Journal of Intelligent Systems*, 37(1):217–263.

Mondal, A., Roy, S. K., and Midya, S. (2021). Intuitionistic fuzzy sustainable multi-objective multi-item multi-choice step fixed-charge solid transportation problem. *Journal of Ambient Intelligence and Humanized Computing*, 14, 6975–6999.

Mondal, A., Roy, S. K., and Pamucar, D. (2023b). Regret-based three-way decision making with possibility dominance and SPA theory in incomplete information system. *Expert Systems with Applications*, 211:118688.

Mondal, A., Roy, S. K., and Zhan, J. (2023c). A reliability-based consensus model and regret theory-based selection process for linguistic hesitant-Z multi-attribute group decision making. *Expert Systems with Applications*, 228:120431.

Mondal, A., Roy, S. K., and Deveci, M. (2023d). Regret-based domination and prospect-based scoring in three-way decision making using q-rung orthopair fuzzy Mahalanobis distance. *Artificial Intelligence Review*, 56(Suppl 2):2311–2348.

Mondal, A., and Roy, S. K. (2023). Behavioral three-way decision making with Fermatean fuzzy Mahalanobis distance: Application to the supply chain management problems. *Applied Soft Computing*, 111182, 151.

Ng, A. Y. and Gujar, G. C. (2009). Government policies, efficiency and competitiveness: The case of dry ports in india. *Transport Policy*, 16(5):232–239.

Ngan, R. T., Cuong, B. C., Ali, M., et al. (2018). H-max distance measure of intuitionistic fuzzy sets in decision making. *Applied Soft Computing*, 69:393–425.

Nguyen, L. C. and Notteboom, T. (2016). A multi-criteria approach to dry port location in developing economies with application to Vietnam. *The Asian Journal of Shipping and Logistics*, 32(1):23–32.

Park, J.-H., Lim, K.-M., and Kwun, Y.-C. (2009). Distance measure between intuitionistic fuzzy sets and its application to pattern recognition. *Journal of the Korean Institute of Intelligent Systems*, 19(4):556–561.

Peng, H.-G., Shen, K.-W., He, S.-S., Zhang, H.-Y., and Wang, J.-Q. (2019). Investment risk evaluation for new energy resources: An integrated decision support model based on regret theory and ELECTRE III. *Energy Conversion and Management*, 183:332–348.

Qi, Z. (2021). An improved similarity measure for generalized trapezoidal fuzzy numbers and its application in the classification of eeg signals. *International Journal of Fuzzy Systems*, 23:890–905.

Roso, V. (2007). Evaluation of the dry port concept from an environmental perspective: A note. *Transportation Research Part D: Transport and Environment*, 12(7):523–527.

Shen, F., Ma, X., Li, Z., Xu, Z., and Cai, D. (2018). An extended intuitionistic fuzzy TOPSIS method based on a new distance measure with an application to credit risk evaluation. *Information Sciences*, 428:105–119.

Szmidt, E. and Kacprzyk, J. (2000). Distances between intuitionistic fuzzy sets. *Fuzzy Sets and Systems*, 114(3):505–518.

Szmidt, E. and Kacprzyk, J. (2004). A similarity measure for intuitionistic fuzzy sets and its application in supporting medical diagnostic reasoning. In *Artificial Intelligence and Soft Computing-ICAISC 2004: 7th International Conference*, Zakopane, Poland, June 7–11, 2004. *Proceedings 7*, pages 388–393. Springer.

Tadi´c, S., Krsti´c, M., Roso, V., and Brnjac, N. (2020). Dry port terminal location selection by applying the hybrid grey MCDM model. *Sustainability*, 12(17):6983.

Wang, W. and Xin, X. (2005). Distance measure between intuitionistic fuzzy sets. *Pattern Recognition Letters*, 26(13):2063–2069.

Xu, Z., Chen, J., and Wu, J. (2008). Clustering algorithm for intuitionistic fuzzy sets. *Information Sciences*, 178(19):3775–3790.

Yang, Y. and Chiclana, F. (2009). Intuitionistic fuzzy sets: Spherical representation and distances. *International Journal of Intelligent Systems*, 24(4):399–420.

Zadeh, L. (1965). Fuzzy sets. *Information Control*, 8:338–353.

Zadeh, L. A. (1975). The concept of a linguistic variable and its application to approximate reasoning—i. *Information Sciences*, 8(3):199–249.

Zeshui, X. (2009). Intuitionistic fuzzy hierarchical clustering algorithms. *Journal of Systems Engineering and Electronics*, 20(1):90–97.

Zhang, Z., Guo, J., Zhang, H., Zhou, L., and Wang, M. (2022). Product selection based on sentiment analysis of online reviews: An intuitionistic fuzzy TODIM method. *Complex & Intelligent Systems*, 8(4):3349–3362.

Chapter 8

Maclaurin symmetric mean operator-based MADM approach for Type-2 intuitionistic fuzzy sets

Kaushik Debnath and Sankar Kumar Roy

8.1 INTRODUCTION

Multi-attribute decision-making (MADM) is a process of ranking a finite set of alternatives based on the attributiv values, which can be tangible or intangible. MADM holds paramount significance within the realm of decision sciences, as it empowers decision-makers to enhance their decision-making prowess by meticulously weighing and incorporating a multitude of crucial factors. Due to its versatility, it has rapidly emerged as one of the most extensively employed methodologies across a diverse range of disciplines in recent years. Aggregation of attribute values is a critical issue in MADM methods, as it influences the overall decision outcome. Additionally, in real decision-making scenarios, the assessments provided by decision-makers are often imprecise and uncertain due to the subjective nature of human thinking. To achieve accurate outcomes, it becomes imperative to effectively address the inherent impreciseness and uncertainty inherent in the information data. To counter it, Zadeh (1965) introduced fuzzy sets (FSs), which consist of membership degree (MD) to tackle uncertainty in the information. After that, Atanassov (1986) extended the concept of FSs to include MD and non-membership degree (NMD). These sets are called intuitionistic fuzzy sets (IFSs). Later, several other extensions have been developed by researchers and utilized in different types of decision-making problems (Debnath and Roy, 2023; Giri and Roy, 2022; Mondal et al., 2023).

In practical situations, decision-makers (DMs) often struggle to accurately determine the membership degree of input data due to various constraints such as limited time or lack of precision. To overcome these challenges, Type-2 fuzzy sets (T2FSs) are employed as an alternative to FSs (Type-1 FSs) in formulating decision-making problems. It was introduced by Zadeh (1975) as an extension of Type-1 FSs. Since its introduction, T2FS has garnered significant attention from researchers, leading to numerous notable works in the field (Das et al., 2020; Ghosh et al., 2023).

DOI: 10.1201/9781003497219-8

Later, Singh and Garg (2017) initiated a new extension of IFSs, Type-2 intuitionistic fuzzy sets (T2IFSs). It incorporates both primary and secondary membership and non-membership functions for each element. This extension allows for the expression of fuzzy preferences of DMs towards their choices under different parameters, capturing both acceptance and non-acceptance perspectives. Thus, it can be treated as an appropriate information expression tool, as it represents the uncertainty of human judgement and impreciseness of information data in a more nuanced way than traditional fuzzy sets.

Information aggregation plays a critical role in the decision-making process, as it involves the amalgamation of information to derive meaningful insights and make well-informed choices. Due to its ability to provide a comprehensive and reliable basis for decision-making, it has gained enormous popularity among researchers. In MADM problems, information aggregation plays a vital role too in combining multiple attributes and DMs' opinions to facilitate effective decision-making and provide a holistic evaluation of alternatives. Till now, various types of aggregation operators have been developed. One such aggregation operators is Maclaurin symmetric mean (MSM) operator. It was first initiated by Maclaurin (1729), and then DeTemple and Robertson (1979) enhanced it. Further, it possesses multifarious advantages over other aggregation operators, including:

- The MSM operator possesses a notable characteristic of not only accounting for the significance of each attribute but also capturing the interdependencies among arguments.
- It is symmetric, meaning that it treats all the numbers in the set equally. It maintains equal importance for all values without prioritizing any particular one.
- Further, it exhibits continuity across a range of inputs, implying that slight changes in the input values lead to minor variations in the output value.
- MSM operator is a generalization of several other mean operators, such as the arithmetic mean, geometric mean, and harmonic mean. By adjusting the parameter of the operator, you can obtain different types of means. This flexibility allows you to choose the appropriate means based on the specific requirements of your problem.

Qin and Liu (2014) introduced an MSM operator under IFSs. Later, Qin and Liu (2015) developed a dual MSM operator. For q-rung orthopair fuzzy information, power MSM was initiated by Liu et al. (2018). Ullah (2021) advanced it for picture FSs. Recently, Ning et al. (2022) expanded it in probabilistic dual hesitant FSs.

To our knowledge, no previous studies have explored the application of the MSM operator in the context of T2IFS. This study aims to bridge this research gap by introducing a novel aggregation operator, namely the "MSM operator with T2IFSs" and applying its applicability to medical waste disposal site selection problem.

The remaining parts of the article are structured as follows: Preliminary concepts are discussed in Section 8.2. In Section 8.3, MSM operators are pioneered under T2IFSs. Based on T2IFS information, a new MADM approach is originated in Section 8.4. A case study of medical waste disposal site selection is illustrated, and solved with the help of the suggested MADM approach in Section 8.5. Finally, Section 8.6 presents the conclusion part.

8.2 PRELIMINARIES

Here, some basic and fundamental concepts of Type-2 fuzzy sets, T2IFSs, and MSM operator are shared.

8.2.1 Type-2 fuzzy sets

Definition 8.1 (Zadeh, 1975) If $\mathcal{U}$ be a fixed universe, then a Type-2 fuzzy set (T2FS) $\mathcal{A} \subset \mathcal{U}$ is defined as

$$\mathcal{A} = \left\{ \left\langle x, \zeta_{\mathcal{A}}(x); \theta_x(\zeta_{\mathcal{A}}) \right\rangle \mid x \in \mathcal{U}, \zeta_{\mathcal{A}} \in [0,1] \right\}, \tag{8.1}$$

where $\zeta_{\mathcal{A}}$ is the primary membership function (PMF) of set $\mathcal{A}$, and $\theta_x(\zeta_{\mathcal{A}}) \in [0,1]$ is the secondary membership function (SMF) of x.

Alternatively, T2FS $\mathcal{A}$ can be expressed as

$$\mathcal{A} = \int_{x \in \mathcal{U}} \frac{\zeta_{\mathcal{A}}(x)}{x} = \int_{x \in \mathcal{U}} \left[\int_{\zeta_{\mathcal{U}} \in [0,1]} \frac{(\theta_x(\zeta_{\mathcal{A}}))}{\zeta_{\mathcal{A}}} \right] / x. \tag{8.2}$$

8.2.2 Type-2 intuitionistic fuzzy sets

Definition 8.2 (Singh and Garg, 2017) A T2IFS $\tilde{\mathcal{A}}$ is a set of membership functions characterized as

$$\tilde{\mathcal{A}} = \left\{ \left\langle x, \zeta_{\mathcal{A}}(x), \eta_{\mathcal{A}}(x); \theta_x(\zeta_{\mathcal{A}}), \vartheta_x(\eta_{\mathcal{A}}) \right\rangle \mid x \in \mathcal{U}, \zeta_{\mathcal{A}}, \eta_{\mathcal{A}} \in [0,1] \right\}, \tag{8.3}$$

where $\zeta_{\mathcal{A}}, \eta_{\mathcal{A}}: \mathcal{U} \to [0,1]$ are the primary membership and non-membership functions, respectively, and $\theta_x(\zeta_{\mathcal{A}}), \vartheta_x(\eta_{\mathcal{A}}) \in [0,1]$ are the secondary membership and non-membership functions. Further, $0 \le \zeta_{\mathcal{A}} + \eta_{\mathcal{A}} \le 1$ and $0 \le \theta_x(\zeta_{\mathcal{A}}) + \vartheta_x(\eta_{\mathcal{A}}) \le 1 \forall x \in \mathcal{U}$.

We entitle quadruplet $(\zeta_A, \eta_A, \theta_A, \vartheta_A)$ as Type-2 intuitionistic fuzzy number (T2IFN).

Example 8.2.1 Let the state government of a particular state brings a bill on the agricultural drainage system. Thereby, the government communicates to local citizens through various activities. Later, the government creates a group of three experts to know public sentiment about the bill. Experts find that in addition to government initiatives, local farmers also play a role in the agricultural drainage system. They believe that these practices are not as important as government-funded projects, but they still play a vital role in it. Suppose one of the experts believes that there is a 20% chance that the bill will be accepted and a 20% chance that it will not be accepted. The expert also believes that there is a 70% chance that the bill will be accepted if the secondary issues are considered, and a 25% chance that it will not be accepted. Then, the assessment can be expressed as ⟨0.75, 0.20; 0.70, 0.25⟩. In a similar way, other experts provide their assessments as ⟨0.80, 0.15; 0.70, 0.30⟩ and ⟨0.80, 0.20; 0.75, 0.20⟩ Therefore, assessments of all experts in terms of a T2IFS can be outlined as $\tilde{A}$ = {⟨0.75, 0.20; 0.70, 0.25⟩, ⟨0.80, 0.15; 0.70, 0.30⟩, ⟨0.80, 0.20; 0.75, 0.20⟩}

Definition 8.3 If $\tilde{A}, \tilde{B}$ be two T2IFNs, then the generalized sum and product of $\tilde{A}$ and $\tilde{B}$ can be describe as

$$\tilde{A} \oplus \tilde{B} = \left\{ \left\langle T_N(\zeta_A, \zeta_B), T_{CN}(\eta_A, \eta_B); T_N(\theta_A, \theta_B), T_{CN}(\vartheta_A, \vartheta_B) \right\rangle \right\}, \quad (8.4)$$

$$\tilde{A} \otimes \tilde{B} = \left\{ \left\langle T_{CN}(\zeta_A, \zeta_B), T_N(\eta_A, \eta_B); T_{CN}(\theta_A, \theta_B), T_N(\vartheta_A, \vartheta_B) \right\rangle \right\}, \quad (8.5)$$

respectively. T_N and T_{CN} indicate *t*-norm and *t*-conorm, respectively.

Definition 8.4 If $\tilde{A}, \tilde{B}$ be two T2IFNs, and λ (>0) be any scalar, then algebraic operations are given below:

Algebraic sum:

$$\tilde{A} \oplus \tilde{B} = \left\{ \left\langle \zeta_A + \zeta_B - \zeta_A \zeta_B, \eta_A \eta_B; \theta_A + \theta_B - \theta_A \theta_B, \vartheta_A \vartheta_B \right\rangle \right\}. \quad (8.6)$$

Algebraic product:

$$\tilde{A} \otimes \tilde{B} = \left\{ \left\langle \zeta_A \zeta_B, \eta_A + \eta_B - \eta_A \eta_B; \theta_A \theta_B, \vartheta_A + \vartheta_B - \vartheta_A \vartheta_B \right\rangle \right\}. \quad (8.7)$$

Algebraic scalar product:

$$\lambda \tilde{A} = \left\{ \left\langle 1 - (1 - \zeta_A)^\lambda, \eta_A^\lambda; 1 - (1 - \theta_A)^\lambda, \vartheta_A^\lambda \right\rangle \right\}. \quad (8.8)$$

Algebraic product:

$$\tilde{\mathcal{A}}^{\lambda} = \left\{ \left\langle \zeta_{\mathcal{A}}^{\lambda}, 1-(1-\eta_{\mathcal{A}})^{\lambda}; \theta_{\mathcal{A}}^{\lambda}, 1-(1-\vartheta_{\mathcal{A}})^{\lambda} \right\rangle \right\}. \tag{8.9}$$

Theorem 8.1 If $\tilde{\mathcal{A}} = \{\langle \zeta_{\mathcal{A}}, \eta_{\mathcal{A}}; \theta_{\mathcal{A}}, \vartheta_{\mathcal{A}} \rangle\}$ and $\tilde{\mathcal{B}} = \{\langle \zeta_{\mathcal{B}}, \eta_{\mathcal{B}}; \theta_{\mathcal{B}}, \vartheta_{\mathcal{B}} \rangle\}$ be two T2IFNs, then algebraic sum and algebraic product operations are commutative, i.e.,

I. $\tilde{\mathcal{A}} \oplus \tilde{\mathcal{B}} = \tilde{\mathcal{B}} \oplus \tilde{\mathcal{A}}$,

II. $\tilde{\mathcal{A}} \otimes \tilde{\mathcal{B}} = \tilde{\mathcal{B}} \otimes \tilde{\mathcal{A}}$.

Theorem 8.2 If $\tilde{\mathcal{A}} = \{\langle \zeta_{\mathcal{A}}, \eta_{\mathcal{A}}; \theta_{\mathcal{A}}, \vartheta_{\mathcal{A}} \rangle\}$, $\tilde{\mathcal{B}} = \{\langle \zeta_{\mathcal{B}}, \eta_{\mathcal{B}}; \theta_{\mathcal{B}}, \vartheta_{\mathcal{B}} \rangle\}$ be two T2IFNs, and λ, λ_1, λ_2 be three non-negative scalars, then the following properties hold.

I. $\lambda\left(\tilde{\mathcal{A}} \oplus \tilde{\mathcal{B}}\right) = \lambda\tilde{\mathcal{A}} \oplus \lambda\tilde{\mathcal{B}},\ \lambda \geq 0;$

II. $(\lambda_1 + \lambda_2)\tilde{\mathcal{A}} = \lambda_1\tilde{\mathcal{A}} \oplus \lambda_2\tilde{\mathcal{A}},\ \lambda_1, \lambda_2 \geq 0;$

III. $\tilde{\mathcal{A}}^{\lambda} \otimes \mathcal{B}^{\lambda} = \left(\tilde{\mathcal{A}} \otimes \tilde{\mathcal{B}}\right)^{\lambda},\ \lambda > 0;$

IV. $\tilde{\mathcal{A}}^{\lambda_1} \otimes \tilde{\mathcal{A}}^{\lambda_2} = \tilde{\mathcal{A}}^{(\lambda_1+\lambda_2)},\ \lambda_1, \lambda_2 > 0.$

8.2.3 MSM operator

The Maclaurin symmetric mean (MSM) operator, initially introduced by Maclaurin (1729) is a powerful tool for modelling interrelationships in decision-making phenomena.

Definition 8.5 For a given collection of real numbers (α_1, α_2, ..., α_n), ($\alpha_i \geq 0$, i=1, 2, ..., n) and a non-negative integer k, the function

$$MSM^{(k)}(\alpha_1, \alpha_2, \ldots, \alpha_n) = \left(\frac{\sum_{1 \le i_1 \le i_2 \le \ldots i_k \le n} \prod_{j=1}^{n} \alpha_{i_j}}{C_n^k} \right)^{\frac{1}{k}}. \tag{8.10}$$

is termed as MSM operator, where (i_1, i_2, ..., i_k) are k-tuple combinations from (1, 2, ..., n) and C_n^k is the binomial coefficient.

MSM operator possesses the following properties:

I. $MSM^{(k)}(0, 0, \ldots, 0)=0$ and $MSM^{(k)}(\alpha, \alpha, \ldots, \alpha)= \alpha$;

II. Whenever $\alpha_i \leq \beta_i \forall j$ $MSM^{(k)}(\alpha_1, \alpha_2, ..., \alpha_n) \leq MSM^{(k)}(\beta_1, \beta_2, ..., \beta_n)$
III. $MSM^{(k)}(\alpha_1,\alpha_2,...,\alpha n) = [min_i(\alpha_i), max_i(\alpha_i)]$

Lemma 8.1 (Maclaurin inequality) For a collection of non-negative real numbers $(\alpha_1, \alpha_2, ..., \alpha_n)$ and for $k = 1, 2, ..., n$

$MSM^{(1)}(\alpha_1, \alpha_2, ..., \alpha_n) \geq MSM^{(2)}(\alpha_1, \alpha_2, ..., \alpha_n) \geq .... \geq MSM^{(n)}(\alpha_1, \alpha_2, ..., \alpha_n)$.

Equality holds if and only if $\alpha_1 = \alpha_2 = ... = \alpha_n$.

Lemma 8.2 For $\alpha_i > 0$, $\varrho_i > 0$ $(i=1, 2, ..., n)$ and $\sum_{i=1}^{n} \varrho_i = 1$, then

$$\prod_{i=1}^{n} (\alpha_i)^{\varrho_i} \leq \sum_{i=1}^{n} \alpha_i \varrho_i.$$

Equality holds if and only if $\alpha_1 = \alpha_2 = ... = \alpha_n$.

8.3 TYPE-2 INTUITIONISTIC FUZZY MSM OPERATORS

In this section, we extend the MSM operator to the T2IFS environment to create the Type-2 intuitionistic fuzzy Maclaurin symmetric mean (T2IFMSM) operator. We also introduce a weighted version of this operator, called the Type-2 intuitionistic fuzzy weighted Maclaurin symmetric mean (T2IFWMSM) operator.

8.3.1 T2IFMSM operator

Definition 8.6 For a given collection of T2IFNs $(\mathcal{F}_1, \mathcal{F}_2, ..., \mathcal{F}_n)$ the mapping

$$T2IFMSM^{(k)}(\mathcal{F}_1, \mathcal{F}_2, ..., \mathcal{F}_n) = \left(\frac{\oplus_{1 \leq i_1 \leq i_2 \leq ... i_k \leq n} \oplus_{j=1}^{n} \mathcal{F}_{i_j}}{C_n^k} \right)^{\frac{1}{k}}. \quad (8.11)$$

is called as T2IFMSM operator, where $(i_1, i_2, ..., i_k)$ are k-tuple combinations from $(i=1, 2, ..., n)$ and C_n^k is the binomial coefficient.

Theorem 8.3 For a given collection of $(\mathcal{F}_1, \mathcal{F}_2, ..., \mathcal{F}_n)$, the aggregated outcome, obtained by utilizing T2IFMSM operator, is also a T2IFN and defined by

$$T2IFMSM^{(k)}\left(\mathcal{F}_1,\mathcal{F}_2,\ldots,\mathcal{F}_n\right)$$

$$=\left\{\begin{array}{l}\left(\left(1-\left(\prod_{\substack{1\le i_1\le i_2\\ \ldots\le i_k\le n}}\left(1-\left(\prod_{j=1}^{k}\zeta_{i_j}\right)\right)\right)^{\frac{1}{C_n^k}}\right)^{\frac{1}{k}},\ \left(1-\left(1-\left(\prod_{\substack{1\le i_1\le i_2\\ \ldots\le i_k\le n}}\left(1-\left(\prod_{j=1}^{k}\left(1-\eta_{i_j}\right)\right)\right)\right)^{\frac{1}{C_n^k}}\right)^{\frac{1}{k}}\right)\right);\\ \left(\left(1-\left(\prod_{\substack{1\le i_1\le i_2\\ \ldots\le i_k\le n}}\left(1-\left(\prod_{j=1}^{k}\theta_{i_j}\right)\right)\right)^{\frac{1}{C_n^k}}\right)^{\frac{1}{k}},\ \left(1-\left(1-\left(\prod_{\substack{1\le i_1\le i_2\\ \ldots\le i_k\le n}}\left(1-\left(\prod_{j=1}^{k}\left(1-\vartheta_{i_j}\right)\right)\right)\right)^{\frac{1}{C_n^k}}\right)^{\frac{1}{k}}\right)\right)\end{array}\right\}. \tag{8.12}$$

Proof. Utilizing algebraic operational laws (described in Definition 8.4), we have

$$\otimes_{j=1}^{k}\mathcal{F}_{i_j}=\left\{\prod_{j=1}^{k}\zeta_{i_j},\left(1-\prod_{j=1}^{k}\left(1-\eta_{i_j}\right)\right);\prod_{j=1}^{k}\theta_{i_j},\left(1-\prod_{j=1}^{k}\left(1-\vartheta_{i_j}\right)\right)\right\}. \tag{8.13}$$

Now,

$$\bigoplus_{\substack{1\le i_1\le i_2\\ \ldots\le i_k\le n}}\left(\bigotimes_{j=1}^{k}\mathcal{F}_{i_j}\right)=\left\{\begin{array}{l}1-\left(\prod_{\substack{1\le i_1\le i_2\\ \ldots\le i_k\le n}}\left(1-\left(\prod_{j=1}^{k}\zeta_{i_j}\right)\right)\right),\quad \prod_{\substack{1\le i_1\le i_2\\ \ldots\le i_k\le n}}\left(1-\prod_{j=1}^{k}\left(1-\eta_{i_j}\right)\right);\\ 1-\left(\prod_{\substack{1\le i_1\le i_2\\ \ldots\le i_k\le n}}\left(1-\left(\prod_{j=1}^{k}\theta_{i_j}\right)\right)\right),\quad \prod_{\substack{1\le i_1\le i_2\\ \ldots\le i_k\le n}}\left(1-\prod_{j=1}^{k}\left(1-\vartheta_{i_j}\right)\right)\end{array}\right\}. \tag{8.14}$$

Again, using algebraic operational laws, we get

$$\frac{1}{C_n^k}\left(\underset{\substack{1\le i_1\le i_2\\ \ldots\le i_k\le n}}{\oplus}\left(\overset{k}{\underset{j=1}{\oplus}}\mathcal{F}_{i_j}\right)\right)$$

$$=\left\{\begin{matrix}1-\left(\prod_{\substack{1\le i_1\le i_2\\ \ldots\le i_k\le n}}\left(1-\left(\prod_{j=1}^{k}\zeta_{i_j}\right)\right)\right)^{\frac{1}{C_n^k}}, & \left(\prod_{\substack{1\le i_1\le i_2\\ \ldots\le i_k\le n}}\left(1-\prod_{j=1}^{k}\left(1-\eta_{i_j}\right)\right)\right)^{\frac{1}{C_n^k}}; \\ 1-\left(\prod_{\substack{1\le i_1\le i_2\\ \ldots\le i_k\le n}}\left(1-\left(\prod_{j=1}^{k}\theta_{i_j}\right)\right)\right)^{\frac{1}{C_n^k}}, & \left(\prod_{\substack{1\le i_1\le i_2\\ \ldots\le i_k\le n}}\left(1-\prod_{j=1}^{k}\left(1-\vartheta_{i_j}\right)\right)\right)^{\frac{1}{C_n^k}}\end{matrix}\right\}. \tag{8.15}$$

Then,

$$T2IFMSM^{(k)}\left(\mathcal{F}_1,\mathcal{F}_2,\ldots,\mathcal{F}_n\right)=\left(\frac{1}{C_n^k}\left(\underset{\substack{1\le i_1\le i_2\\ \ldots\le i_k\le n}}{\oplus}\left(\overset{k}{\underset{j=1}{\oplus}}\mathcal{F}_{i_j}\right)\right)\right)^{\frac{1}{k}}$$

$$=\left\{\begin{matrix}\left(1-\left(\prod_{\substack{1\le i_1\le i_2\\ \ldots\le i_k\le n}}\left(1-\left(\prod_{j=1}^{k}\zeta_{i_j}\right)\right)\right)^{\frac{1}{C_n^k}}\right)^{\frac{1}{k}}, & 1-\left(1-\left(\prod_{\substack{1\le i_1\le i_2\\ \ldots\le i_k\le n}}\left(1-\left(\prod_{j=1}^{k}\left(1-\eta_{i_j}\right)\right)\right)\right)^{\frac{1}{C_n^k}}\right)^{\frac{1}{k}}; \\ \left(1-\left(\prod_{\substack{1\le i_1\le i_2\\ \ldots\le i_k\le n}}\left(1-\left(\prod_{j=1}^{k}\theta_{i_j}\right)\right)\right)^{\frac{1}{C_n^k}}\right)^{\frac{1}{k}}, & 1-\left(1-\left(\prod_{\substack{1\le i_1\le i_2\\ \ldots\le i_k\le n}}\left(1-\left(\prod_{j=1}^{k}\left(1-\vartheta_{i_j}\right)\right)\right)\right)^{\frac{1}{C_n^k}}\right)^{\frac{1}{k}}\end{matrix}\right\}. \tag{8.16}$$

Property 8.1 (Idempotency) For all equal $\mathcal{F}_i$ ($i=1, 2, \ldots, n$), i.e., $\mathcal{F}_i = \mathcal{F}\ \ \forall i$,

$$T2IMSM^{(k)}\ \left(\mathcal{F}_1, \mathcal{F}_2, \ldots, \mathcal{F}_n\right) = \mathcal{F}. \tag{8.17}$$

Proof. Let $\mathcal{F}=\langle\zeta,\eta;\theta,\vartheta\rangle$.
Now, according to Eq. (8.12), we obtain

$$T2IFMSM^{(k)}\left(\mathcal{F},\mathcal{F},\ldots,\mathcal{F}\right)$$

$$=\left\{\begin{array}{l}\left(\left(1-\left(\prod_{\substack{1\le i_1\le i_2\\ \ldots\le i_k\le n}}\left(1-\left(\prod_{j=1}^{k}\zeta\right)\right)\right)^{\frac{1}{C_n^k}}\right)^{\frac{1}{k}},\quad\left(1-\left(1-\left(\prod_{\substack{1\le i_1\le i_2\\ \ldots\le i_k\le n}}\left(1-\left(\prod_{j=1}^{k}(1-\eta)\right)\right)\right)^{\frac{1}{C_n^k}}\right)^{\frac{1}{k}}\right)\right);\\ \left(\left(1-\left(\prod_{\substack{1\le i_1\le i_2\\ \ldots\le i_k\le n}}\left(1-\left(\prod_{j=1}^{k}\theta\right)\right)\right)^{\frac{1}{C_n^k}}\right)^{\frac{1}{k}},\quad\left(1-\left(1-\left(\prod_{\substack{1\le i_1\le i_2\\ \ldots\le i_k\le n}}\left(1-\left(\prod_{j=1}^{k}(1-\vartheta)\right)\right)\right)^{\frac{1}{C_n^k}}\right)^{\frac{1}{k}}\right)\right)\end{array}\right\}$$

$$=\left\{\begin{array}{l}\left(\left(1-\left(\prod_{\substack{1\le i_1\le i_2\\ \ldots\le i_k\le n}}\left(1-\left(\zeta^{k}\right)\right)\right)^{\frac{1}{C_n^k}}\right)^{\frac{1}{k}},\quad\left(1-\left(1-\left(\prod_{\substack{1\le i_1\le i_2\\ \ldots\le i_k\le n}}\left(1-(1-\eta)^{k}\right)\right)^{\frac{1}{C_n^k}}\right)^{\frac{1}{k}}\right)\right);\\ \left(\left(1-\left(\prod_{\substack{1\le i_1\le i_2\\ \ldots\le i_k\le n}}\left(1-\left(\theta^{k}\right)\right)\right)^{\frac{1}{C_n^k}}\right)^{\frac{1}{k}},\quad\left(1-\left(1-\left(\prod_{\substack{1\le i_1\le i_2\\ \ldots\le i_k\le n}}\left(1-(1-\vartheta)^{k}\right)\right)^{\frac{1}{C_n^k}}\right)^{\frac{1}{k}}\right)\right)\end{array}\right\}$$

$$\left\{\begin{array}{l}\left(\left(1-\left(1-\left(\zeta^{k}\right)\right)\right)^{\frac{1}{k}},\quad\left(1-\left(1-\left(1-(1-\eta)^{k}\right)\right)^{\frac{1}{k}}\right)\right);\\ \left(\left(1-\left(1-\left(\theta^{k}\right)\right)\right)^{\frac{1}{k}},\quad\left(1-\left(1-\left(1-(1-\vartheta)^{k}\right)\right)^{\frac{1}{k}}\right)\right)\end{array}\right\}$$

$$=\left\{\left(\zeta^{k}\right)^{\frac{1}{k}},\left(1-(1-\eta)^{k\times\frac{1}{k}}\right);\left(\theta^{k}\right)^{\frac{1}{k}},\left(1-(1-\vartheta)^{k\times\frac{1}{k}}\right)\right\}$$

$$=\{\zeta,\eta;\theta,\vartheta\}=\mathcal{F}.$$

Property 8.2 (Commutativity) Let $\mathcal{F}_i$ $(i=1, 2, \ldots, n)$ *and* $\mathcal{F}_i^{'}\left(i=1,2,\ldots,n\right)$ be two collections of T2IFNs that are commutative, then

$$T2IFMSM^{(k)}\left(\mathcal{F}_1,\mathcal{F}_2,\ldots,\mathcal{F}_n\right)=T2IFMSM^{(k)}\left(\mathcal{F}_1^{'},\mathcal{F}_2^{'},\ldots,\mathcal{F}_n^{'}\right). \quad (8.18)$$

Proof. As $\mathcal{F}_i$ $(i=1, 2,\ldots,n)$ and $\mathcal{F}_i^{'}\left(i=1,2,\ldots,n\right)$ are commutative, then we have

$$T2IFMSM^{(k)}\left(\mathcal{F}_1,\mathcal{F}_2,\ldots,\mathcal{F}_n\right)=\left(\frac{\oplus_{1\le i_1\le i_2\le\ldots\le i_k\le n}\left(\oplus_{j=1}^{k}\mathcal{F}_{i_j}\right)}{C_n^k}\right)^{\frac{1}{k}}$$

$$=\left(\frac{\oplus_{1\le i_1\le i_2\le\ldots\le i_k\le n}\left(\oplus_{j=1}^{k}\mathcal{F}_{i_j}^{'}\right)}{C_n^k}\right)^{\frac{1}{k}}$$

$$=\mathrm{T2IFMSM}^{(k)}\left(\mathcal{F}_1^{'},\mathcal{F}_2^{'},\ldots,\mathcal{F}_n^{'}\right).$$

This is the conclusion of the theorem.

Property 8.3 (Monotonicity) If $\mathcal{F}_i=\langle\zeta_i,\eta_i;\theta_i,\vartheta_i\rangle$ (i = 1,2,…,n) and $\mathcal{F}i=\langle\zeta_i',\eta_i';\theta_i',\vartheta_i'\rangle$ 1,2,…,n) are two collections of T2IFNs, such that

$$\zeta_i\ge\zeta_i^{'},\ \eta_i\le\eta_i^{'},\ \theta_i\ge\theta_i^{'},\ \vartheta_i\le\vartheta_i^{'},\quad\forall i,$$

then

$$T2IFMSM^{(k)}\left(\mathcal{F}_1,\mathcal{F}_2,\ldots,\mathcal{F}_n\right)\ge T2IFMSM^{(k)}\left(\mathcal{F}_1^{'},\mathcal{F}_2^{'},\ldots,\mathcal{F}_n^{'}\right). \quad (8.19)$$

Proof. Since, $\zeta_{i_j}\ge\zeta_{i_j}^{'}$, we have

$$\prod_{j=1}^{k}\zeta_{i_j}\ge\prod_{j=1}^{k}\zeta_{i_j}^{'}\Rightarrow1-\prod_{j=1}^{k}\zeta_{i_j}\le1-\prod_{j=1}^{k}\zeta_{i_j}^{'},$$

$$\therefore\prod_{\substack{1\le i_1\le i_2\le\\\ldots\le i_k\le n}}\left(1-\prod_{j=1}^{k}\zeta_{i_j}\right)\le\prod_{\substack{1\le i_1\le i_2\le\\\ldots\le i_k\le n}}\left(1-\prod_{j=1}^{k}\zeta_{i_j}^{'}\right)$$

$$\Rightarrow\left(\prod_{\substack{1\le i_1\le i_2\le\\\ldots\le i_k\le n}}\left(1-\prod_{j=1}^{k}\zeta_{i_j}\right)\right)^{\frac{1}{C_n^k}}\ge\left(\prod_{\substack{1\le i_1\le i_2\le\\\ldots\le i_k\le n}}\left(1-\prod_{j=1}^{k}\zeta_{i_j}^{'}\right)\right)^{\frac{1}{C_n^k}}$$

$$\Rightarrow 1-\left(\prod_{\substack{1\le i_1\le i_2\le \\ \ldots\le i_k\le n}}\left(1-\prod_{j=1}^{k}\zeta_{i_j}\right)\right)^{\frac{1}{C_n^k}} \le 1-\left(\prod_{\substack{1\le i_1\le i_2\le \\ \ldots\le i_k\le n}}\left(1-\prod_{j=1}^{k}\zeta'_{i_j}\right)\right)^{\frac{1}{C_n^k}}$$

$$\Rightarrow \left(1-\left(\prod_{\substack{1\le i_1\le i_2\le \\ \ldots\le i_k\le n}}\left(1-\prod_{j=1}^{k}\zeta_{i_j}\right)\right)^{\frac{1}{C_n^k}}\right)^{\frac{1}{k}} \ge \left(1-\left(\prod_{\substack{1\le i_1\le i_2\le \\ \ldots\le i_k\le n}}\left(1-\prod_{j=1}^{k}\zeta'_{i_j}\right)\right)^{\frac{1}{C_n^k}}\right)^{\frac{1}{k}}.$$

Now, when $\eta_{i_j} \le \eta'_{i_j}$, we have

$$\left(1-\eta_{i_j}\right) \ge \left(1-\eta'_{i_j}\right) \Rightarrow \prod_{j=1}^{k}\left(1-\eta_{i_j}\right) \ge \prod_{j=1}^{k}\left(1-\eta'_{i_j}\right)$$

$$\therefore \left(1-\prod_{j=1}^{k}\left(1-\eta_{i_j}\right)\right) \le \left(1-\prod_{j=1}^{k}\left(1-\eta'_{i_j}\right)\right)$$

$$\Rightarrow \prod_{\substack{1\le i_1\le i_2 \\ \ldots i_k\le n}}\left(1-\prod_{j=1}^{k}\left(1-\eta_{i_j}\right)\right) \le \prod_{\substack{1\le i_1\le i_2 \\ \ldots i_k\le n}}\left(1-\prod_{j=1}^{k}\left(1-\eta'_{i_j}\right)\right)$$

$$\Rightarrow \left(\prod_{\substack{1\le i_1\le i_2 \\ \ldots i_k\le n}}\left(1-\prod_{j=1}^{k}\left(1-\eta_{i_j}\right)\right)\right)^{\frac{1}{C_n^k}} \ge \left(\prod_{\substack{1\le i_1\le i_2 \\ \ldots i_k\le n}}\left(1-\prod_{j=1}^{k}\left(1-\eta'_{i_j}\right)\right)\right)^{\frac{1}{C_n^k}}$$

$$\Rightarrow \left(1-\left(\prod_{\substack{1\le i_1\le i_2 \\ \ldots i_k\le n}}\left(1-\prod_{j=1}^{k}\left(1-\eta_{i_j}\right)\right)\right)^{\frac{1}{C_n^k}}\right) \le \left(1-\left(\prod_{\substack{1\le i_1\le i_2 \\ \ldots i_k\le n}}\left(1-\prod_{j=1}^{k}\left(1-\eta'_{i_j}\right)\right)\right)^{\frac{1}{C_n^k}}\right)$$

$$\Rightarrow \left(1-\left(\prod_{\substack{1\le i_1\le i_2 \\ \ldots i_k\le n}}\left(1-\prod_{j=1}^{k}\left(1-\eta_{i_j}\right)\right)\right)^{\frac{1}{C_n^k}}\right)^{\frac{1}{k}} \ge \left(1-\left(\prod_{\substack{1\le i_1\le i_2 \\ \ldots i_k\le n}}\left(1-\prod_{j=1}^{k}\left(1-\eta'_{i_j}\right)\right)\right)^{\frac{1}{C_n^k}}\right)^{\frac{1}{k}}$$

$$\Rightarrow\left(1-\left(1-\left(\prod_{\substack{1\le i_1\le i_2\\ \ldots i_k\le n}}\left(1-\prod_{j=1}^{k}\left(1-\eta_{i_j}\right)\right)\right)^{\frac{1}{C_n^k}}\right)^{\frac{1}{k}}\right)\le\left(1-\left(1-\left(\prod_{\substack{1\le i_1\le i_2\\ \ldots i_k\le n}}\left(1-\prod_{j=1}^{k}\left(1-\eta_{i_j}'\right)\right)\right)\right)^{\frac{1}{k}}\right).$$

Similarly, for the other two terms, we have the same.

This ends the proof.

Property 8.4 (Boundedness) If $\mathcal{F}_i$ $(i=1, 2, \ldots, n)$ is a collection of T2IFNs, and

$$\mathcal{F}^- = \min_i \mathcal{F}_i = \left\langle \min_i \zeta_i, \max_i \eta_i; \min_i \theta_i, \max_i \vartheta_i \right\rangle,$$

$$\mathcal{F}^+ = \max_i \mathcal{F}_i = \left\langle \max_i \zeta_i, \min_i \eta_i; \max_i \theta_i, \min_i \vartheta_i \right\rangle,$$

then

$$\mathcal{F}^- \le T2IFMSM^{(k)}\left(\mathcal{F}_1, \mathcal{F}_2, \ldots, \mathcal{F}_n\right) \le \mathcal{F}^+.$$

Proof. The proof is similar to the proof of Property 8.3.

Theorem 8.4 (Shift-invariance) If $\mathcal{F}_i\left(i=1,2,\ldots,n\right)$ be a collection of T2IFNs and $\mathcal{E}$ be any other T2IFN, then

$$T2IFMSM^{(k)}\left(\mathcal{F}_1 \oplus \mathcal{E}, \mathcal{F}_2 \oplus \mathcal{E}, \ldots, \mathcal{F}_n \oplus \mathcal{E}\right) = T2IFMSM^{(k)}\left(\mathcal{F}_1, \mathcal{F}_2, \ldots, \mathcal{F}_n\right) \oplus \mathcal{E}.$$

Theorem 8.5 (Homogeneity) If $\mathcal{F}_i\left(i=1,2,\ldots,n\right)$ is a collection of T2IFNs and λ is a non-negative real number, then

$$T2IFMSM^{(k)}\left(\lambda \otimes \mathcal{F}_1, \lambda \otimes \mathcal{F}_2, \ldots, \lambda \otimes \mathcal{F}_n\right) = \lambda \otimes T2IFMSM^{(k)}\left(\mathcal{F}_1, \mathcal{F}_2, \ldots, \mathcal{F}_n\right).$$

Theorem 8.6 If $\mathcal{F}_i\left(i=1,2,\ldots,n\right)$ be a collection of T2IFNs and $k = 1, 2, \ldots, n$, then the T2IFMSM operator monotonically decreases as the value of k increases.

8.3.2 T2IFWMSM operator

The importance of attributes is a crucial factor in resolving practical decision-making issues. But the T2IFMSM operator does not consider the importance of the aggregated arguments. In practical MADM, the weights of attributes are always taken into account in the aggregation process.

To address the limitations of the T2IFMSM operator, we propose the T2IFWMSM operator, which is defined as follows:

Definition 8.7 If $(\mathcal{F}_1, \mathcal{F}_2, \ldots, \mathcal{F}_n)$ be a collection of T2IFNs, and $\mathcal{W} = (w_1, w_2, \ldots, w_n)$ be the weight vector, where w_i representing the weight of argument $\mathcal{F}_i$ with the condition $w_i \in [0,1]$ and $\sum w_i = 1$, then the mapping

$$T2IFWMSM_{\mathcal{W}}^{(k)}(\mathcal{F}_1, \mathcal{F}_2, \ldots, \mathcal{F}_n) = \left(\frac{\oplus_{1 \le i_1 \le i_2 \le \ldots \le i_k \le n} \left(\oplus_{j=1}^{n} \left(\mathcal{F}_{i_j} \right)^{w_{i_j}} \right)}{C_n^k} \right)^{\frac{1}{k}}. \tag{8.20}$$

is called as T2IFWMSM operator, where $(i_1, i_2, \ldots, i_k)$ are k-tuple combinations from $(1, 2, \ldots, n)$ and C_n^k is the binomial coefficient.

Theorem 8.7 For a given collection of T2IFNs $(\mathcal{F}_1, \mathcal{F}_2, \ldots, \mathcal{F}_n)$, and a given corresponding weight vector $\mathcal{W} = (w_1, w_2, \ldots, w_n)$, satisfying the conditions $w_i \in [0,1]$ and $\sum w_i = 1$, the aggregated outcome, obtained by utilizing T2IFWMSM operator, is also a T2IFN and defined by

$$T2IFWMSM_{\mathcal{W}}^{(k)}(\mathcal{F}_1, \mathcal{F}_2, \ldots, \mathcal{F}_n)$$
$$= \left\{ \begin{array}{l} \left(\left(1 - \left(\prod_{\substack{1 \le i_1 \le i_2 \\ \ldots \le i_k \le n}} \left(1 - \left(\prod_{j=1}^{k} \left(\theta_{i_j} \right)^{w_{i_j}} \right) \right) \right)^{\frac{1}{C_n^k}} \right)^{\frac{1}{k}}, \left(1 - \left(1 - \left(\prod_{\substack{1 \le i_1 \le i_2 \\ \ldots \le i_k \le n}} \left(1 - \left(\prod_{j=1}^{k} \left(1 - \eta_{i_j} \right)^{w_{i_j}} \right) \right) \right)^{\frac{1}{C_n^k}} \right)^{\frac{1}{k}} \right) \right); \\ \left(\left(1 - \left(\prod_{\substack{1 \le i_1 \le i_2 \\ \ldots \le i_k \le n}} \left(1 - \left(\prod_{j=1}^{k} \left(\theta_{i_j} \right)^{w_{i_j}} \right) \right) \right)^{\frac{1}{C_n^k}} \right)^{\frac{1}{k}}, \left(1 - \left(1 - \left(\prod_{\substack{1 \le i_1 \le i_2 \\ \ldots \le i_k \le n}} \left(1 - \left(\prod_{j=1}^{k} \left(1 - \vartheta_{i_j} \right)^{w_{i_j}} \right) \right) \right)^{\frac{1}{C_n^k}} \right)^{\frac{1}{k}} \right) \right) \end{array} \right\}. \tag{8.21}$$

Proof. The proof is similar to Theorem 8.3.

For different values of parameter k, the following remarks can be outlined from Theorem 8.5.

Remark 8.1 If $k = 1$, the T2IFWMSM takes the following form:

$$T2IFWMSM_W^{(1)}\left(\mathcal{F}_1,\mathcal{F}_2,\ldots,\mathcal{F}_n\right)=\begin{pmatrix}\prod_{i=1}^{n}\zeta_i^{w_i},\prod_{i=1}^{n}\left(1-(1-\eta_i)^{w_i}\right);\\ \prod_{i=1}^{n}\theta_i^{w_i},\prod_{i=1}^{n}\left(1-(1-\vartheta_i)^{w_i}\right)\end{pmatrix}.$$

Remark 8.2 If $k = 2$, the T2IFWMSM takes the following form:

$$T2IFWMSM_W^{(2)}\left(\mathcal{F}_1,\mathcal{F}_2,\ldots,\mathcal{F}_n\right)$$

$$=\begin{Bmatrix}\left(\left(1-\left(\prod_{1\le i_1\le i_2\le n}\left(1-(\zeta_{i_1})^{w_{i1}}(\zeta_{i_2})^{w_{i2}}\right)\right)^{\frac{1}{C_n^2}}\right)^{\frac{1}{2}},\left(1-\left(1-\left(\prod_{1\le i_1\le i_2\le n}\left(1-(1-\eta_{i_1})^{w_{i1}}(1-\eta_{i_2})^{w_{i2}}\right)\right)^{\frac{1}{C_n^2}}\right)^{\frac{1}{2}}\right)\right);\\ \left(\left(1-\left(\prod_{1\le i_1\le i_2\le n}\left(1-(\theta_{i_1})^{w_{i1}}(\theta_{i_2})^{w_{i2}}\right)\right)^{\frac{1}{C_n^2}}\right)^{\frac{1}{2}},\left(1-\left(1-\left(\prod_{1\le i_1\le i_2\le n}\left(1-(1-\vartheta_{i_1})^{w_{i1}}(1-\vartheta_{i_2})^{w_{i2}}\right)\right)^{\frac{1}{C_n^2}}\right)^{\frac{1}{2}}\right)\right)\end{Bmatrix}$$

Remark 8.3 If $k = n$, the T2IFWMSM takes the following form:

$$T2IFWMSM_W^{(n)}\left(\mathcal{F}_1,\mathcal{F}_2,\ldots,\mathcal{F}_n\right)=\begin{pmatrix}\prod_{i=1}^{n}\left(1-(1-\zeta_i)^{w_i}\right),\prod_{i=1}^{n}\eta_i^{w_i};\\ \prod_{i=1}^{n}\left(1-(1-\theta_i)^{w_i}\right),\prod_{i=1}^{n}\vartheta_i^{w_i}\end{pmatrix}.$$

8.4 MADM APPROACH WITH T2IFS INFORMATION

In this section, we propose a new MADM framework that uses the developed aggregation operator to solve decision-making problems. For that, let us consider an MADM problem under T2IFS information. Let there are a finite set of alternatives $\mathcal{P} = \{\mathcal{P}_1, \mathcal{P}_2, \ldots, \mathcal{P}_m\}$ and a finite set of attributes, $\mathfrak{C} = \{\mathfrak{C}_1, \mathfrak{C}_2, \ldots, \mathfrak{C}_n\}$. Further, $\mathcal{W} = \{w_1, w_2, \ldots, w_n\}$ be the weight vector corresponding to the attributes, which satisfies the conditions $w_i > 0$ and $\sum w_i = 1$. Aiming to improve the assessment process, let us consider a finite collection of decisionmakers (DMs), $\mathfrak{D} = \{\mathfrak{D}_1, \mathfrak{D}_2, \ldots, \mathfrak{D}_p\}$. The weights in $\mathcal{W}_{\mathcal{D}} = \{\lambda_1, \lambda_2, \ldots, \lambda_p\}$ represent the relative importance of each DM. Furthermore, in a T2IFS environment, the assessment of attribute $\mathfrak{C}_j$ for alternative $\mathcal{P}_i$ provided by expert $\mathfrak{D}_l$ can be expressed as $\alpha_{ij}^l = \langle \zeta_{ij}^l, \eta_{ij}^l; \theta_{ij}^l, \vartheta_{ij}^l \rangle$, $(l = 1, 2, \ldots, p;\ i = 1, 2, \ldots, m;\ j = 1, 2, \ldots, n)$. Decision-matrices can be obtained based on the above information as follows:

$$\mathcal{A}^{\ell} = \langle \alpha_{ij} \rangle_{m \times n}^{l} = \begin{matrix} \\ \mathcal{P}_1 \\ \mathcal{P}_2 \\ \vdots \\ \mathcal{P}_m \end{matrix} \begin{matrix} \begin{matrix} \mathfrak{C}_1 & \mathfrak{C}_2 & \cdots & \mathfrak{C}_n \end{matrix} \\ \begin{pmatrix} \alpha_{11}^l & \alpha_{12}^l & \cdots & \alpha_{1n}^l \\ \alpha_{21}^l & \alpha_{22}^l & \cdots & \alpha_{2n}^l \\ \vdots & \vdots & \ddots & \vdots \\ \alpha_{m1}^l & \alpha_{m2}^l & \cdots & \alpha_{mn}^l \end{pmatrix} \end{matrix} \tag{8.22}$$

Attributes are homogeneous and all attributes are correlated to each other. To address this type of problem, we have developed a new algorithm based on the proposed operator. The algorithm follows the steps outlined below:

Step I. In MADM problems, attributes are typically divided into two types: beneficial attributes and non-beneficial attributes. The distinction between beneficial and non-beneficial attributes is important as it affects the way that the attributes are evaluated. Normalization ensures that all the attributes are on a comparable scale. The elements of the normalized decision matrices are obtained by

$$\beta_{ij}^l = \begin{cases} \langle \zeta_{ij}^l, \eta_{ij}^l; \theta_{ij}^l, \vartheta_{ij}^l \rangle, & \text{if attribute is of beneficial type} \\ \langle \eta_{ij}^l, \zeta_{ij}^l; \vartheta_{ij}^l, \theta_{ij}^l \rangle, & \text{if attribute is of non-beneficial type.} \end{cases} \tag{8.23}$$

Step II. The decision-matrices provided by the experts are aggregated into a single decision matrix, $\mathfrak{M} = [\gamma_{ij}]_{m \times n}$ using the T2IFWMSM operator, and given by

$$\gamma_{ij} = T2IFWMSM_{\mathcal{W}_{\mathcal{D}}}^{(k)} \left(\beta_{ij}^1, \beta_{ij}^2, \ldots, \beta_{ij}^p \right), \tag{8.24}$$

where $\mathcal{W}_{\mathcal{D}} = (\lambda_1, \lambda_2, \ldots, \lambda_p)$ represents the weight vector of the DMs.

Step III. With the help of T2IFWMSM operator, aggregated value δ_i $(i = 1, 2, ..., m)$ are formulated as follows:

$$\delta_i = T2IFWMSM_{\mathcal{W}}^{(k)}\left(\gamma_{i1}, \gamma_{i2}, \ldots, \gamma_{in}\right), \tag{8.25}$$

where $\mathcal{W} = \left(w_1, w_2, \ldots, w_n\right)$ represents the weight vector of attributes.

Step IV. The ranking value of each alternative is articulated using the following equation:

$$\mathcal{R}\left(\delta_i\right) = {}^{1+(\zeta \times \theta)-(\eta \times \vartheta)}\!\big/_{2}. \tag{8.26}$$

Step V. Using the obtained ranking values, the alternatives are ranked accordingly.

8.5 ILLUSTRATIVE EXAMPLE

Here, with help from the developed MADM approach, a hypothetical case study is solved. A comparison is then conducted between the proposed methodology and other existing studies. Furthermore, a detailed analysis is provided to identify the strengths and advantages of the proposed approach.

8.5.1 A case study: Medical waste disposal site selection

With growing time, medical waste disposal has become a major problem for humankind. It is a critical process that ensures the safe and proper management of waste generated from healthcare facilities to protect public health and the environment. Proper disposal of medical waste is important to prevent the spread of infections, protect the environment, and ensure the safety of healthcare workers, waste management personnel, and the general public. Improper handling or disposal of medical waste can have serious health and environmental consequences.

Effective medical waste disposal requires collaboration among healthcare facilities, waste management companies, regulatory bodies, and local communities. Public awareness campaigns and training programs also play a significant role in promoting proper waste segregation, handling, and disposal practices. So, the importance of medical waste disposal site selection cannot be overstated as it directly impacts public health, worker safety, and environmental protection. Selection of an appropriate site for medical waste disposal is crucial to ensure the effective and safe management of hazardous waste materials. A well-selected disposal site plays a critical role in minimizing contamination risks, preventing the spread of infections, and mitigating potential impacts on nearby communities. It also facilitates efficient transportation logistics, compliance with regulations, and the implementation of proper waste treatment methods, thereby protecting public health and maintaining environmental sustainability (Aung et al., 2019; Birpınar et al., 2009; Tirkolaee et al., 2021).

Recognizing the significance of the matter, the Midnapore municipality corporation in West Bengal, India, seeks to identify an optimal site for the disposal of medical waste. Given the limited availability of precise information or knowledge, assessing the exact attributes essential for identifying the optimal location for establishing a medical waste disposal site is an extremely challenging task. The process entails complexity and requires a significant amount of time. In this study, we conduct a hypothetical survey at Midnapore, a small town of West Bengal, to identify the optimal location for the establishment of the medical waste disposal site, utilizing the proposed MADM algorithm. To achieve this, a group of DMs $\{\mathfrak{D}_1, \mathfrak{D}_2, \mathfrak{D}_3\}$ is invited to provide their expertise and insights on the topic at hand. Decision-maker, $\mathfrak{D}_1$, is a waste management professional. Whereas decision-maker,

Table 8.1 Attributes identified by the experts

Attributes	*Description*	*Type*
Distance from residential areas $(\mathfrak{C}_1)$	By locating medical waste disposal sites far away from residential areas, we can help mitigate the risk of exposure to these hazardous materials.	Beneficial
Accessibility of roads $(\mathfrak{C}_2)$	Accessibility of roads is an important criterion in medical waste disposal site selection, ensuring convenient, and efficient transportation of hazardous materials.	Beneficial
Distance from open water $(\mathfrak{C}_3)$	Distance from open water is a key criterion in medical waste disposal site selection to prevent potential contamination and protect water resources from hazardous waste materials.	Beneficial
Land area $(\mathfrak{C}_4)$	Medical waste can be ponderous and hazardous, so it is important to have enough space to properly dispose of it.	Beneficial
Soil condition $(\mathfrak{C}_5)$	Evaluating the soil condition is crucial in medical waste disposal site selection to ensure the suitability of the land for proper waste containment and prevent potential contamination of surrounding soil and groundwater.	Beneficial
Social acceptance $(\mathfrak{C}_6)$	Considering social acceptance is important to ensure that the community is supportive of the project.	Beneficial
Sensitivity for environmental aspects $(\mathfrak{C}_7)$	Refers to the fact that does not have a negative impact on the ecosystem.	Beneficial

Table 8.2 Rating values, provided by the DMS

		$\mathfrak{C}_1$	$\mathfrak{C}_2$	$\mathfrak{C}_3$	$\mathfrak{C}_4$	$\mathfrak{C}_5$	$\mathfrak{C}_6$	$\mathfrak{C}_7$
$\mathcal{P}_1$	$\mathfrak{D}_1$	5	4	6	3	4	2	3
	$\mathfrak{D}_2$	5	5	6	4	4	3	3
	$\mathfrak{D}_3$	6	6	7	5	3	4	5
$\mathcal{P}_2$	$\mathfrak{D}_1$	4	5	4	6	3	2	4
	$\mathfrak{D}_2$	4	6	5	7	2	2	3
	$\mathfrak{D}_3$	5	6	4	6	4	3	3
$\mathcal{P}_3$	$\mathfrak{D}_1$	6	4	5	4	4	3	4
	$\mathfrak{D}_2$	6	4	5	3	3	4	5
	$\mathfrak{D}_3$	7	5	6	4	4	4	4
$\mathcal{P}_4$	$\mathfrak{D}_1$	6	6	5	5	3	4	5
	$\mathfrak{D}_2$	5	7	7	5	4	3	6
	$\mathfrak{D}_3$	5	6	6	5	4	5	4

$\mathfrak{D}_2$, is a public health official, and decision-maker, $\mathfrak{D}_3$, is a geotechnical engineer cum environmental consultant.

Four sites in the Midnapore town have been selected as alternatives by the DMs. Selected options are as follows: (i) Dharme $(\mathcal{P}_1)$, (ii) Keranitola $(\mathcal{P}_2)$, (iii) Judge Court $(\mathcal{P}_3)$, and (iv) Sepoyi Bazaar $(\mathcal{P}_4)$. All three DMs choose ten attributes each, seven common attributes are taken in this study. Details of the attributes are presented in Table 8.1. The crisp weights of the DMs are {0.4,0.35,0.25}, reflecting their respective opinions and expertise. The weight elements of the attributes are {0.18,0.15,0.17,0.16,0.11,0.09,0.1 4}, assigned by the group of DMs. All three DMs provide their assessments

Table 8.3 Rating scale values and corresponding T2IFNs

Rating scale value	*T2IFN*
9	$\langle 0.9, 0.1; 0.9, 0.1 \rangle$
8	$\langle 0.8, 0.2; 0.8, 0.2 \rangle$
7	$\langle 0.7, 0.3; 0.7, 0.3 \rangle$
6	$\langle 0.6, 0.4; 0.6, 0.4 \rangle$
5	$\langle 0.5, 0.5; 0.5, 0.5 \rangle$
4	$\langle 0.4, 0.6; 0.4, 0.6 \rangle$
3	$\langle 0.3, 0.7; 0.3, 0.7 \rangle$
2	$\langle 0.2, 0.8; 0.2, 0.8 \rangle$
1	$\langle 0.1, 0.9; 0.1, 0.9 \rangle$

Table 8.4 Decision-matrix, provided by $\mathfrak{D}_1$

	$\mathfrak{C}_1$	$\mathfrak{C}_2$	$\mathfrak{C}_3$	$\mathfrak{C}_4$	$\mathfrak{C}_5$	$\mathfrak{C}_6$	$\mathfrak{C}_7$
$\mathcal{P}_1$	$\langle 0.5,0.5;0.5,0.5\rangle$	$\langle 0.4,0.6;0.4,0.6\rangle$	$\langle 0.6,0.4;0.6,0.4\rangle$	$\langle 0.3,0.7;0.3,0.7\rangle$	$\langle 0.4,0.6;0.4,0.6\rangle$	$\langle 0.2,0.8;0.2,0.8\rangle$	$\langle 0.3,0.7;0.3,0.7\rangle$
$\mathcal{P}_2$	$\langle 0.4,0.6;0.4,0.6\rangle$	$\langle 0.5,0.5;0.5,0.5\rangle$	$\langle 0.4,0.6;0.4,0.6\rangle$	$\langle 0.6,0.4;0.6,0.4\rangle$	$\langle 0.3,0.7;0.3,0.7\rangle$	$\langle 0.2,0.8;0.2,0.8\rangle$	$\langle 0.4,0.6;0.4,0.6\rangle$
$\mathcal{P}_3$	$\langle 0.6,0.4;0.6,0.4\rangle$	$\langle 0.4,0.6;0.4,0.6\rangle$	$\langle 0.5,0.5;0.5,0.5\rangle$	$\langle 0.4,0.6;0.4,0.6\rangle$	$\langle 0.4,0.6;0.4,0.6\rangle$	$\langle 0.3,0.7;0.3,0.7\rangle$	$\langle 0.4,0.6;0.4,0.6\rangle$
$\mathcal{P}_4$	$\langle 0.6,0.4;0.6,0.4\rangle$	$\langle 0.6,0.4;0.6,0.4\rangle$	$\langle 0.5,0.5;0.5,0.5\rangle$	$\langle 0.5,0.5;0.5,0.5\rangle$	$\langle 0.3,0.7;0.3,0.7\rangle$	$\langle 0.4,0.6;0.4,0.6\rangle$	$\langle 0.5,0.5;0.5,0.5\rangle$

Table 8.5 Decision-matrix, provided by $\mathfrak{D}_2$

	$\mathfrak{C}_1$	$\mathfrak{C}_2$	$\mathfrak{C}_3$	$\mathfrak{C}_4$	$\mathfrak{C}_5$	$\mathfrak{C}_6$	$\mathfrak{C}_7$
$\mathcal{P}_1$	$\langle 0.5,0.5;0.5,0.5\rangle$	$\langle 0.5,0.5;0.5,0.5\rangle$	$\langle 0.6,0.4;0.6,0.4\rangle$	$\langle 0.4,0.6;0.4,0.6\rangle$	$\langle 0.4,0.6;0.4,0.6\rangle$	$\langle 0.3,0.7;0.3,0.7\rangle$	$\langle 0.3,0.7;0.3,0.7\rangle$
$\mathcal{P}_2$	$\langle 0.4,0.6;0.4,0.6\rangle$	$\langle 0.6,0.4;0.6,0.4\rangle$	$\langle 0.5,0.5;0.5,0.5\rangle$	$\langle 0.7,0.3;0.7,0.3\rangle$	$\langle 0.2,0.8;0.2,0.8\rangle$	$\langle 0.2,0.8;0.2,0.8\rangle$	$\langle 0.3,0.7;0.3,0.7\rangle$
$\mathcal{P}_3$	$\langle 0.6,0.4;0.6,0.4\rangle$	$\langle 0.4,0.6;0.4,0.6\rangle$	$\langle 0.5,0.5;0.5,0.5\rangle$	$\langle 0.3,0.7;0.3,0.7\rangle$	$\langle 0.3,0.7;0.3,0.7\rangle$	$\langle 0.4,0.6;0.4,0.6\rangle$	$\langle 0.5,0.5;0.5,0.5\rangle$
$\mathcal{P}_4$	$\langle 0.5,0.5;0.5,0.5\rangle$	$\langle 0.7,0.3;0.7,0.3\rangle$	$\langle 0.7,0.3;0.7,0.3\rangle$	$\langle 0.5,0.5;0.5,0.5\rangle$	$\langle 0.4,0.6;0.4,0.6\rangle$	$\langle 0.3,0.7;0.3,0.7\rangle$	$\langle 0.6,0.4;0.6,0.4\rangle$

Table 8.6 Decision-matrix, provided by $\mathfrak{D}_3$

	$\mathfrak{C}_1$	$\mathfrak{C}_2$	$\mathfrak{C}_3$	$\mathfrak{C}_4$	$\mathfrak{C}_5$	$\mathfrak{C}_6$	$\mathfrak{C}_7$
$\mathcal{P}_1$	$\langle 0.6,0.4;0.6,0.4\rangle$	$\langle 0.6,0.4;0.6,0.4\rangle$	$\langle 0.7,0.3;0.7,0.3\rangle$	$\langle 0.5,0.5;0.5,0.5\rangle$	$\langle 0.3,0.7;0.3,0.7\rangle$	$\langle 0.4,0.6;0.4,0.6\rangle$	$\langle 0.5,0.5;0.5,0.5\rangle$
$\mathcal{P}_2$	$\langle 0.5,0.5;0.5,0.5\rangle$	$\langle 0.6,0.4;0.6,0.4\rangle$	$\langle 0.4,0.6;0.4,0.6\rangle$	$\langle 0.6,0.4;0.6,0.4\rangle$	$\langle 0.4,0.6;0.4,0.6\rangle$	$\langle 0.3,0.7;0.3,0.7\rangle$	$\langle 0.3,0.7;0.3,0.7\rangle$
$\mathcal{P}_3$	$\langle 0.7,0.3;0.7,0.3\rangle$	$\langle 0.5,0.5;0.5,0.5\rangle$	$\langle 0.6,0.4;0.6,0.4\rangle$	$\langle 0.4,0.6;0.4,0.6\rangle$	$\langle 0.4,0.6;0.4,0.6\rangle$	$\langle 0.4,0.6;0.4,0.6\rangle$	$\langle 0.4,0.6;0.4,0.6\rangle$
$\mathcal{P}_4$	$\langle 0.5,0.5;0.5,0.5\rangle$	$\langle 0.6,0.4;0.6,0.4\rangle$	$\langle 0.6,0.4;0.6,0.4\rangle$	$\langle 0.5,0.5;0.5,0.5\rangle$	$\langle 0.4,0.6;0.4,0.6\rangle$	$\langle 0.5,0.5;0.5,0.5\rangle$	$\langle 0.4,0.6;0.4,0.6\rangle$

in rating scale values, which is shown in Table 8.2. Conversion of rating scale values to T2IFNs is presented in Table 8.3. So, the converted decision-matrices in terms of T2IFNs are presented in Tables 8.4–8.6.

For testing the applicability and effectiveness of the proposed algorithm, we conducted a case study to determine the optimal choice using the proposed methodology. The calculation process is outlined below:

Step I. Since all the attributes are of the beneficial type, normalization is not required.

Step II. All three decision-matrices are aggregated using Eq. (8.24). Before starting the calculation process, we fix the value of parameter k to 3.

For instance,

$$\gamma_{11} = T2IFWMSM_{W_D}^{(k)}\left(\langle 0.5,0.5;0.5,0.5\rangle, \langle 0.5,0.5;0.5,0.5\rangle, \langle 0.6,0.4;0.6,0.4\rangle\right),$$

where W_D =(0.40,0.35,0.25) is the weight vector of DMs.
Similarly, $\gamma_{ij}\left(i = 1,2,\ldots,m;\ j = 1,2,\ldots,n\right)$ are derived.

Aggregated decision-matrix is put forward in Table 8.7.

Step III. Again, using the T2IFWMSM operator, we aggregate values for each alternative. Let us fix the value of parameter k=2. The aggregated values are formulated as follows:

$$\delta_1 = T2IFWMSM_W^{(k)}\left(\gamma_{11}, \gamma_{12}, \gamma_{13}, \gamma_{14}, \gamma_{15}, \gamma_{16}, \gamma_{17}\right),$$

$$\delta_2 = T2IFWMSM_W^{(k)}\left(\gamma_{21}, \gamma_{22}, \gamma_{23}, \gamma_{24}, \gamma_{25}, \gamma_{26}, \gamma_{27}\right),$$

$$\delta_3 = T2IFWMSM_W^{(k)}\left(\gamma_{31}, \gamma_{32}, \gamma_{33}, \gamma_{34}, \gamma_{35}, \gamma_{36}, \gamma_{37}\right),$$

$$\delta_4 = T2IFWMSM_W^{(k)}\left(\gamma_{41}, \gamma_{42}, \gamma_{43}, \gamma_{44}, \gamma_{45}, \gamma_{46}, \gamma_{47}\right),$$

where $\mathcal{W} = \left(0.18,0.15,0.17,0.16,0.11,0.09,0.14\right)$ is the weight vector of attributes.

Step IV. The rating value of each alternative is determined utilizing Eq. (8.26), and the values are given below:

$$R(\delta_1) = 0.7775, \quad R(\delta_2) = 0.7743, \quad R(\delta_3) = 0.7839, \quad R(\delta_4) = 0.8022.$$

Step V. So, the final ranking is $\mathcal{P}_4 \succ \mathcal{P}_3 \succ \mathcal{P}_1 \succ \mathcal{P}_2$, where "$\succ$" indicates "superiority." Hence, $\mathcal{P}_4$, i.e., Sepoyi Bazaar is the best alternative for medical waste disposal site.

Table 8.7 Aggregated decision-matrix

	$\mathfrak{C}_1$	$\mathfrak{C}_2$	$\mathfrak{C}_3$	$\mathfrak{C}_4$
$\mathcal{P}_1$	$\langle 0.4678,0.5322;0.4678,0.5322\rangle$	$\langle 0.4418,0.5582;0.4418,0.5582\rangle$	$\langle 0.5272,0.4728;0.5272,0.4728\rangle$	$\langle 0.3546,0.6454;0.3546,0.6454\rangle$
$\mathcal{P}_2$	$\langle 0.4092,0.5908;0.4092,0.5908\rangle$	$\langle 0.4880,0.5120;0.4880,0.5120\rangle$	$\langle 0.4148,0.5852;0.4148,0.5852\rangle$	$\langle 0.5331,0.4669;0.5331,0.4669\rangle$
$\mathcal{P}_3$	$\langle 0.5272,0.4728;0.5272,0.4728\rangle$	$\langle 0.4092,0.5908;0.4092,0.5908\rangle$	$\langle 0.4678,0.5322;0.4678,0.5322\rangle$	$\langle 0.3728,0.6272;0.3728,0.6272\rangle$
$\mathcal{P}_4$	$\langle 0.4763,0.5237;0.4763,0.5237\rangle$	$\langle 0.5331,0.4669;0.5331,0.4669\rangle$	$\langle 0.5063,0.4937;0.5063,0.4937\rangle$	$\langle 0.4542,0.5458;0.4842,0.5458\rangle$

Table 8.7 Contd.

	$\mathfrak{C}_5$	$\mathfrak{C}_6$	$\mathfrak{C}_7$
$\mathcal{P}_1$	$\langle 0.3792,0.6208;0.3792,0.6208\rangle$	$\langle 0.3182,0.6818;0.3182,0.6818\rangle$	$\langle 0.3602,0.6398;0.3602,0.6398\rangle$
$\mathcal{P}_2$	$\langle 0.3219,0.6781;0.3219,0.6781\rangle$	$\langle 0.2829,0.7171;0.2829,0.7171\rangle$	$\langle 0.3576,0.6424;0.3576,0.6424\rangle$
$\mathcal{P}_3$	$\langle 0.3728,0.6272;0.3728,0.6272\rangle$	$\langle 0.3697,0.6303;0.3697,0.6303\rangle$	$\langle 0.4148,0.5852;0.4148,0.5852\rangle$
$\mathcal{P}_4$	$\langle 0.3697,0.6303;0.3697,0.6303\rangle$	$\langle 0.3852,0.6148;0.3852,0.6148\rangle$	$\langle 0.4565,0.5435;0.4565,0.5435\rangle$

8.5.2 Importance of the parameter *k*

Developed MADM approach consists of parameter k, which plays a crucial role in the outcomes of results. Mainly, there is a single parameter k, which reflects the interrelationship among arguments. Therefore, it is essential to figure out the impact of the parameter on the decision-making procedure.

Initially, we fix the value of $k = 3$ for decision-matrices aggregation process, which corresponds to a more pessimistic nature. However, for aggregated values of each alternative, we fix the value of $k = 2$. It shows a more balanced nature of assessment. Here, we take different values of parameter$k$ and find the changes in the ranking outcomes.

We can conclude from Table 8.8 that $\mathcal{P}_4$ i.e., "Sepoyi Bazaar" is the best choice. On the other hand, for $k = 1, 2$, and 3, $\mathcal{P}_2$ is the worst choice. Again, for $k = 4, 5, 6$, and 3, $\mathcal{P}_1$ is the worst choice.

In general, the T2IFMSM is a versatile tool that can be used to smooth data in a variety of applications. The choice of the parameter k depends on the specific application and the desired level of smoothing. Further, it is also able to take into consideration of DMs' confidence in decision-making procedures. Along with that, it is capable of removing outliers from data.

8.5.3 Comparison analysis and discussion

In this section, we examine the comparability aspect of the suggested MADM approach with existing methods based on operators to demonstrate the superior efficiency of the proposed algorithm. Due to the limited availability of studies (T2IFS information) in the literature, we compare the developed approach with few existing approaches. Existing approaches include different types of information, such as intuitionistic fuzzy sets. Comparison outcomes are shown in Table 8.9. In Table 8.10, characteristics of developed operators are compared with corresponding existing operators. From Table 8.9, it is clear that all the methods yield similar outcomes. In each method, $\mathcal{P}_4$ is the best alternative. Despite the outcomes of all the MADM approaches being the same, the proposed MADM approach stands out as superior to the compared methods. Compared approaches are based on extensions of type-1 fuzzy sets. However, the developed MADM approach is based on the T2IFS environment. It demonstrates robustness and a greater ability to handle uncertainties, offering distinct advantages over existing type-1 environments. Hence, we make a conclusion that the developed MADM approach is versatile and effective. At the same time, it is suitable for large-scale data. Further, it can be used to model more complex problems than those existing MADM approach that can be modelled by IFSs.

The designed operators offer several key advantages, providing a more comprehensive perspective for DMs. Additionally, the developed operator yields several other advantages, including the following:

Table 8.8 Impact of parameter *k* on ranking results

Value of k	*Ranking*
k = 1	$\mathcal{P}_4 \succ \mathcal{P}_3 \succ \mathcal{P}_1 \succ \mathcal{P}_2$
k = 2	$\mathcal{P}_4 \succ \mathcal{P}_3 \succ \mathcal{P}_1 \succ \mathcal{P}_2$
k = 3	$\mathcal{P}_4 \succ \mathcal{P}_3 \succ \mathcal{P}_1 \succ \mathcal{P}_2$
k = 4	$\mathcal{P}_4 \succ \mathcal{P}_3 \succ \mathcal{P}_2 \succ \mathcal{P}_1$
k = 5	$\mathcal{P}_4 \succ \mathcal{P}_3 \succ \mathcal{P}_2 \succ \mathcal{P}_1$
k = 6	$\mathcal{P}_4 \succ \mathcal{P}_3 \succ \mathcal{P}_2 \succ \mathcal{P}_1$
k = 7	$\mathcal{P}_4 \succ \mathcal{P}_3 \succ \mathcal{P}_2 \succ \mathcal{P}_1$

Table 8.9 Comparison study with existing studies

Method	*Ranking result*
AWHS operator-based method (Karmakar et al., 2021)	$\mathcal{P}_4 \succ \mathcal{P}_3 \succ \mathcal{P}_1 \succ \mathcal{P}_2$
TFNIFWG operator-based method (Chen et al., 2010)	$\mathcal{P}_4 \succ \mathcal{P}_3 \succ \mathcal{P}_1 \succ \mathcal{P}_2$
TFNIFOWG operator-based method (Chen et al., 2010)	$\mathcal{P}_4 \succ \mathcal{P}_3 \succ \mathcal{P}_2 \succ \mathcal{P}_1$
GFNIFWBM operator-based method (Verma, 2015)	$\mathcal{P}_4 \succ \mathcal{P}_3 \succ \mathcal{P}_2 \succ \mathcal{P}_1$
IFGWHM operator-based method (Yu, 2013)	$\mathcal{P}_4 \succ \mathcal{P}_3 \succ \mathcal{P}_2 \succ \mathcal{P}_1$

Abbreviations: AWHS: Aggregated weighted Hamacher sum; TFNIFWG: Triangle fuzzy number intuitionistic fuzzy weighted geometric; TFNIFOWG: Triangle fuzzy number intuitionistic fuzzy ordered weighted geometric; GFNIFWBM: Generalized fuzzy number intuitionistic fuzzy weighted Bonferroni mean; IFGWHM: Intuitionistic fuzzy geometric weighted Heronian mean.

Table 8.10 Characteristics comparison

Method	*Proficiency of correlation among multiple arguments*	*Ability to analyze DMs' attitudes*	*Inclusion of secondary membership degrees*
AWHS	No	Yes	Yes
TFNIWG	No	No	Yes
TFNIOWG	No	No	Yes
GFNIFWBM	No	Yes	No
IFGWHM	No	Yes	No
Proposed operator	Yes	Yes	Yes

Abbreviations: AWHS: Aggregated weighted Hamacher sum; TFNIFWG: Triangle fuzzy number intuitionistic fuzzy weighted geometric; TFNIFOWG: Triangle fuzzy number intuitionistic fuzzy ordered weighted geometric; GFNIFWBM: Generalized fuzzy number intuitionistic fuzzy weighted Bonferroni mean; IFGWHM: Intuitionistic fuzzy geometric weighted Heronian mean.

- The formulated operator enables the incorporation of interdependencies among multiple input arguments.
- The designed approach considers the symmetric nature of arguments, which ensures a balanced treatment of uncertainty, leading to more accurate and reliable results.
- The developed operator explores secondary membership degrees, providing a better representation of uncertainty. It allows for the fusion of uncertain information at different levels, providing a more comprehensive representation of uncertainty. This flexibility enables capturing and representing uncertain and vague information more effectively.
- Compared to other operators, the proposed operator has lesser computational complexity. This efficiency makes it practical for real-world applications where computational resources and time constraints are considerations.
- The developed MADM approach ensures consistency and reliability in the aggregation process.
- The proposed operator involves handling a single parameter, whereas other existing operators require dealing with multiple parameters.

8.6 CONCLUSION

Information aggregation is the process of collecting and merging information to make comprehensive and reliable decisions. The MSM operator plays a crucial role in information aggregation by capturing the interrelationships among different criteria and effectively considering their relative importance. Whereas, T2IFSs serve as an effective uncertainty-handling mechanism, allowing for a more nuanced representation and management of uncertainty in decision-making processes. This study has proposed a novel integrated structure that combines the MSM operator with T2IFS information, resulting in the development of the T2IFMSM operator. Furthermore, a new MADM algorithm has been proposed by utilizing the T2IFWMSM operator. Selection of medical waste disposal sites is essential for ensuring public health, and preventing environmental contamination. To find the best possible medical waste disposal site, a case study has been performed using the developed MADM algorithm. Furthermore, a comprehensive comparative analysis has been conducted to compare the proposed algorithm with existing methods, providing valuable insights into its effectiveness and advantages.

In the future, the T2IFMSM operator-based MADM algorithm can be extended to other fuzzy environments, opening up new possibilities for their application and further advancing the field of decision-making. Furthermore, the developed algorithm can be implemented in various fields.

For instance, the developed operator can be utilized in the image processing sector to aggregate attribute information, enabling more effective analysis and processing of images.

FUNDING:

The author, Kaushik Debnath, expresses gratitude to the Council of Scientific & Industrial Research, India for their support in funding his research activities under the JRF (CSIR) scheme. The award letter is numbered [F. No. 09/0599(12846)/2021-EMR-I] and is dated 01/12/2021.

REFERENCES

Atanassov, K. T. (1986). Intuitionistic fuzzy sets. *Fuzzy Sets and Systems*, 20(1), 137–142.

Aung, T. S., Luan, S., and Xu, Q. (2019). Application of multi-criteria-decision approach for the analysis of medical waste management systems in Myanmar. *Journal of Cleaner Production*, 222, 733–745.

Birpınar, M. E., Bilgili, M. S., and Erdogan, T. (2009). Medical waste management in Turkey: A case study of Istanbul. *Waste Management*, 29(1), 445–448.

Chen, D., Zhang, L., and Jiao, J. (2010). Triangle fuzzy number intuitionistic fuzzy aggregation operators and their application to group decision making. In *Artificial Intelligence and Computational Intelligence: International Conference*, AICI 2010, Sanya, China, October 23–24, 2010, Proceedings, Part II 2, pages 350–357. Springer.

Das, S. K., Roy, S. K., and Weber, G. W. (2020). Application of type-2 fuzzy logic to a multiobjective green solid transportation–location problem with dwell time under carbon tax, cap, and offset policy: Fuzzy versus nonfuzzy techniques. *IEEE Transactions on Fuzzy Systems*, 28(11), 2711–2725.

Debnath, K., and Roy, S. K. (2023). Power partitioned neutral aggregation operators for Tspherical fuzzy sets: An application to H2 refuelling site selection. *Expert Systems with Applications*, 216, 119470.

DeTemple, D. W., and Robertson, J. M. (1979). On generalized symmetric means of two variables. *Publikacije Elektrotehnickog fakulteta. Serija Matematika i fizika*, (634/677), 236–238.

Ghosh, S., Kufer, K. H., Roy, S. K., and Weber, G. W. (2023). Type-2 zigzag uncertain multi-objective fixed-charge solid transportation problem: Time window vs. preservation technology. *Central European Journal of Operations Research*, 31(1), 337–362.

Giri, B. K., and Roy, S. K. (2022). Neutrosophic multi-objective green four-dimensional fixed-charge transportation problem. *International Journal of Machine Learning and Cybernetics*, 13(10), 3089–3112.

Karmakar, S., Seikh, M. R., and Castillo, O. (2021). Type-2 intuitionistic fuzzy matrix games based on a new distance measure: Application to biogas-plant implementation problem. *Applied Soft Computing*, 106, 107357.

Liu, P., Chen, S. M., and Wang, P. (2018). Multiple-attribute group decision-making based on q-rung orthopair fuzzy power Maclaurin symmetric mean operators. *IEEE Transactions on Systems, Man, and Cybernetics: Systems*, 50(10), 3741–3756.

Maclaurin, C. (1729). A second letter to Martin Folkes, Esq.; concerning the roots of equations, with demonstration of other rules of algebra. *Philosophical Transactions of the Royal Society of London Series A*, 1729(36), 59–96.

Mondal, A., Roy, S. K., and Pamucar, D. (2023). Regret-based three-way decision making with possibility dominance and SPA theory in incomplete information system. *Expert Systems with Applications*, 211, 118688.

Ning, B., Wei, G., Lin, R., and Guo, Y. (2022). A novel MADM technique based on extended power generalized Maclaurin symmetric mean operators under probabilistic dual hesitant fuzzy setting and its application to sustainable suppliers selection. *Expert Systems with Applications*, 204, 117419.

Qin, J., and Liu, X. (2014). An approach to intuitionistic fuzzy multiple attribute decision making based on Maclaurin symmetric mean operators. *Journal of Intelligent & Fuzzy Systems*, 27(5), 2177–2190.

Qin, J., and Liu, X. (2015). Approaches to uncertain linguistic multiple attribute decision making based on dual Maclaurin symmetric mean. *Journal of Intelligent & Fuzzy Systems*, 29(1), 171–186.

Singh, S., and Garg, H. (2017). Distance measures between type-2 intuitionistic fuzzy sets and their application to multicriteria decision-making process. *Applied Intelligence*, 46, 788–799.

Tirkolaee, E. B., Abbasian, P., and Weber, G. W. (2021). Sustainable fuzzy multi-trip location-routing problem for medical waste management during the COVID-19 outbreak. *Science of the Total Environment*, 756, 143607.

Ullah, K. (2021). Picture fuzzy Maclaurin symmetric mean operators and their applications in solving multiattribute decision-making problems. *Mathematical Problems in Engineering*, 2021, 1–13.

Verma, R. (2015). Generalized Bonferroni mean operator for fuzzy number intuitionistic fuzzy sets and its application to multiattribute decision making. *International Journal of Intelligent Systems*, 30(5), 499–519.

Yu, D. (2013). Intuitionistic fuzzy geometric Heronian mean aggregation operators. *Applied Soft Computing*, 13(2), 1235–1246.

Zadeh, L. (1965). Fuzzy sets. *Information and Control*, 8(3), 338–353.

Zadeh, L. A. (1975). The concept of a linguistic variable and its application to approximate reasoning—I. *Information Sciences*, 8(3), 199–249.

Chapter 9

A new correlation coefficient of linguistic intuitionistic fuzzy sets and its use in medical diagnosis problems

Ritu Malik, Chirag Dhankhar, Devansh Sharma, Vijay Kumar, and Pravesh Kumar

9.1 INTRODUCTION

Intuitionistic fuzzy set (IFS), an extension of fuzzy set (FS) (Zadeh, 1965), was generalized by Atanassov (1986). IFSs offer information in quantitative form on the given data, including membership degree and non-membership degree. For processing uncertain data, IFSs have a variety of uses, including decision-making (Szmidt and Kacprzyk, 1996), image processing (Jurio et al., 2010), pattern recognition (Hwang et al., 2012), and medical diagnosis (Thao et al., 2019). Firstly, the correlation coefficient for IFS was studied by Gerstenkorn and Manko (1991). Xu (2006) developed a correlation coefficient measure for the IFS environment. Garg (2016) defined the idea of correlation and correlation coefficient for Pythagorean fuzzy sets (PFSs). Lin et al. (2021) presented a directional correlation coefficient for Pythagorean fuzzy sets (PFS) by considering four parameters of the PFSs, which are membership degree of PFS, non-membership degree of PFS, strength of commitment of PFS, and direction of commitment of PFS. Du (2019) developed a correlation and correlation coefficient for q-rung orthopair fuzzy sets. Liao et al. (2015) developed a correlation coefficient for hesitant fuzzy sets (HFSs), and it lies between [–1, 1]. Researchers later presented several correlation coefficients (Liu et al., 2016; Thao et al., 2019; Xu, 2006; Xu et al., 2008) to illustrate the connection between two items. The correlation coefficient is favourable in the field of medical diagnosis (Thao et al., 2019; Szmidt and Kacprzyk, 2001; Xuan, 2018) since it can demonstrate a relationship between two random variables. Even though numerous studies have reported the correlation coefficient for various types of fuzzy data, these results cannot be used to determine the connection between any two linguistic intuitionistic fuzzy sets. The fact that the correlation coefficient in our literature is only determined for IFS and its extensions, not for linguistic intuitionistic fuzzy sets. LIFSs are fusions of intuitionistic fuzzy sets (IFSs) and linguistic term sets (LTSs) (Chen et al., 2015). Linguistic intuitionistic fuzzy sets (LIFSs) (Chen et al., 2015) provide a better scope for decision-makers to express vague and imprecise information in a qualitative environment. Motivated by this, we studied the correlation coefficient

DOI: 10.1201/9781003497219-9

of linguistic intuitionistic fuzzy sets in this chapter. Further, we practiced the proposed correlation coefficient of LIFSs for medical diagnosis. For the verification of the developed formula, numerical analysis is done. The contribution of this chapter is to provide a correlation coefficient and medical diagnosis for the LIFS environment, which have not yet been developed to our best knowledge. The rest of the chapter is arranged as follows: Section 9.2 describes elementary concepts relevant to this article. In Section 9.3, a new correlation coefficient is constructed for LIFSs, and the necessary properties along with numerical examples are proved. In Section 9.4, a few case studies are done for medical diagnosis based on the proposed correlation coefficient. In Section 9.5, we provide the conclusion.

9.2 PRELIMINARIES

Definition 9.1 (Chen et al., 2015) A finite linguistic term set (LTS) $S = \{s_0, s_1, \ldots, s_h\}$ with odd cardinality, where s_k express the feasible value of a linguistic variable. For example, when evaluating the location of a playground area, we usually prefer a linguistic variable over a numeric value. We might use linguistic terms such as $S = \{s_0 =$ "Not Good," $s_1 =$ "Good," $s_2 =$ "Favorable," $s_3 =$ "Great," $s_4 =$ "Superb," $s_5 =$ "Marvelous"$\}$.

An LTS satisfies the following properties:

i. $s_j < s_k \Leftrightarrow j \leq k$;
ii. $Neg(s_j) = s_{h-k}$;
iii. $max(s_{j,} s_k) = s_j \Leftrightarrow s_j \geq s_k$;
iv. $min(s_j, s_k) = s_k \Leftrightarrow s_j \geq s_k$.

Later, Xu (2004) proposes the extended continuous LTS.

$$S_{[0,h]} = \{s_p \setminus s_0 \leq s_p \leq s_h\}.$$

Definition 9.2 (Chen et al., 2015) A linguistic intuitionistic fuzzy set (LIFS) in a universal set F is an expression G stated as:

$$G = \left\{ \left\langle f_i, s_{\alpha(f_i)}, s_{\beta(f_i)} \right\rangle \middle| f_i \in F \right\}. \tag{9.1}$$

where $s_{\alpha(f_i)} \in S_{[0,h]}$ and $s_{\beta(f_i)} \in S_{[0,h]}$ notify the membership degree (MD) and non-membership degree (NMD) of the element $f_i \in F$ to G, respectively, $0 \leq \alpha(f_i) \leq h$, $0 \leq \beta(f_i) \leq h$ and $0 \leq \alpha(f_i) + \beta(f_i) \leq h.s_{\pi(f_i)} = s_{h-\alpha(f_i)-\beta(f_i)}$ is known as hesitancy degree (HD) of f_i to G where $0 \leq \pi(f_i) \leq h$, $f_i \in F$.

Here, the pair $\langle s_\alpha, s_\beta \rangle$ represent a linguistic intuitionistic fuzzy number (LIFN) where $0 \leq \alpha \leq h$, $0 \leq \beta \leq h$, and $0 \leq \alpha + \beta \leq h$.

Let $\Gamma_{[0,h]}$ be the class of the LIFSs.

9.3 CORRELATION COEFFICIENT FOR LIFS

Let $F = \{f_1, f_2, \ldots, f_n\}$ be a finite universal set and a continuous linguistic term set $S = \{s_\gamma | \gamma \in [0,h]\}$, if $U = \{\langle f_i, s_{\alpha U(f_i)}, s_{\beta U(f_i)} \rangle | f_i \in F\}$ is an LIFS on F, then its information energy $T(U)$ is defined as:

$$T(U) = \sum_{i=1}^{n} \alpha_U^2(f_i) + \beta_U^2(f_i) + \pi_U^2(f_i). \tag{9.2}$$

Assume that two LIFSs U and V in the universe of discourse $F = \{f_1, f_2, \ldots, f_n\}$ are denoted by $U = \{\langle f_i, s_{\alpha U(f_i)}, s_{\beta U(f_i)} \rangle | f_i \in F\}$ and $V = \{\langle f_i, s_{\alpha V(f_i)}, s_{\beta V(f_i)} \rangle | f_i \in F\}$, where $\alpha_U(f_i), \beta_U(f_i), \alpha_V(f_i), \beta_V(f_i) \in [0,h]$ for every $f_i \in F$. The correlation between two LIFSs U and V is defined as:

$$C(U,V) = \sum_{i=1}^{n} \left(\alpha_U(f_i).\alpha_V(f_i) + \beta_U(f_i).\beta_V(f_i) + \pi_U(f_i).\pi_V(f_i) \right). \tag{9.3}$$

The correlation of LIFSs satisfies the listed properties:

(i) $C(U,U) = T(U)$;
(ii) $C(U,V) = C(V,U)$.

Definition 9.3 Let $U = \{\langle f_i, s_{\alpha U(f_i)}, s_{\beta U(f_i)} \rangle\} | f_i \in F\}$ and $V = \{\langle f_i, s_{\alpha V(f_i)}, s_{\beta V(f_i)} \rangle | f_i \in F\}$ be two LIFSs in the universal set $F = \{f_1, f_2, \ldots, f_n\}$ and a continuous LTS $S = \{s_\gamma \backslash \gamma \in [0,h]\}$. Then the correlation coefficient between them is expressed as:

$$K(U,V) = \frac{C(U,V)}{[T(U).T(V)]^{1/2}}$$

$$= \frac{\sum_{i=1}^{n} \alpha_U(f_i).\alpha_V(f_i) + \beta_U(f_i).\beta_V(f_i) + \pi_U(f_i).\pi_V(f_i)]}{\sqrt{\sum_{i=1}^{n} \alpha_U^2(f_i) + \beta_U^2(f_i) + \pi_U^2(f_i)}.\sqrt{\sum_{i=1}^{n} \alpha_V^2(f_i) + \beta_V^2(f_i) + \pi_V^2(f_i)}}. \tag{9.4}$$

Theorem 9.1 For any two LIFSs $U = \{\langle f_i, s_{\alpha U(f_i)}, s_{\beta U(f_i)} \rangle | f_i \in F\}$ and $V = \{\langle f_i, s_{\alpha V(f_i)}, s_{\beta V(f_i)} \rangle | f_i \in F\}$, in the universe of discourse $F = \{f_1, f_2, \ldots, f_n\}$, the correlation coefficient of LIFSs satisfies the following conditions:

(P1) $K(U,V) = K(V,U)$;
(P2) $0 \leq K(U,V) \leq 1$;
(P3) $U = V \Rightarrow K(U,V) = 1$.

Proof.

(P1) $K(U,V) = \frac{C(U,V)}{[T(U).T(V)]^{1/2}}$

$$= \frac{\sum_{i=1}^{n}\left[\alpha_U(f_i).\alpha_V(f_i) + \beta_U(f_i).\beta_V(f_i) + \pi_U(f_i).\pi_V(f_i)\right]}{\sqrt{\sum_{i=1}^{n}\alpha_U^2(f_i) + \beta_U^2(f_i) + \pi_U^2(f_i)}.\sqrt{\sum_{i=1}^{n}\alpha_V^2(f_i) + \beta_V^2(f_i) + \pi_V^2(f_i)}} = K(V,U).$$

(P2) It is obvious $K(U,V) \geq 0$ .Then, we will prove $K(U,V) \leq 1$.

$$C(U,V) = \sum_{i=1}^{n}\left(\alpha_U(f_i).\alpha_V(f_i) + \beta_U(f_i).\beta_V(f_i) + \pi_U(f_i).\pi_V(f_i)\right)$$

$$= \left(\alpha_U(f_1).\alpha_V(f_1) + \beta_U(f_1).\beta_V(f_1) + \pi_U(f_1).\pi_V(f_1)\right)$$

$$+\left(\alpha_U(f_2).\alpha_V(f_2) + \beta_U(f_2).\beta_V(f_2) + \pi_U(f_2).\pi_V(f_2)\right)\ldots$$

$$+\left(\alpha_U(f_n).\alpha_V(f_n) + \beta_U(f_n).\beta_V(f_n) + \pi_U(f_n).\pi_V(f_n)\right).$$

By Cauchy–Schwarz inequality,
$(f_1z_1 + f_2z_2 + \ldots + f_nz_n)^2 \leq \left(f_1^2 + f_2^2 + \ldots + f_n^2\right).\left(z_1^2 + z_2^2 + \ldots + z_n^2\right)$, where, $(f_1 + f_2 + \ldots + f_n) \in R^n$ and $\left(z_1 + z_2 + \ldots + z_n\right) \in R^n$, we get

$$\left(C(U,V)\right)^2 \leq \begin{bmatrix}\left((\alpha_U f_1) + \beta_U(f_1) + \pi_U(f_1)\right) + \left((\alpha_U f_2) + \beta_U(f_2) + \pi_U(f_2)\right) + \ldots \\ +\left((\alpha_U f_n) + \beta_U(f_n) + \pi_U(f_n)\right)\end{bmatrix}$$

$$+\begin{bmatrix}\left((\alpha_V f_1) + \beta_V(f_1) + \pi_V(f_1)\right) + \left((\alpha_V f_2) + \beta_V(f_2) + \pi_V(f_2)\right) + \ldots \\ +\left((\alpha_V f_n) + \beta_V(f_n) + \pi_V(f_n)\right)\end{bmatrix}$$

$$= \sum_{i=1}^{n}\alpha_U^2(f_i) + \beta_U^2(f_i) + \pi_U^2(f_i) \times \sum_{i=1}^{n}\alpha_V^2(f_i) + \beta_V^2(f_i) + \pi_V^2(f_i)$$

$$= T(U).T(V)$$

Therefore,$(C\ (U,V))^2 \leq T(U).T(V)$ and thus $K(U,V) \leq 1$. Hence, we obtain:

$$0 \leq K(U,V) \leq 1.$$

(P3) As $U = V$ implies that $\alpha_U(f_i) = \alpha_V(f_i)$ and $\beta_U(f_i) = \beta_V(f_i)$ for $f_i \in$ F. Thus, $K(U,V) = 1$.

Example 9.1 Consider U and V be two LIFS in the universal set $F = \{f_1, f_2, f_3\}$, and $U = \{\langle f_1, s_6, s_2\rangle, \langle f_2, s_2, s_4\rangle, \langle f_3, s_4, s_3\rangle\}$ *and* $V = \{\langle f_1, s_7, s_1\rangle, \langle f_2, s_2, s_3\rangle, \langle f_3, s_5, s_2\rangle\}$ where $U,V \in s_{[0,h]}$. By using Eq. (9.2), The information energy $T(U) = 90$ and $T(V) = 102$. By using Eq. (9.3), the correlation between the linguistic intuitionistic fuzzy sets U and V is $C(U,V) = 93$. Hence, the correlation coefficient between LIFSs U and V is computed by using Eq. (9.4), $K(U,V) = \frac{93}{\sqrt{90 \times 102}} = 0.9706$.

Example 9.2 Consider U and V be two LIFSs in the universal set $F = \{f_1, f_2, f_3\}$ and $U = \{\langle f_1, s_3, s_2\rangle, \langle f_2, s_4, s_3\rangle, \langle f_3, s_6, s_1\rangle\}$ and $V = \{\langle f_1, s_2, s_2\rangle, \langle f_2, s_5, s_3\rangle, \langle f_3, s_7, s_1\rangle\}$ where $U,V \in s_{[0,8]}$. By using Eq. (9.2), the information energy $T(U) = 86$ and $T(V) = 108$. By using Eq. (9.3), the correlation between the linguistic intuitionistic fuzzy sets U and V is $C(U,V) = 94$. Hence, the correlation coefficient between LIFSs U and V is calculated by using Eq. (9.4), $K(U,V) = \frac{94}{\sqrt{86 \times 108}} = 0.9754$.

9.4 CASE STUDY

Example 9.3 In this case study, the developed correlation coefficient has been used for medical diagnosis. The information comes from a study conducted by Talukdar et al. (2020). In this example, the dataset is containing four patients P = {Jim, Sam, Tom, Dan}, five symptoms E = {E_1(Temperature), E_2 (Headache), E_3 (Stomach Pain), E_4 (Cough), E_5 (Chest Pain)} and five diseases: W = W_1(Viral fever), W_2 (Malaria), W_3 (Typhoid), W_4 (Stomach problem), W_5 (chest problem)}. The relations between the patients and the symptoms, denoted by PE and relations between the symptoms and the diseases, denoted by EW, are displayed in the form of LIFNs in Tables 9.1 and 9.2. Our goal is to make the final decision of diagnosis for each patient, based on the collection of symptoms for every disease.

Table 9.1 Patients–symptoms linguistic intuitionistic fuzzy relation

PE =	E_1	E_2	E_3	E_4	E_5
Jim	$\langle s_{0.8}, s_{0.1}\rangle$	$\langle s_{0.6}, s_{0.1}\rangle$	$\langle s_{0.2}, s_{0.8}\rangle$	$\langle s_{0.6}, s_{0.1}\rangle$	$\langle s_{0.1}, s_{0.6}\rangle$
Sam	$\langle s_{0.0}, s_{0.8}\rangle$	$\langle s_{0.4}, s_{0.4}\rangle$	$\langle s_{0.6}, s_{0.1}\rangle$	$\langle s_{0.1}, s_{0.7}\rangle$	$\langle s_{0.1}, s_{0.8}\rangle$
Tom	$\langle s_{0.8}, s_{0.1}\rangle$	$\langle s_{0.8}, s_{0.1}\rangle$	$\langle s_{0.0}, s_{0.6}\rangle$	$\langle s_{0.2}, s_{0.7}\rangle$	$\langle s_{0.0}, s_{0.5}\rangle$
Dan	$\langle s_{0.6}, s_{0.1}\rangle$	$\langle s_{0.5}, s_{0.4}\rangle$	$\langle s_{0.3}, s_{0.4}\rangle$	$\langle s_{0.7}, s_{0.2}\rangle$	$\langle s_{0.3}, s_{0.4}\rangle$

Note: Data from Talukdar et al. (2020).

Talukdar, P., Goala, S., Dutta, P., & Limboo, B. (2020). Fuzzy multicriteria decision making in medical diagnosis using an advanced distance measure on linguistic Pythagorean fuzzy sets. *Annals of Optimization Theory and Practice*, 3(4), 113–131.

Table 9.2 Symptoms–diseases linguistic intuitionistic fuzzy relation

$$EW = \begin{array}{c} \\ E_1 \\ E_2 \\ E_3 \\ E_4 \\ E_5 \end{array} \begin{array}{ccccc} W_1 & W_2 & W_3 & W_4 & W_5 \\ \left[\begin{array}{ccccc} \langle s_{0.4}, s_{0.0} \rangle & \langle s_{0.7}, s_{0.0} \rangle & \langle s_{0.3}, s_{0.3} \rangle & \langle s_{0.1}, s_{0.7} \rangle & \langle s_{0.1}, s_{0.8} \rangle \\ \langle s_{0.3}, s_{0.5} \rangle & \langle s_{0.2}, s_{0.6} \rangle & \langle s_{0.6}, s_{0.1} \rangle & \langle s_{0.2}, s_{0.4} \rangle & \langle s_{0.0}, s_{0.8} \rangle \\ \langle s_{0.1}, s_{0.7} \rangle & \langle s_{0.0}, s_{0.9} \rangle & \langle s_{0.2}, s_{0.7} \rangle & \langle s_{0.8}, s_{0.0} \rangle & \langle s_{0.2}, s_{0.8} \rangle \\ \langle s_{0.4}, s_{0.3} \rangle & \langle s_{0.7}, s_{0.0} \rangle & \langle s_{0.2}, s_{0.6} \rangle & \langle s_{0.2}, s_{0.7} \rangle & \langle s_{0.2}, s_{0.8} \rangle \\ \langle s_{0.1}, s_{0.7} \rangle & \langle s_{0.1}, s_{0.8} \rangle & \langle s_{0.1}, s_{0.9} \rangle & \langle s_{0.2}, s_{0.7} \rangle & \langle s_{0.8}, s_{0.1} \rangle \end{array}\right] \end{array}$$

Note: Data from Talukdar et al. (2020).[1]

To apply the suggested method to the considered data, firstly, the information energy of the patients with respect to the symptoms is determined by using Eq. (9.2) in Table 9.3. The information energy of symptoms with respect to the diseases (EW) is calculated by using Eq. (9.2) in Table 9.4. The correlation of two LIFSs PE and EW is computed by using Eq. (9.3) in Table 9.5.

The proposed correlation coefficient presented in Eq. (9.4) is utilized for the disease's diagnosis in all patients.

From Table 9.6, it is observed that the correlation coefficient of Jim is the highest with viral fever and malaria. So, it is obvious that Jim is suffering from viral fever and malaria diseases. The correlation coefficient between Sam and stomach problem is the maximum. So, it is easy to decide that Sam is suffering from stomach problems. We will take the same path for Tom

Table 9.3 Information energy of PE

$$T(PE) = \begin{bmatrix} s_{51.0600} & s_{53.6600} & s_{49.6800} & s_{53.6600} & s_{53.6600} \\ s_{52.4800} & s_{52.1600} & s_{53.6600} & s_{52.3400} & s_{51.0600} \\ s_{51.0600} & s_{51.0600} & s_{55.1200} & s_{50.9400} & s_{56.5000} \\ s_{53.6600} & s_{50.8200} & s_{53.5400} & s_{50.9400} & s_{53.5400} \end{bmatrix}$$

Table 9.4 Information energy of EW

$$(EW) = \begin{bmatrix} s_{57.9200} & s_{53.7800} & s_{54.9400} & s_{52.3400} & s_{51.0600} \\ s_{52.1800} & s_{52.2400} & s_{53.6600} & s_{54.9600} & s_{52.4800} \\ s_{52.3400} & s_{51.2200} & s_{50.9400} & s_{52.4800} & s_{49.6800} \\ s_{53.5400} & s_{53.7800} & s_{52.2400} & s_{50.9400} & s_{49.6800} \\ s_{52.3400} & s_{51.0600} & s_{49.8200} & s_{50.9400} & s_{51.0600} \end{bmatrix}$$

Table 9.5 Correlation between *PE* and *EW*

$$C(PE,EW)=\begin{bmatrix} s_{264.6000} & s_{261.5800} & s_{261.1600} & s_{260.3000} & s_{256.1600} \\ s_{264.0700} & s_{260.3300} & s_{261.0900} & s_{261.5800} & s_{256.7700} \\ s_{256.8600} & s_{262.4300} & s_{262.6200} & s_{261.8300} & s_{257.6500} \\ s_{265.1000} & s_{261.8600} & s_{261.4400} & s_{261.1500} & s_{257.0000} \end{bmatrix}$$

Table 9.6 Correlation coefficient between patients and diseases

	W_1	W_2	W_3	W_4	W_5
Jim	$s_{0.9989}$	$s_{0.9989}$	$s_{0.9984}$	$s_{0.9949}$	$s_{0.9937}$
Sam	$s_{0.9967}$	$s_{0.9941}$	$s_{0.9980}$	$s_{0.9997}$	$s_{0.9961}$
Tom	$s_{0.9983}$	$s_{0.9969}$	$s_{0.9988}$	$s_{0.9954}$	$s_{0.9941}$
Dan	$s_{0.9991}$	$s_{0.9986}$	$s_{0.9980}$	$s_{0.9967}$	$s_{0.9955}$

and Dan's medical diagnosis. Thus, it is concluded that Tom is suffering from typhoid and Dan is suffering from viral fever.

9.5 CONCLUSION

Linguistic intuitionistic fuzzy set (LIFS) is a unique concept that was introduced by Chen et al. (2015) to deal with many real-life issues in uncertain situations. In this chapter, correlation coefficient for LIFSs is introduced and axioms for the same are verified. The correlation coefficient is a significant information measure in many fields. The proposed correlation coefficient is used for decision-making in medical diagnosis problems. We have illustrated numerical examples of medical diagnosis in support of the presented formula. The proposed method gives a flexible and easy solution for medical diagnosis problems in linguistic intuitionistic fuzzy environment. It will be advantageous to expand the proposed correlation coefficient to the linguistic interval-valued intuitionistic fuzzy sets (LIVIFS) environment because LIVIFSs are more accurate to accomplish vague information.

REFERENCES

Atanassov, K. T. (1986). Intuitionistic fuzzy sets. *Fuzzy Sets and Systems*, 20(1), 87–96.

Chen, Z., Liu, P., & Pei, Z. (2015). An approach to multiple attribute group decision making based on linguistic intuitionistic fuzzy numbers. *International Journal of Computational Intelligence Systems*, *8*(4), 747–760.

Du, W. S. (2019). Correlation and correlation coefficient of generalized orthopair fuzzy sets. *International Journal of Intelligent Systems*, *34*(4), 564–583.

Garg, H. (2016). A novel correlation coefficient between Pythagorean fuzzy sets and its applications to decision-making processes. *International Journal of Intelligent Systems*, *31*(12), 1234–1252.

Gerstenkorn, T., & Mańko, J. (1991). Correlation of intuitionistic fuzzy sets. *Fuzzy Sets and Systems*, *44*(1), 39–43.

Hwang, C. M., Yang, M. S., Hung, W. L., & Lee, M. G. (2012). A similarity measure of intuitionistic fuzzy sets based on the Sugeno integral with its application to pattern recognition. *Information Sciences*, *189*, 93–109.

Jurio, A., Paternain, D., Bustince, H., Guerra, C., & Beliakov, G. (2010). A construction method of Atanassov's intuitionistic fuzzy sets for image processing. In *2010 5th IEEE International Conference Intelligent Systems* (pp. 337–342). IEEE.

Liao, H., Xu, Z., & Zeng, X. J. (2015). Novel correlation coefficients between hesitant fuzzy sets and their application in decision making. *Knowledge-Based Systems*, *82*, 115–127.

Lin, M., Huang, C., Chen, R., Fujita, H., & Wang, X. (2021). Directional correlation coefficient measures for Pythagorean fuzzy sets: Their applications to medical diagnosis and cluster analysis. *Complex & Intelligent Systems*, *7*, 1025–1043.

Liu, B., Shen, Y., Mu, L., Chen, X., & Chen, L. (2016). A new correlation measure of the intuitionistic fuzzy sets. *Journal of Intelligent & Fuzzy Systems*, *30*(2), 1019–1028.

Szmidt, E., & Kacprzyk, J. (1996). Intuitionistic fuzzy sets in group decision making. *Notes on IFS*, *2*(1), 15–32.

Szmidt, E., & Kacprzyk, J. (2001). Intuitionistic fuzzy sets in some medical applications. In *International Conference on Computational Intelligence* (pp. 148–151). Berlin, Heidelberg: Springer Berlin Heidelberg.

Talukdar, P., Goala, S., Dutta, P., & Limboo, B. (2020). Fuzzy multicriteria decision making in medical diagnosis using an advanced distance measure on linguistic Pythagorean fuzzy sets. *Annals of Optimization Theory and Practice*, *3*(4), 113–131.

Thao, N. X., Ali, M., & Smarandache, F. (2019). An intuitionistic fuzzy clustering algorithm based on a new correlation coefficient with application in medical diagnosis. *Journal of Intelligent & Fuzzy Systems*, *36*(1), 189–198.

Xu, Z. (2006). On correlation measures of intuitionistic fuzzy sets. In *International Conference on Intelligent Data Engineering and Automated Learning* (pp. 16–24). Berlin, Heidelberg: Springer Berlin Heidelberg.

Xu, Z. (2004) A method based on linguistic aggregation operators for group decision making linguistic preference relations. *Inform Sci,* 166(1), 19–30.

Xu, Z., Chen, J., & Wu, J. (2008). Clustering algorithm for intuitionistic fuzzy sets. *Information Sciences*, *178*(19), 3775–3790.

Xuan Thao, N. (2018). A new correlation coefficient of the intuitionistic fuzzy sets and its application. *Journal of Intelligent & Fuzzy Systems*, *35*(2), 1959–1968.

Zadeh, L. A. (1965). Fuzzy sets. *Information and Control*, *8*(3), 338–353.

Chapter 10

Power Einstein aggregation operators of intuitionistic fuzzy sets and their application in MADM

Amit Sharma, Naveen Mani, Rishu Arora, and Reeta Bhardwaj

10.1 INTRODUCTION

In real life, the most common multi-attribute decision-making (MADM) challenges, such as staff selection, supplier selection, and contractor selection for building development, rely on the consideration of multiple elements simultaneously. These factors contribute to the increased complexity and uncertainty of the situation for decision-makers when evaluating qualities. Decision-makers are unable to provide precise numerical estimates. To remove such types of difficulties of the decision-makers, Zadeh (1965) introduced the concept of the fuzzy sets (FSs), and after that extensions of it like intuitionistic FS (IFS) (Atannasov and Gargov, 1989) and interval-valued IFS (IVIFS) (Atannasov, 1986) have been proposed. Recently, these theories have been used widely by decision-makers to address MADM problems in real life. The aggregating method and attribute weight are used to determine the ranking of alternatives in the MADM problem-solving process. The weight of an attribute is important during the decision-making process because weight fluctuations can modify how the alternatives are ranked. Decision-makers assign attribute weights in certain MADM problems, but in other cases, the fuzziness of the data prevents decision-makers from assigning attribute weights. To handle it, the entropy measure is an effective tool that can depict the fuzziness of the data and we can get the weights of the attribute. Garg et al. (2017) proposed a generalized entropy measure of degree β and order α. Garg and Kaur (2018) proposed (R, S)-Norm novel entropy measure for IFS. Bhardwaj et al. (2022) defined the entropy measures for the IFSs and their applications in the MADM problem. Neelam et al. (2023) defined the entropy measures for the linguistic q-rung orthopair fuzzy number.

In the aggregating stage, under the IFS and IVIFS, several types of AOs have been developed by the researchers to fuse the decision-maker(s) data. For example, Xu and Yager (2006) and Xu (2007) developed geometric and averaging AOs correspondingly for aggregating the IFS environment. Xu and Chen (2007) defined various geometric and averaging AOs for the IVIFS environment. Garg (2016) presented intuitionistic fuzzy interactive

DOI: 10.1201/9781003497219-10

averaging (IFIA) AOs by adding the hesitance degree into these existing operators. Later on, Garg (2017) improved Einstein's operational laws and defined AOs based on it for aggregating IFNs. Garg (2019) defined the Hamacher AOs for handling the MADM issues for IF information with the entropy weights. Seikh and Mandal (2021) defined the Dombi AOs for the IFNs environment. Senapati et al. (2022) presented the AO based on the Aczel–Alsina norm and the MADM method based on it. Kumar and Chen (2022) defined the Heronian mean AO for the IFNs

In every one of these works, it is expected that the inclination given by decision-maker(s) is an autonomous and steady connection among the contention isn't considered in any way. But in some situations, there are certain actual issues in which a few degrees of interrelationship exist among the contention and it is basic to require under consideration this interdependency during the aggregation process in order to make an optimal decision. For handling such issues, Yager (2001) presented an instrument to supply more flexibility in the information aggregation process, i.e., defined as power-average (PA) and power-ordered weighted averaging (POWA) operators. Based on the PA and geometric mean, Xu (2011) presented power geometric (PG), weighted PG, and power-ordered weighted geometric (POWG) operators. Xu and Yager (2009) developed an approach based on PA AOs for handling MADM issues in the IF environment. Garg and Kumar (2019) defined the power AOs for linguistic IFNs. Garg and Kumar (2020) proposed the power AOs for aggregating the IFNs environment. Dhankar and Kumar (2023) defined the power AO based on the Yager's norm for the q-rung orthopair fuzzy values.

Based on the above analysis, it seems that IFS theory is extremely used by researchers and we find out that there are some drawbacks in existing entropy measure that are explained by taking some counterexamples in this article. Therefore, in this article, firstly, we construct a new entropy measure that depicts the fuzziness of a set in the IFS environment and presents the advantages of this measure by comparing it with existing entropy measures of IFSs. Certain properties of the proposed entropy measure have been demonstrated to show legality and validity. The main purpose of introducing a new entropy measure is to evaluate the weights of the attributes under the characteristics of the weights of the attributes which are either partly known or entirely unknown. Secondly, by the motivated and improved Einstein operations proposed by Garg (2017) and by adding the features of power AO, we proposed some new power Einstein aggregating operators for aggregating the IFNs information. The proposed operators take into account how individual input argument relates to the others. Because the related weights of the attributes in the power operator are derived directly from their support steps, in accordance with the theory that more support value is given to all the other values, more weight can be derived, and some effect of the unreasonable details provided by the experts can be moderated.

Therefore, based on the above analysis, the main targets of this chapter are summarized as

- (i) We present new power Einstein aggregating operators, namely, intuitionistic fuzzy power Einstein averaging (IFPEA) and intuitionistic fuzzy weighted power Einstein averaging (IFWPEA) operators.
- (ii) The different alluring properties and extraordinary cases of the proposed operators have been presented in detail.
- (iii) An MADM approach has been developed for IF information.
- (iv) To evaluate the developed approach, an actual illustrative illustration has been taken and the results of the proposed method are compared with the results of existing methods to show the advantages of the proposed method.

In order to achieve the aforementioned goals, this chapter is organized as follows: Section 10.2, briefly introduces the basics of IFS theory and also reviews the drawbacks of the existing entropy measures. Some power Einstein AOs for aggregating the IF information are developed in Section 10.3. In Section 10.4, an MADM approach based on the proposed AOs under the IFNs environment has been developed. In Section 10.5, an actual case has been discussed to illustrate the effectiveness of the presented DM approach. Finally, Section 10.6 concludes the chapter.

10.2 PRELIMINARIES

In this segment, a few elementary concepts related to the IFS theory on finite universal set X are characterized as follows:

Definition 10.1 (Atannasov and Gargove, 1989) An IFS H is defined as

$$H = \{\langle x, \tau_H(x), \theta_H(x)\rangle \mid x \in X\},$$

where $\tau_H(x), \theta_H(x) \in [0,1]$ represents membership and non-membership degrees of x to H, respectively, such that $0 \le \tau_H(x) + \theta_H(x) \le 1$ holds for $\forall\, x$. The intuitionistic index of x to H is defined as $\pi_H(x) = 1 - \tau_H(x) - \theta_H(x)$. Usually, the pair $\langle \tau, \theta \rangle$ is called an IFN.

Definition 10.2 (Hong and Choi, 2000) The score function for IFN $\gamma = \langle \tau, \theta \rangle$ is defined as

$$S(\gamma) = \tau - \theta, \tag{10.1}$$

and accuracy function

$$H(\gamma) = \tau + \theta. \tag{10.2}$$

Definition 10.3 (Garg, 2017) If $\gamma_1 = \langle \tau_1, \theta_1 \rangle, \gamma_2 = \langle \tau_2, \theta_2 \rangle$ and $\gamma = \langle \tau, \theta \rangle$ be three IFNs, then Einstein operations are defined as:

(i) $$\lambda\gamma = \left\langle \frac{(1+\tau)^{\lambda} - (1-\tau)^{\lambda}}{(1+\tau)^{\lambda} + (1-\tau)^{\lambda}}, \frac{2((1-\tau)^{\lambda} - (1-\tau-\theta)^{\lambda})}{(1+\tau)^{\lambda} + (1-\tau)^{\lambda}} \right\rangle, \quad \lambda > 0.$$

Theorem 10.1 (Liu and Ren, 2014) For two IFSs $H_1 = \{\langle x, \tau_1(x), \theta_1(x) \rangle \mid x \in X\}$ and $H_2 = \{\langle x, \tau_2(x), \theta_2(x) \rangle \mid x \in X\}$, we have

1. $H_1 \subseteq H_2$ if $\tau_1(x) \geq \tau_2(x)$ and $\theta_1(x) \leq \theta_2(x)$.
2. $H_1 = H_2 \Leftrightarrow \tau_1(x) = \tau_2(x)$ and $\theta_1(x) = \theta_2(x)$.
3. $H^c = \langle x, \theta_H(x), \tau_H(x) \mid x \in X \rangle$.

Definition 10.4 (Yager, 2001) For aggregating the real numbers γ_t, $t = 1, 2, 3, \ldots, n$, the power averaging (PA) operator is defined as

$$\mathrm{PA}(\gamma_1, \gamma_2, \ldots, \gamma_n) = \sum_{t=1}^{n} \frac{(1 + T(\gamma_t))}{\sum_{t=1}^{n} (1 + T(\gamma_t))} \gamma_t, \tag{10.3}$$

where $T(\gamma_t) = \sum_{\substack{l=1 \\ l \neq t}}^{n} Sup(\gamma_t, \gamma_l)$ and $Sup(\gamma_t, \gamma_l)$ means support degree to which γ_t supports γ_l which satisfies the following rules:

(*i*). $Sup(\gamma_t, \gamma_l) \in [0,1]$;
(*ii*). $Sup(\gamma_t, \gamma_l) = Sup(\gamma_l, \gamma_t)$
(*iii*). $Sup(\gamma_t, \gamma_l) \geq Sup(\gamma_m, \gamma_n)$ if $|\gamma_t - \gamma_l| \leq |\gamma_m - \gamma_n|$

10.3 PROPOSED INTUITIONISTIC FUZZY POWER EINSTEIN AVERAGING (IFPEA) AGGREGATION OPERATOR

Definition 10.5 For IFNs γ_t, $t = 1, 2, 3, \ldots, n$, $\mathrm{IFPEA} : \Omega^n \to \Omega$ operator defined as

$$\mathrm{IFPEA}(\gamma_1, \gamma_2, \ldots \gamma_n) = \bigoplus_{t=1}^{n} \frac{(1 + T(\gamma_t))}{\sum_{t=1}^{n} (1 + T(\gamma_t))} \gamma_t,$$

$$= \left\langle \frac{\prod_{t=1}^{n}(1+\tau_t)^{\frac{(1+T(\gamma_t))}{\sum_{t=1}^{n}(1+T(\gamma_t))}} - \prod_{t=1}^{n}(1-\tau_t)^{\frac{(1+T(\gamma_t))}{\sum_{t=1}^{n}(1+T(\gamma_t))}}}{\prod_{t=1}^{n}(1+\tau_t)^{\frac{(1+T(\gamma_t))}{\sum_{t=1}^{n}(1+T(\gamma_t))}} + \prod_{t=1}^{n}(1-\tau_t)^{\frac{(1+T(\gamma_t))}{\sum_{t=1}^{n}(1+T(\gamma_t))}}}, \frac{2\left(\prod_{t=1}^{n}(1-\tau_t)^{\frac{(1+T(\gamma_t))}{\sum_{t=1}^{n}(1+T(\gamma_t))}} - \prod_{t=1}^{n}(1-\tau_t-\theta_t)^{\frac{(1+T(\gamma_t))}{\sum_{t=1}^{n}(1+T(\gamma_t))}}\right)}{\prod_{t=1}^{n}(1+\tau_t)^{\frac{(1+T(\gamma_t))}{\sum_{t=1}^{n}(1+T(\gamma_t))}} + \prod_{t=1}^{n}(1-\tau_t)^{\frac{(1+T(\gamma_t))}{\sum_{t=1}^{n}(1+T(\gamma_t))}}} \right\rangle \quad (10.4)$$

where $T(\gamma_t) = \sum_{\substack{l=1 \\ l \neq t}}^{n} Sup(\gamma_t, \gamma_l)$ and $Sup(\gamma_t, \gamma_l) = S(\gamma_t, \gamma_l)$, means support degree to which γ_1 supports γ_t which satisfies the following rules:

(i). $Sup(\gamma_t, \gamma_l) \in [0,1]$;

(ii). $Sup(\gamma_t, \gamma_l) = Sup(\gamma_l, \gamma_t)$;

(iii). $Sup(\gamma_t, \gamma_l) \geq Sup(\gamma_m, \gamma_n)$ if $S(\gamma_t, \gamma_l) \geq S(\gamma_m, \gamma_n)$

Theorem 10.2 For a collection of IFNs $\gamma_t = \langle \tau_t, \theta_t \rangle$, $t = 1, 2, \ldots, n$, the IFPEA operator satisfies the following properties:

(P1) **Idempotency:** If all γ_t are equal, i.e, $\gamma_t = \gamma$, for all t, then

$\text{IFPEA}(\gamma_1, \gamma_2, \ldots \gamma_n) = \gamma$.

(P2) **Boundedness:** $\gamma^- = \langle \tau^-, \theta^- \rangle$ and $\gamma^+ = \langle \tau^+, \theta^+ \rangle$ and where $\tau^- = \min_t \{\tau_t\}, \theta^- = \max_t \{\theta_t\}, \tau^+ = \max_t$ and $\theta^+ = \min_t \{\theta_t\}$, then we have

$\gamma^- \leq \text{IFPEA}(\gamma_1, \gamma_2, \ldots, \gamma_n) \leq \gamma^+$.

(P3) **Commutativity:** If γ_t, γ_t^*, $t = 1,2, \ldots, n$, be two collection of IFNs, then

$$\text{IFPEA}(\gamma_1,\gamma_2,\ldots,\gamma_n) = \text{IFPEA}(\gamma_1^*,\gamma_2^*,\ldots\gamma_n^*),$$

where $(\gamma_1^*,\gamma_2^*,\ldots\gamma_n^*)$ is any permutation of $(\gamma_1,\gamma_2,\ldots,\gamma_n)$.

Proof. For a collection of IFNs $\gamma_t = \langle \tau_t, \theta_t \rangle$, $t = 1, 2, \ldots, n$,, where $T(\gamma_t) = \sum_{\substack{l=1 \\ l \neq t}}^{n} Sup(\gamma_t, \gamma_l)$, and $Sup(\gamma_t, \gamma_l) = S(\gamma_t, \gamma_l)$, we have

(P1) Since, $\gamma_t = \gamma = \langle \tau, \theta \rangle$, $t = 1,2,\ldots,n$, then

$$\text{IFPEA}(\gamma_1,\gamma_2,\ldots,\gamma_n)$$

$$= \left\langle \frac{\prod_{t=1}^{n}(1+\tau)^{\frac{(1+T(\gamma))}{\sum_{t=1}^{n}(1+T(\gamma))}} - \prod_{t=1}^{n}(1-\tau)^{\frac{(1+T(\gamma))}{\sum_{t=1}^{n}(1+T(\gamma))}}}{\prod_{t=1}^{n}(1+\tau)^{\frac{(1+T(\gamma))}{\sum_{t=1}^{n}(1+T(\gamma))}} + \prod_{t=1}^{n}(1-\tau)^{\frac{(1+T(\gamma))}{\sum_{t=1}^{n}(1+T(\gamma))}}}, \frac{2\left(\prod_{t=1}^{n}(1-\tau)^{\frac{(1+T(\gamma))}{\sum_{t=1}^{n}(1+T(\gamma))}} - \prod_{t=1}^{n}(1-\tau-\theta)^{\frac{(1+T(\gamma))}{\sum_{t=1}^{n}(1+T(\gamma))}}\right)}{\prod_{t=1}^{n}(1+\tau)^{\frac{(1+T(\gamma))}{\sum_{t=1}^{n}(1+T(\gamma))}} + \prod_{t=1}^{n}(1-\tau)^{\frac{(1+T(\gamma))}{\sum_{t=1}^{n}(1+T(\gamma))}}} \right\rangle$$

$$= \left\langle \frac{(1+\tau)^{\sum_{t=1}^{n}\frac{(1+T(\gamma))}{\sum_{t=1}^{n}(1+T(\gamma))}} - (1-\tau)^{\sum_{t=1}^{n}\frac{(1+T(\gamma))}{\sum_{t=1}^{n}(1+T(\gamma))}}}{(1+\tau)^{\sum_{t=1}^{n}\frac{(1+T(\gamma))}{\sum_{t=1}^{n}(1+T(\gamma))}} + (1-\tau)^{\sum_{t=1}^{n}\frac{(1+T(\gamma))}{\sum_{t=1}^{n}(1+T(\gamma))}}}, \frac{2\left((1-\tau)^{\sum_{t=1}^{n}\frac{(1+T(\gamma))}{\sum_{t=1}^{n}(1+T(\gamma))}} - (1-\tau-\theta)^{\sum_{t=1}^{n}\frac{(1+T(\gamma))}{\sum_{t=1}^{n}(1+T(\gamma))}}\right)}{(1+\tau)^{\sum_{t=1}^{n}\frac{(1+T(\gamma))}{\sum_{t=1}^{n}(1+T(\gamma))}} + (1-\tau)^{\sum_{t=1}^{n}\frac{(1+T(\gamma))}{\sum_{t=1}^{n}(1+T(\gamma))}}} \right\rangle$$

$$= \left\langle \frac{(1+\tau)-(1-\tau)}{(1+\tau)+(1-\tau)}, \frac{2\{(1-\tau)-(1-\tau-\theta)\}}{(1+\tau)+(1-\tau)} \right\rangle = \langle \tau, \theta \rangle = \gamma \cdot$$

(P2) Let $g(x) = \frac{1-x}{1+x}, x \in (0,1]$, then $g'(x) = \frac{-2}{(1+x)^2} < 0$, which implies $g(x)$ is decreasing function on [0, 1]. Since $\tau^- \le \tau_t \le \tau^+, \forall t$, then $g(\tau^-) \le g(\tau_t) \le g(\tau^+)$, $t = 1, 2, \ldots, n$. Therefore, we have

$$\left(\frac{1-\tau^+}{1+\tau^+}\right) \le \left(\frac{1-\tau_t}{1+\tau_t}\right) \le \left(\frac{1-\tau^-}{1+\tau^-}\right)$$

$$\Leftrightarrow \prod_{t=1}^{n} \left(\frac{1-\tau^+}{1+\tau^+}\right)^{\frac{(1+T(\gamma))}{\sum_{t=1}^{n}(1+T(\gamma))}} \le \prod_{t=1}^{n} \left(\frac{1-\tau_t}{1+\tau_t}\right)^{\frac{(1+T(\gamma))}{\sum_{t=1}^{n}(1+T(\gamma))}} \le \prod_{t=1}^{n} \left(\frac{1-\tau^-}{1+\tau^-}\right)^{\frac{(1+T(\gamma))}{\sum_{t=1}^{n}(1+T(\gamma))}}$$

$$\Leftrightarrow \left(\frac{1-\tau^+}{1+\tau^+}\right)^{\sum_{t=1}^{n}\frac{(1+T(\gamma))}{\sum_{t=1}^{n}(1+T(\gamma))}} \le \prod_{t=1}^{n} \left(\frac{1-\tau_t}{1+\tau_t}\right)^{\frac{(1+T(\gamma))}{\sum_{t=1}^{n}(1+T(\gamma))}} \le \left(\frac{1-\tau^-}{1+\tau^-}\right)^{\sum_{t=1}^{n}\frac{(1+T(\gamma))}{\sum_{t=1}^{n}(1+T(\gamma))}}$$

$$\Leftrightarrow \left(\frac{1-\tau^+}{1+\tau^+}\right) \le \prod_{t=1}^{n} \left(\frac{1-\tau_t}{1+\tau_t}\right)^{\frac{(1+T(\gamma))}{\sum_{t=1}^{n}(1+T(\gamma))}} \le \left(\frac{1-\tau^-}{1+\tau^-}\right)$$

$$\Leftrightarrow \left(\frac{2}{1+\tau^+}\right) \le 1 + \prod_{t=1}^{n} \left(\frac{1-\tau_t}{1+\tau_t}\right)^{\frac{(1+T(\gamma))}{\sum_{t=1}^{n}(1+T(\gamma))}} \le \left(\frac{2}{1+\tau^-}\right)$$

$$\Leftrightarrow \left(\frac{1+\tau^-}{2}\right) \le \frac{1}{1 + \prod_{t=1}^{n} \left(\frac{1-\tau_t}{1+\tau_t}\right)^{\frac{(1+T(\gamma))}{\sum_{t=1}^{n}(1+T(\gamma))}}} \le \left(\frac{1+\tau^+}{2}\right)$$

$$\Leftrightarrow \tau^- \le \frac{2}{1 + \prod_{t=1}^{n} \left(\frac{1-\tau_t}{1+\tau_t}\right)^{\frac{(1+T(\gamma))}{\sum_{t=1}^{n}(1+T(\gamma))}}} - 1 \le \tau^+$$

Thus,

$$\Leftrightarrow \tau^- \le \frac{\prod_{t=1}^{n}(1+\tau_t)^{\frac{(1+T(\gamma))}{\sum_{t=1}^{n}(1+T(\gamma))}} - \prod_{t=1}^{n}(1-\tau_t)^{\frac{(1+T(\gamma))}{\sum_{t=1}^{n}(1+T(\gamma))}}}{\prod_{t=1}^{n}(1+\tau_t)^{\frac{(1+T(\gamma))}{\sum_{t=1}^{n}(1+T(\gamma))}} + \prod_{t=1}^{n}(1-\tau_t)^{\frac{(1+T(\gamma))}{\sum_{t=1}^{n}(1+T(\gamma))}}} \le \tau^+ \tag{10.5}$$

Again,

$$\left(\frac{1+\tau^+}{1-\tau^+}\right) \le \left(\frac{1+\tau_t}{1-\tau_t}\right) \le \left(\frac{1+\tau^-}{1-\tau^-}\right)$$

$$\Leftrightarrow \left(\frac{1+\tau^+}{1-\tau^+}\right) \le \prod_{t=1}^{n}\left(\frac{1+\tau_t}{1-\tau_t}\right)^{\frac{(1+T(\gamma))}{\sum_{t=1}^{n}(1+T(\gamma))}} \le \left(\frac{1+\tau^-}{1-\tau^-}\right)$$

$$\Leftrightarrow \left(\frac{1-\tau^-}{2}\right) \le \frac{1}{1+\prod_{t=1}^{n}\left(\frac{1+\tau_t}{1-\tau_t}\right)^{\frac{(1+T(\gamma))}{\sum_{t=1}^{n}(1+T(\gamma))}}} \le \left(\frac{2}{1-\tau^+}\right)$$

$$\Leftrightarrow 1-\tau^- \le \frac{2}{1+\prod_{t=1}^{n}\left(\frac{1+\tau_t}{1-\tau_t}\right)^{\frac{(1+T(\gamma))}{\sum_{t=1}^{n}(1+T(\gamma))}}} \le 1-\tau^+$$

Also, $\left(\frac{1-\tau^+-\theta^-}{1-\tau^-}\right) \le \left(\frac{1-\tau_t-\theta_t}{1-\tau_t}\right) \le \left(\frac{1-\tau^--\theta^+}{1-\tau^+}\right)$.

$$\Leftrightarrow 1-\left(\frac{1-\tau^--\theta^+}{1-\tau^+}\right) \le 1-\prod_{t=1}^{n}\left(\frac{1-\tau_t-\theta_t}{1-\tau_t}\right)^{\frac{(1+T(\gamma))}{\sum_{t=1}^{n}(1+T(\gamma))}} \le \left(\frac{1-\tau^+-\theta^-}{1-\tau^-}\right)$$

$$\Leftrightarrow \left(\frac{-\tau^-+\theta^+-\tau^-}{1-\tau^+}\right) \le 1-\prod_{t=1}^{n}\left(\frac{1-\tau_t-\theta_t}{1-\tau_t}\right)^{\frac{(1+T(\gamma))}{\sum_{t=1}^{n}(1+T(\gamma))}} \le \left(\frac{-\tau^-+\theta^-+\tau^+}{1-\tau^-}\right)$$

$$\Leftrightarrow -\tau^-+\theta^+-\tau^- \le \frac{2\left\{1-\prod_{t=1}^{n}\left(\frac{1-\tau_t-\theta_t}{1-\tau_t}\right)^{\frac{(1+T(\gamma))}{\sum_{t=1}^{n}(1+T(\gamma))}}\right\}}{1+\prod_{t=1}^{n}\left(\frac{1+\tau_t}{1-\tau_t}\right)^{\frac{(1+T(\gamma))}{\sum_{t=1}^{n}(1+T(\gamma))}}} \le -\tau^-+\theta^-+\tau^+$$

$$\Leftrightarrow \theta^{+} \le \frac{2\left\{1-\prod_{t=1}^{n}\left(\frac{1-\tau_t-\theta_t}{1-\tau_t}\right)^{\frac{(1+T(\gamma))}{\sum_{t=1}^{n}(1+T(\gamma))}}\right\}}{1+\prod_{t=1}^{n}\left(\frac{1+\tau_t}{1-\tau_t}\right)^{\frac{(1+T(\gamma))}{\sum_{t=1}^{n}(1+T(\gamma))}}} \le \theta^{-}$$

$$\theta^{+} \le \frac{2\left(\prod_{t=1}^{n}(1-\tau_t)^{\frac{(1+T(\gamma_t))}{\sum_{t=1}^{n}(1+T(\gamma_t))}}-\prod_{t=1}^{n}(1-\tau_t-\theta_t)^{\frac{(1+T(\gamma_t))}{\sum_{t=1}^{n}(1+T(\gamma_t))}}\right)}{\prod_{t=1}^{n}(1+\tau_t)^{\frac{(1+T(\gamma_t))}{\sum_{t=1}^{n}(1+T(\gamma_t))}}+\prod_{t=1}^{n}(1-\tau_t)^{\frac{(1+T(\gamma_t))}{\sum_{t=1}^{n}(1+T(\gamma_t))}}} \le \theta^{-} \tag{10.6}$$

Let $\text{IFPEA}(\gamma_1,\gamma_2,\ldots,\gamma_n)=\langle\tau_\gamma,\theta_\gamma\rangle$, then by Eqs. (10.5) and (10.6), we have

$$\tau^{-} \le \tau_\gamma \le \tau^{+}, \qquad \theta^{+} \le \theta_\gamma \le \theta^{-}$$

Thus, $\tau^{-}-\theta^{-} \le \tau_\gamma-\theta_\gamma \le \tau^{+}-\theta^{+}$. Therefore $S(\gamma^{-}) \le S(\gamma) \le S(\gamma^{+})$ and hence $\gamma^{-} \le \text{IFPEA}(\gamma_1,\gamma_2,\ldots,\gamma_n) \le \gamma^{+}$.

(P3) Let γ_t, γ_t^{*}, $t = 1, 2, \ldots, n$ be any two collections of IFNs, then we have

$$\text{IFPEA}(\gamma_1,\gamma_2,\ldots,\gamma_n)=\bigoplus_{t=1}^{n}\frac{(1+T(\gamma_t))}{\sum_{t=1}^{n}(1+T(\gamma_t))}\gamma_t,$$

$$\text{IFPEA}(\gamma_1^{*},\gamma_2^{*},\ldots\gamma_n^{*})=\bigoplus_{t=1}^{n}\frac{(1+T(\gamma_t))}{\sum_{t=1}^{n}(1+T(\gamma_t))}\gamma_t.$$

Since $(\gamma_1^{*},\gamma_2^{*},\ldots\gamma_n^{*})$ is any permutation of $(\gamma_1,\gamma_2,\ldots\gamma_n)$, therefore we have

$$\sum_{t=1}^{n}(1+T(\gamma_t))=\sum_{t=1}^{n}(1+T(\gamma_t^*)),$$

$$\text{and } \bigoplus_{t=1}^{n}(1+T(\mu_t))\gamma_t=\bigoplus_{t=1}^{n}(1+T(\mu_t^*))\gamma_t^*$$

Thus, $\text{IFEPA}(\gamma_1,\gamma_2,\ldots,\gamma_n)=\text{IFEPA}(\gamma_1^*,\gamma_2^*,\ldots,\gamma_n^*)$.

All aggregating objects are of equal value in the IFPEA operator. However, the alternatives typically do not have equal significance in MADM problems, and the weights of the alternatives should be taken into account. To resolve this problem, here we present the IFWPEA AO as:

Definition 10.6 For IFNs γ_t, t = 1, 2, 3, …, n, $\text{IFWPEA}:\Omega^n\to\Omega$ operator defined as:

$$\text{IFWPEA}(\gamma_1,\gamma_2,\ldots\gamma_n)=\bigoplus_{t=1}^{n}\frac{\omega_t(1+T(\gamma_t))}{\sum_{t=1}^{n}\omega_t(1+T(\gamma_t))}\gamma_t,$$

$$=\left\langle \frac{\prod_{t=1}^{n}(1+\tau_t)^{\frac{\omega_t(1+T(\gamma_t))}{\sum_{t=1}^{n}\omega_t(1+T(\gamma_t))}}-\prod_{t=1}^{n}(1-\tau_t)^{\frac{\omega_t(1+T(\gamma_t))}{\sum_{t=1}^{n}\omega_t(1+T(\gamma_t))}}}{\prod_{t=1}^{n}(1+\tau_t)^{\frac{\omega_t(1+T(\gamma_t))}{\sum_{t=1}^{n}\omega_t(1+T(\gamma_t))}}+\prod_{t=1}^{n}(1-\tau_t)^{\frac{\omega_t(1+T(\gamma_t))}{\sum_{t=1}^{n}\omega_t(1+T(\gamma_t))}}}, \frac{2\left(\prod_{t=1}^{n}(1-\tau_t)^{\frac{\omega_t(1+T(\gamma_t))}{\sum_{t=1}^{n}\omega_t(1+T(\gamma_t))}}-\prod_{t=1}^{n}(1-\tau_t-\theta_t)^{\frac{\omega_t(1+T(\gamma_t))}{\sum_{t=1}^{n}\omega_t(1+T(\gamma_t))}}\right)}{\prod_{t=1}^{n}(1+\tau_t)^{\frac{\omega_t(1+T(\gamma_t))}{\sum_{t=1}^{n}\omega_t(1+T(\gamma_t))}}+\prod_{t=1}^{n}(1-\tau_t)^{\frac{\omega_t(1+T(\gamma_t))}{\sum_{t=1}^{n}\omega_t(1+T(\gamma_t))}}}\right\rangle \quad (10.7)$$

where $T(\gamma_t) = \sum_{\substack{l=1 \\ l \neq t}}^{n} Sup(\gamma_t, \gamma_l)$, ω_t is the weight of γ_t, $t = 1, 2, \ldots, n$ such that

$\omega_t > 0$ and $\sum_{t=1}^{n} \omega_t = 1$.

Especially, if $\omega_t = \frac{1}{n}$, $t = 1, 2, \ldots, n$, then IFWPEA reduces to the IFPEA.

Theorem 10.3 For a collection of IFNs $\gamma_t = \langle \tau_t, \theta_t \rangle$, $t = 1, 2, \ldots, n$, IFWPEA operator satisfies the following properties:

(P1) **Idempotency:** If all γ_t are equal, i.e., $\gamma_t = \gamma$, for all t, then

$$\text{IFWPEA}(\gamma_1, \gamma_2, \ldots \gamma_n) = \gamma.$$

(P2) **Boundedness:** $\gamma^- = \langle \tau^-, \theta^- \rangle$ and $\gamma^+ = \langle \tau^+, \theta^+ \rangle$ where $\tau^- = \min_t \{\tau_t\}$, $\theta^- = \max_t \{\theta_t\}$ $\tau^+ = \max_t \{\tau_t\}$ and $\theta^+ = \min_t \{\theta_t\}$, then we have

$$\gamma^- \leq \text{IFWPEA}(\gamma_1, \gamma_2, \ldots, \gamma_n) \leq \gamma^+.$$

(P3) **Commutative:** If γ_t, γ_t^*, $t = 1, 2, \ldots, n$, be two collections of IFNs, then

$$\text{IFWPEA}(\gamma_1, \gamma_2, \ldots, \gamma_n) = \text{IFWPEA}(\gamma_1^*, \gamma_2^*, \ldots \gamma_n^*),$$

where $(\gamma_1^*, \gamma_2^*, \ldots \gamma_n^*)$ is any permutation of $(\gamma_1, \gamma_2, \ldots \gamma_n)$.

Proof. We can prove this similar to **Theorem 10.2.**

10.4 PROPOSED MULTIATTRIBUTE DECISION-MAKING METHOD BASED ON THE PROPOSED AGGREGATION OPERATOR

Here, in the IFNs context, we have developed an MADM approach based on proposed AOs.

Consider an MADM issue in which there are m' alternatives $H = \{H_1, H_2, \ldots, H_m\}$ which are assessed under the set $G = \{G_1, G_2, \ldots, G_n\}$ of different attributes. Experts prefer to evaluate the H_k alternatives under the G_t attributes in terms of IFN $\tilde{\gamma}_{kt} = \langle \tilde{\tau}_{kt}, \tilde{\theta}_{kt} \rangle$. Then, we summarized the following steps of the developed approach as follows:

(Step 1): Arrange the collective information in the form of the decision matrix $\tilde{R} = \left(\tilde{\gamma}_{kt}\right)_{m\times n}$ as follows:

$$\tilde{R} = \begin{array}{c} \\ H_1 \\ H_2 \\ \vdots \\ H_m \end{array} \begin{array}{c} \begin{array}{cccc} G_1 & G_2 & \cdots & G_n \end{array} \\ \begin{pmatrix} \tilde{\gamma}_{11} & \tilde{\gamma}_{12} & \cdots & \tilde{\gamma}_{1n} \\ \tilde{\gamma}_{21} & \tilde{\gamma}_{22} & \cdots & \tilde{\gamma}_{2n} \\ \vdots & \vdots & \ddots & \vdots \\ \tilde{\gamma}_{m1} & \tilde{\gamma}_{m2} & \cdots & \tilde{\gamma}_{mn} \end{pmatrix} \end{array} \tag{10.8}$$

(Step 2): Decision matrix $\tilde{R}$ is changed over into its normalized matrix, indicated by $R = (\gamma_{kt})_{m\times n}$, to remove the impact of physical dimension, as follows:

$$\gamma_{kt} = \begin{cases} \langle \tilde{\tau}_{kt}, \tilde{\theta}_{kt} \rangle : \text{If } G_t \text{ is benefit type attribute} \\ \langle \tilde{\theta}_{kt}, \tilde{\tau}_{kt} \rangle : \text{If } G_t \text{ is cost type attribute} \end{cases} \tag{10.9}$$

(Step 3): Calculate the weight ω_t of the attribute G_t, $t = 1, 2, \ldots, n$ as:

$$\omega_t = \frac{1 - e_t}{n - \sum_{t=n}^{n} e_t}, \tag{10.10}$$

where $e_t = \frac{1}{m}\sum_{k=1}^{m} E(\gamma_{kt})$ and $E(\gamma_{kt}) = \frac{1}{3}\left(4\sqrt{\tau_{kt}\theta_{kt}} + \pi_{kt} + 2\sqrt{(1-\theta_{kt})(1-\theta_{kt})}\right)$ is the entropy measure for $\gamma_{kt} = \langle \tau_{kt}, \theta_{kt} \rangle$.

(Step 4): Supports $Sup(\gamma_{kt}, \gamma_{kl})$, $k = 1,2,\ldots,m; t,l = 1,2,\ldots,n, l \neq t$, are calculated as

$$Sup(\gamma_{kt}, \gamma_{kl}) = S(\gamma_{kt}, \gamma_{kl})$$

$$= \frac{1}{3}\left(\begin{array}{c} 2\sqrt{\tau_{kt}\tau_{kl}} + 2\sqrt{\theta_{kt}\theta_{kl}} + \sqrt{\pi_{kt}\pi_{kl}} \\ + \sqrt{(1-\tau_{kt})(1-\tau_{kl})} + \sqrt{(1-\theta_{kt})(1-\theta_{kl})} \end{array}\right) \tag{10.11}$$

(Step 5): Support $T(\gamma_{kt})$ and weights $w_{kt}, k = 1,2,\ldots,m, t = 1,2,\ldots,n,$, are calculated as

$$T(\gamma_{kt}) = \sum_{\substack{l=1 \\ t\neq k}}^{n} Sup(\gamma_{kt}, \gamma_{kl}) \tag{10.12}$$

and

$$w_{kt}=\frac{\omega_t(1+T(\gamma_{kt}))}{\sum_{t=1}^{n}\omega_t(1+T(\gamma_{kt}))} \tag{10.13}$$

where $w_{kt}>0$ and $\sum_{t=1}^{n} w_{kt}=1$.

(Step 6): Calculate the overall IFN γ_k for the alternative H_k, $k = 1, 2, 3, \ldots, m$ by using the IFWPEA operator as follows:

$$\gamma_k=\text{IFWPEA}(\gamma_{k1},\gamma_{k2},\ldots,\gamma_{kn})$$

$$=\left\langle \frac{\prod_{t=1}^{n}(1+\tau_{kt})^{w_{kt}}-\prod_{t=1}^{n}(1-\tau_{kt})^{w_{kt}}}{\prod_{t=1}^{n}(1+\tau_{kt})^{w_{kt}}+\prod_{t=1}^{n}(1-\tau_{kt})^{w_{kt}}}, \frac{2\left(\prod_{t=1}^{n}(1-\tau_{kt})^{w_{kt}}-\prod_{t=1}^{n}(1-\tau_{t}-\theta_{kt})^{w_{kt}}\right)}{\prod_{t=1}^{n}(1+\tau_{kt})^{w_{kt}}+\prod_{t=1}^{n}(1-\tau_{kt})^{w_{kt}}} \right\rangle. \tag{10.14}$$

(Step 7): For ranking the alternatives H_k, we compute the score value $S(\gamma_k)$ and the accuracy value $H(\gamma_k)$ of the overall IFN, γ_k , $k = 1, 2, 3, \ldots, m$, as follows:

$$S(\gamma_k)=\tau_k-\theta_k, \tag{10.15}$$

and

$$H(\gamma_k)=\tau_k+\theta_k. \tag{10.16}$$

(Step 8): According to the value of $S(\gamma_k)$, $t = 1, 2, \ldots, m$ rank the alternative H_k, and subsequently the best one is picked on a superior value. If score values are equal for two alternatives, then we will calculate the accuracy values for those alternatives and choose the best alternative based on the greater value.

10.5 ILLUSTRATIVE EXAMPLE

Here, we illustrate a real-life case to demonstrate the legitimacy and adequacy of the proposed DM approach.

Metro cities in India are facing serious traffic problems in which the Meerut City is one of New Delhi's national capital region city. Traffic delays are one of this city's main problems in peak hours. Meerut Development Authority (MDA) wants to build the flyover on the heavy-duty crossing point to diminish this issue. For this reason, MDA invites the construction company to set up five attributes to pick the best company for this work as "project cost" (G_1), "completion time" (G_2), "technical capability" (G_3), "financial status"' (G_4), and "company background" (G_5), The four construction firms took the form of alternatives, namely "PNC Infratech Ltd." (H_1), "Hindustan Construction Company" (H_2), "J.P." (H_3), and "Gammon India Ltd." (H_4) are involved in these projects. Currently, the main aim is to select the best organization for the task among them.

10.5.1 By proposed approach

To choose the best firm for the job, the following are the steps according to the developed approach.

(Step 1:) The expert(s) give the assessment for the construction firms under the given attributes in the form of IFNs and are summarized as follows:

$$\begin{array}{c} \\ H_1 \\ \tilde{R} = H_2 \\ H_3 \\ H_4 \end{array} \begin{array}{c} \begin{array}{ccccc} G_1 & G_2 & G_3 & G_4 & G_5 \end{array} \\ \begin{pmatrix} \langle 0.3,0.6\rangle & \langle 0.5,0.4\rangle & \langle 0.7,0.2\rangle & \langle 0.5,0.2\rangle & \langle 0.7,0.1\rangle \\ \langle 0.5,0.3\rangle & \langle 0.6,0.2\rangle & \langle 0.5,0.4\rangle & \langle 0.6,0.3\rangle & \langle 0.4,0.2\rangle \\ \langle 0.5,0.4\rangle & \langle 0.7,0.2\rangle & \langle 0.8,0.1\rangle & \langle 0.6,0.2\rangle & \langle 0.5,0.3\rangle \\ \langle 0.6,0.2\rangle & \langle 0.4,0.3\rangle & \langle 0.7,0.2\rangle & \langle 0.4,0.4\rangle & \langle 0.2,0.8\rangle \end{pmatrix} \end{array}$$

(Step 2:) Normalized the values of the cost attributes G_1 and G_2 by using Eq. (10.9) and the results are summarized as:

$$\begin{array}{c} \\ H_1 \\ R = H_2 \\ H_3 \\ H_4 \end{array} \begin{array}{c} \begin{array}{ccccc} G_1 & G_2 & G_3 & G_4 & G_5 \end{array} \\ \begin{pmatrix} \langle 0.6,0.3\rangle & \langle 0.4,0.5\rangle & \langle 0.7,0.2\rangle & \langle 0.5,02\rangle & \langle 0.7,0.1\rangle \\ \langle 0.3,0.5\rangle & \langle 0.2,0.6\rangle & \langle 0.5,0.4\rangle & \langle 0.6,0.3\rangle & \langle 0.4,0.2\rangle \\ \langle 0.4,0.5\rangle & \langle 0.2,0.7\rangle & \langle 0.8,0.1\rangle & \langle 0.6,0.2\rangle & \langle 0.5,0.3\rangle \\ \langle 0.2,0.6\rangle & \langle 0.3,0.4\rangle & \langle 0.7,0.2\rangle & \langle 0.4,0.4\rangle & \langle 0.2,0.8\rangle \end{pmatrix} \end{array}$$

(Step 3:) Estimate the weight ω_t of the attribute G_t, $t = 1, 2, \ldots, n$ by using Eq. (10.10) and the obtained results are summarized as:

$$\omega = (\omega_1, \omega_2, \omega_3, \omega_4, \omega_5)^T = (0.1005, 0.1456, 0.3506, 0.1176, 0.2857)^T$$

(Step 4:) Compute the $S_{kt,kl} = Sup(\gamma_{kt}, \gamma_{kl})$ by using Eq. (10.11) and obtain the values as:

$$S_{11,12} = S_{12,11} = 0.9786, S_{11,13} = S_{13,11} = 0.9936, S_{11,14} = S_{14,11} = 0.9847,$$
$$S_{11,15} = S_{15,11} = 0.9747, S_{12,13} = S_{13,12} = 0.9492, S_{12,14} = S_{14,12} = 0.9601,$$
$$S_{12,15} = S_{15,12} = 0.9140, S_{13,14} = S_{14,13} = 0.9812, S_{13,15} = S_{15,13} = 0.9909,$$
$$S_{14,15} = S_{15,14} = 0.9823, S_{21,22} = S_{22,21} = 0.9936, S_{21,23} = S_{23,21} = 0.9833,$$
$$S_{21,24} = S_{24,21} = 0.9618, S_{21,25} = S_{25,21} = 0.9629, S_{22,23} = S_{23,22} = 0.9587,$$
$$S_{22,24} = S_{24,22} = 0.9259, S_{22,25} = S_{25,22} = 0.9333, S_{23,24} = S_{24,23} = 0.9945,$$
$$S_{23,25} = S_{25,23} = 0.9669, S_{24,25} = S_{25,24} = 0.9693, S_{31,32} = S_{32,31} = 0.9763,$$
$$S_{31,33} = S_{33,31} = 0.8986, S_{31,34} = S_{34,31} = 0.9587, S_{31,35} = S_{35,31} = 0.9833,$$
$$S_{32,33} = S_{33,32} = 0.7829, S_{32,34} = S_{34,32} = 0.8794, S_{32,35} = S_{35,32} = 0.9270,$$
$$S_{33,34} = S_{34,33} = 0.9804, S_{33,35} = S_{35,33} = 0.9542, S_{34,35} = S_{35,34} = 0.9936,$$
$$S_{41,42} = S_{42,41} = 0.9843, S_{41,43} = S_{43,41} = 0.8794, S_{41,44} = S_{44,41} = 0.9761,$$
$$S_{41,45} = S_{45,41} = 0.9562, S_{42,43} = S_{43,42} = 0.9355, S_{42,44} = S_{44,42} = 0.9953,$$
$$S_{42,45} = S_{45,42} = 0.9053, S_{43,44} = S_{44,43} = 0.9608, S_{43,45} = S_{45,43} = 0.8127,$$
$$S_{44,45} = S_{45,44} = 0.9121.$$

(Step 5:) Support $T(\gamma_{kt})$, $k = 1, 2, 3, 4$, $t = 1, 2, 3, 4, 5$ are calculated by using Eq. (10.12) and obtained results are as

$$T_{11} = 3.9316, T_{12} = 3.8019, T_{13} = 3.9149, T_{14} = 3.9083, T_{15} = 3.8619,$$
$$T_{21} = 3.9015, T_{22} = 3.8115, T_{23} = 3.9033, T_{24} = 3.8515, T_{15} = 3.8324,$$
$$T_{31} = 3.8169, T_{32} = 3.5657, T_{33} = 3.6162, T_{34} = 3.8121, T_{15} = 3.8582,$$
$$T_{41} = 3.7959, T_{42} = 3.8204, T_{43} = 3.5885, T_{44} = 3.8443, T_{15} = 3.5863$$

Now, weights are calculated by using Eq. (10.13) and get

$$w_{11} = 0.1014, w_{12} = 0.1432, w_{13} = 0.3528, w_{14} = 0.1181, w_{15} = 0.2844,$$
$$w_{21} = 0.1013, w_{22} = 0.1441, w_{23} = 0.3535, w_{24} = 0.1173, w_{25} = 0.2839,$$
$$w_{31} = 0.1025, w_{32} = 0.1408, w_{33} = 0.3428, w_{34} = 0.1198, w_{35} = 0.2940,$$
$$w_{41} = 0.1031, w_{42} = 0.1502, w_{43} = 0.3443, w_{44} = 0.1219, w_{45} = 0.2805$$

(Step 6:) Overall collective IFNs γ_k for the alternatives H_k, $k = 1, 2, 3, 4$ are computed by using Eq. (10.14) and the obtained results are given as follows:

$$\gamma_1 = \langle 0.6343, 0.2257 \rangle, \qquad \gamma_2 = \langle 0.4271, 0.3957 \rangle,$$

$$\gamma_3 = \langle 0.6000, 0.2623 \rangle, \qquad \gamma_4 = \langle 0.4418, 0.5582 \rangle$$

(Step 7:) Score values $S(\gamma_k)$, $k = 1, 2, 3, 4$ are calculated by using Eq. (10.15) and obtained results are as:

$$S(\gamma_1) = 0.4085, S(\gamma_2) = 0.0314, S(\gamma_3) = 0.3377, \ S(\gamma_4) = -0.1163$$

(Step 8:) Since $S(\gamma_1) > S(\gamma_3) > S(\gamma_2) > S(\gamma_4)$. Therefore $H_1 \succ H_3 \succ H_2 \succ H_4$ and hence H_1 is the finest company for the task.

10.5.2 Comparative study with existing decision-making methods

To demonstrate the effectiveness of the proposed MADM aproach, we use the existing MADM approaches by utilizing the weight vector $\omega = (0.1005, 0.1456, 0.3506, 0.1176, 0.2857)^T$ of attributes on the above-considered illustrative example., and results are presented in Table 10.1. Table 10.2 compares the characteristics of the proposed MADM approach with the existing MADM approaches. From Table 10.1 and 10.2, we can see that the proposed MADM approach is superior to the existing MADM appaoches.

Table 10.1 Results comparison with existing approaches

	Rating Values				
Author	H_1	H_2	H_3	H_4	*Ranking*
Xu (2007)	0.4065	0.0538	0.3771	0.1076	$H_1 \succ H_3 \succ H_4 \succ H_2$
He et al. (2014)	0.3412	0.0102	0.1501	0.0221	$H_1 \succ H_3 \succ H_4 \succ H_2$
Garg (2016)	0.3678	−0.0016	0.3130	−0.0957	$H_1 \succ H_3 \succ H_4 \succ H_2$
Garg (2017)	0.3711	0.0044	0.3264	−0.0773	$H_1 \succ H_3 \succ H_2 \succ H_4$
Ye (2017)	0.3670	0.0125	0.2495	−0.0087	$H_1 \succ H_3 \succ H_2 \succ H_4$
Joshi and Kumar (2014)	0.8101	0.3700	0.6550	0.2923	$H_1 \succ H_3 \succ H_2 \succ H_4$
Kumar and Garg (2018)	0.9632	0.1679	0.7300	0.1901	$H_1 \succ H_3 \succ H_4 \succ H_2$
Proposed method	0.4085	0.0314	0.3377	−0.1163	$H_1 \succ H_3 \succ H_2 \succ H_4$

Table 10.2 Characteristic comparison with existing approaches

	Aggregating the evaluation information	*Consider the interrelationship between input argument*	*Describe the IFNs information*	*Determine the attributes weights*
Xu (2007)	Yes	No	Yes	No
He et al. (2014)	Yes	No	Yes	No
Joshi and Kumar (2014)	No	No	Yes	No
Garg (2016)	Yes	No	Yes	No
Garg (2017)	Yes	No	Yes	No
Ye (2017)	Yes	No	Yes	No
Kumar and Garg (2018)	No	No	Yes	No
Proposed method	Yes	Yes	Yes	Yes

10.6 CONCLUSION

In this article, some aggregating operators have been proposed by combining the features of Einstein operations and power aggregating operator, namely, IFPEA and IFWPEA for aggregating the IFNs information. By utilizing the proposed entropy measure and aggregating operators, we have defined a DM approach for fathoming the MADM issues beneath the IFNs context. At last, a real-life case of the MADM issue is explained to demonstrate the effectiveness of the presented DM approach and the obtained results are compared with the results obtained from other existing approaches. The comparison analysis and computed results indicate that the suggested DM strategy is reasonable and practicable, providing an additional means of addressing MADM concerns in the context of IFNs. In the future, we will apply the proposed DM approach to the other field.

REFERENCES

Atanassov, K. T. (1986). Intuitionistic fuzzy sets. *Fuzzy Sets and Systems*, *20*(1), 87–96.

Atanassov, K. T., & Gargov, G. (1989). Interval valued intuitionistic fuzzy sets. *Fuzzy Sets and System*, *31*(3), 343–349.

Bhardwaj, R., Sharma, A., Mani, N., & Kumar, K. (2022). An intuitionistic fuzzy entropy measure and its application in multi-attribute decision making with incomplete weights information. In *Handbook of research on advances and applications of fuzzy sets and logic* (pp. 324–338). IGI Global.

Dhankhar, C., & Kumar, K. (2023). Multi-attribute decision making based on the q-rung orthopair fuzzy Yager power weighted geometric aggregation operator of q-rung orthopair fuzzy values. *Granular Computing*, *8*(1), 1–13.

Garg, H. (2016). Some series of intuitionistic fuzzy interactive averaging aggregation operators. *Springer Plus*, *5*(1), 999.

Garg, H. (2017). Novel intuitionistic fuzzy decision making method based on an improved operation laws and its application. *Engineering Applications of Artificial Intelligence*, *60*, 164–174.

Garg, H. (2019). Intuitionistic fuzzy Hamacher aggregation operators with entropy weight and their applications to multi-criteria decision-making problems. *Iranian Journal of Science and Technology, Transactions of Electrical Engineering*, *43*(3), 597–613.

Garg, H., Agarwal, N., & Tripathi, A. (2017). Generalized intuitionistic fuzzy entropy measure of order α and degree β and its applications to multi-criteria decision making problem. *International Journal of Fuzzy System Applications (IJFSA)*, *6*(1), 86–107.

Garg, H., & Kaur, J. (2018). A novel (R, S)-norm entropy measure of intuitionistic fuzzy sets and its applications in multi-attribute decision-making. *Mathematics*, *6*(6), 92.

Garg, H., & Kumar, K. (2019). Multiattribute decision making based on power operators for linguistic intuitionistic fuzzy set using set pair analysis. *Expert Systems*, *36*(4), e12428.

Garg, H., & Kumar, K. (2020). Power geometric aggregation operators based on connection number of set pair analysis under intuitionistic fuzzy environment. *Arabian Journal for Science and Engineering*, *45*(3), 2049–2063.

He, Y., Chen, H., Zhou, L., Liu, J., & Tao, Z. (2014). Intuitionistic fuzzy geometric interaction averaging operators and their application to multi-criteria decision making. *Information Sciences*, *259*, 142–159.

Hong, D. H., & Choi, C. H. (2000). Multicriteria fuzzy decision-making problems based on vague set theory. *Fuzzy Sets and Systems*, *114*(1), 103–113.

Joshi, D., & Kumar, S. (2014). Intuitionistic fuzzy entropy and distance measure based TOPSIS method for multi-criteria decision making. *Egyptian Informatics Journal*, *15*(2), 97–104.

Kumar, K., & Chen, S. M. (2022). Group decision making based on advanced intuitionistic fuzzy weighted Heronian mean aggregation operator of intuitionistic fuzzy values. *Information Sciences*, *601*, 306–322.

Kumar, K., & Garg, H. (2018). Connection number of set pair analysis based TOPSIS method on intuitionistic fuzzy sets and their application to decision making. *Applied Intelligence*, *48*, 2112–2119.

Liu, M., & Ren, H. (2014). A new intuitionistic fuzzy entropy and application in multi-attribute decision making. *Information*, *5*(4), 587–601.

Neelam, Kumar, K., & Bhardwaj, R. (2023). Entropy measure for the linguistic q-Rung orthopair fuzzy set. In *Soft computing: Theories and applications: Proceedings of SOCTA 2022* (pp. 161–171). Singapore: Springer Nature Singapore.

Seikh, M. R., & Mandal, U. (2021). Intuitionistic fuzzy Dombi aggregation operators and their application to multiple attribute decision-making. *Granular Computing*, *6*, 473–488.

Senapati, T., Chen, G., & Yager, R. R. (2022). Aczel–Alsina aggregation operators and their application to intuitionistic fuzzy multiple attribute decision making. *International Journal of Intelligent Systems*, *37*(2), 1529–1551.

Xu, Z. (2007). Intuitionistic fuzzy aggregation operators. *IEEE Transactions on Fuzzy Systems*, *15*(6), 1179–1187.

Xu, Z. (2011). Approaches to multiple attribute group decision making based on intuitionistic fuzzy power aggregation operators. *Knowledge-Based Systems*, *24*(6), 749–760.

Xu, Z., & Chen, J. (2007). On geometric aggregation over interval-valued intuitionistic fuzzy information. In *Fourth international conference on fuzzy systems and knowledge discovery (FSKD 2007)* (Vol. 2, pp. 466–471). IEEE.

Xu, Z., & Yager, R. R. (2006). Some geometric aggregation operators based on intuitionistic fuzzy sets. *International Journal of General Systems*, *35*(4), 417–433.

Xu, Z., & Yager, R. R. (2009). Power-geometric operators and their use in group decision making. *IEEE Transactions on Fuzzy Systems*, *18*(1), 94–105.

Yager, R. R. (2001). The power average operator. *IEEE Transactions on Systems, Man, and Cybernetics-Part A: Systems and Humans*, *31*(6), 724–731.

Ye, J. (2017). Intuitionistic fuzzy hybrid arithmetic and geometric aggregation operators for the decision-making of mechanical design schemes. *Applied Intelligence*, *47*, 743–751.

Zadeh, L. A. (1965). Fuzzy sets. *Information and Control*, *8*(3), 338–353.

Chapter 11

Effectiveness of MADM in *q*-rung orthopair fuzzy sets

A novel entropy measure-based TOPSIS method

Binoy Krishna Giri and Sankar Kumar Roy

11.1 INTRODUCTION

Multi-criteria decision-making (MCDM) represents to make decisions on multiple and basically conflicting criteria. Based on the alternative's domain, MCDM is categorized as discrete or continuous. Hwang and Yoon (1981) divide them such as multi-attribute decision making (MADM) and multiple objective decision-making (MODM). MADM is generally limited with specific number of alternatives (Debnath and Roy, 2023). In MODM, the decision variable is to be evaluated on a continuous domain of large-scale data for best satisfying the constraints of decision makers (DMs) with priorities and preferences. In other words, finding the most preferable alternative(s) from a discrete collection of possible alternatives with respect to a finite set of attributes is known as MADM. MADM has been a popular research subject for the past few decades and has been widely used in a variety of fields, including society, economics, the military, management, supplier selection, etc.

The efficiency of the supplier selection procedure is restricted on how the DMs execute the supplier accurately. The selection of the suppliers mainly focuses on product price, product quality, previous performance of the suppliers, and delivery time. Also, some other factors such as various risk factors, safety factors, and social, political, environmental, and economical factors can be taken during the selection process (Giri and Roy, 2022; Mondal et al., 2023). Therefore, sustainable supplier selection (S^3) becomes a rising new pattern in enterprises and industries. However, it includes several uncontrolled and unexpected factors that need to be addressed, otherwise an inappropriate selection can affect the states of the enterprises. Therefore, S^3 is to be taken as an MADM process on various conflicting criteria that are to be executed for selecting the appropriate supplier. Major portions of DMs prefer the most economical issues, but due to carbon zero emission mission, increasing of global warming, climate change, health implication and so forth, S^3 is essential day by day (Giri and Roy, 2024). Sometimes, organizations should focus on social criteria to tackle very critical situation like, COVID-19 shutdown, migrant workers, joblessness over

 DOI: 10.1201/9781003497219-11

population, and so forth. Due to these reasons, an S^3 is most important activity for today's competitive market.

Generally, the real parameters of S^3 in an MADM are inefficient to describe the information accurately. Practically, most of the cases, the information on S^3 is vague, hesitant, and hazy. For this situation, researchers used several uncertainties to represent the parameter in various decision-making systems (DMS). To tackle imprecise information, Zadeh (1965) introduced fuzzy set (FS) with membership function for the first time. Thereafter, intuitionistic fuzzy sets (IFSs) along with membership and non-membership degrees were proposed by Atanassov (1986). For some situation, the addition of membership and non-membership degrees of a parameter that satisfies an attribute may be bigger than one, but the addition of the squares is smaller than or equal to one. Therefore, IFS do not operate this situation. Then Yager and Abbasov (2013) developed Pythagorean fuzzy set (PFS) according to great mathematician Pythagoras. Suppose a decision-maker intends to achieve the level of satisfaction in any DMS with the membership degree 0.80, and wants to reduce the level of dissatisfaction with non-membership degree 0.70. Upon adding the membership and non-membership degrees, $0.80+0.70 \geq 1$. So, IFS cannot determine the situation. Again, in Pythagorean fuzzy sense, $0.80^2 + 0.70^2 \geq 1$. It was also unable to handle this situation. To overcome this critical situation, Yager (2016) introduced q-rung orthopair fuzzy (q-ROF) set which satisfy the condition $0.80^3 + 0.70^3 < 1$. Therefore, the concept of q-ROF set is utilized extensively for solving the DMS, namely, S^3, supply chain management, inventory management, medical diagnosis, engineering aspects, and water resources (Giri and Roy, 2023). In S^3, the values of the attributes with respect to different alternatives are uncertain due to various situations; therefore, we assume q-rung orthopair fuzzy number for the proposed study.

Hwang and Yoon (1981) introduced technique for order of preference by similarity to ideal solution (TOPSIS) method which is an efficient method for evaluating an MADM problem with finite attributes of definite alternatives. The core principle of the method is to select the best alternatives and rank the alternatives by evaluating the Euclidean distance of the alternatives from the positive ideal solution (PIS) and the negative ideal solution (NIS) for DMs. Nowadays, TOPSIS method is widely utilized in several fields like S^3, risk management, renewable energy (Giri et al., 2023; Ghosh et al., 2023), sustainability assessment due to its easy calculation, simple understanding, and mostly corrected outcomes. In recent times, TOPSIS method is expanded with various fuzzy environments. In this study, we utilized TOPSIS method under q-ROF environment. In the determination steps of the TOPSIS method, it is essential to calculate the weight of the attributes. To determine the weight of the attributes, many researchers used various weight determination methods like analytic hierarchy process (AHP), indifference threshold-based attribute ratio analysis (ITARA), criteria importance through intercriteria correlation (CRITIC), etc. (Jana and

Roy, 2023; Mondal and Roy, 2021). Furthermore, to rank of the alternatives, several methods are existed such as multi-objective optimization on the basis of ratio analysis (MOORA) (Dincer et al., 2019), complex proportional assessment (COPRAS) (Amoozad et al., 2018), etc. Here the entropy method is applied to evaluate the weight of the attributes in the steps of the TOPSIS method. After using the entropy method in TOPSIS, the proposed determination process becomes entropy measure-based TOPSIS method. It is an objective method that is totally dependent on objective value of the attributes but does not involve the subjective preferences. The main investigation topics related to this research initiative are as follows:

(i) The subjective evaluation of the weighting of various attributes by decision-makers is a downside of TOPSIS. Therefore, how should we address the issue of DMs arbitrary weighting in the TOPSIS method when they are in a q-ROF environment?
(ii) The evaluation of q-ROF entropy differs from the traditional entropy weight measurement. How can we use the TOPSIS method in addition to q-ROF entropy to produce calculated weight values?
(iii) The selection of sustainable supplier by various methods is a hot issue in recent times. Which is a better sustainable supplier out of listed suppliers? To validate and efficiency of the developed method, we consider two numerical examples based on sustainable supplier selection.

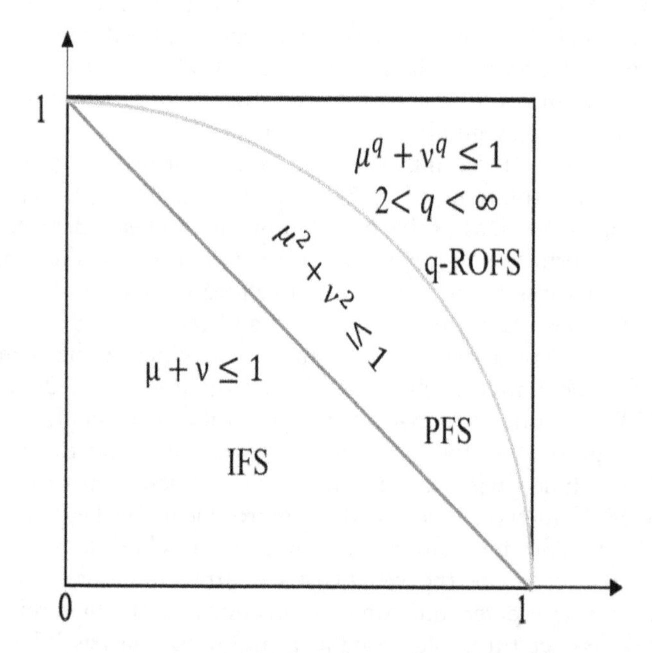

Figure 11.1 Comparison among q-ROF set, IFS, and PFS.

In this study, we develop entropy measure-based TOPSIS method under q-ROF environment for selecting sustainable suppliers. In the sustainable supplier selection process, the entropy based TOPSIS method based on q-ROF set has a very high possibility of success.

The remainder of this study is divided as follows: Section 11.2 represents the basic definition for designing the proposed model. Problem description, entropy measure, and TOPSIS method are discussed to formulate the proposed q-ROF entropy-based TOPSIS method in Section 11.3. Section 11.4 illustrates two examples and the outcomes for selecting sustainable supplier. Comparative study with other methods is represented in Section 11.5. Lastly, conclusion and future study are explained in Section 11.6.

11.2 PRELIMINARIES

In this section, the definitions of entropy measure, IFS, PFS, and q-ROF set and its basic operations are discussed. The score and accuracy functions are also illustrated.

Definition 11.1 (Atanassov, 1986) An IFS S in universe of discourse Z is defined as $S = \{z_i, (\mu_S(z_i), \nu_S(z_i)) \mid z_i \in Z\}$, where $\mu_S(z_i) \in [0,1]$ and $\nu_S(z_i) \in [0,1]$ are the membership function and nonmembership function of $z_i \in Z$, respectively, with satisfying the condition $0 \leq \mu_S(z_i) + \nu_S(z_i) \leq 1, \forall z_i \in Z$.

Definition 11.2 (Yager and Abbasov, 2013) A PFS P in universe of discourse R is defined as $P = \{r_i, (\mu_P(r_i), \nu_P(r_i)) \mid r_i \in R\}$, where $\mu_P(r_i) \in [0,1]$ and $\nu_P(r_i) \in [0,1]$ are the membership function and nonmembership function of $r_i \in R$, respectively, with satisfying the condition $0 \leq \mu_P^2(r_i) + \nu_P^2(r_i) \leq 1, \forall r_i \in R$.

Definition 11.3 (Yager, 2016) Let $Y = \{y_1, y_2, \ldots, y_n\}$ be a set, then q-ROF set on Y is represented as $A = \{\langle y_i, (\mu_A(y_i), \nu_A(y_i))\rangle_q \mid y_i \in Y\}, q \geq 1$, where $\mu_A(y_i)$ and $\nu_A(y_i) \in [0,1]$ are the membership function and nonmembership function of $y_i \in Y$, respectively, with satisfying the condition $0 \leq \mu_A^q(y_i) + \nu_A^q(y_i) \leq 1, \forall y_i \in Y$. Comparison among q-ROF set, IFS, and PFS is depicted in Figure 11.1. According to Figure 11.1, DMs are more comfortable to take decision in q-ROF environment.

In case, the set $Y = \{y_1, y_2, \ldots, y_n\}$ contains only single element, such as $Y = \{y\}$, then the set A is converted to $A = (\mu_A(y), \nu_A(y))_q = (\mu_A, \nu_A)_q$. For simplicity, $A = (\mu_A, \nu_A)_q$ is called a q-rung orthopair fuzzy number (q-ROFN).

Definition 11.4 (Yager, 2016) Let $A_1 = (\mu_{A_1}, \nu_{A_1})_q$ and $A_2 = (\mu_{A_2}, \nu_{A_2})_q$ be two q-ROFNs, then the basic operations are described as

(i) $A_1 \cup A_2 = (\mu_{A_1} \vee \mu_{A_2}, \nu_{A_1} \wedge \nu_{A_2})_q$;
(ii) $A_1 \cap A_2 = (\mu_{A_1} \wedge \mu_{A_2}, \nu_{A_1} \vee \nu_{A_2})_q$;
(iii) $A_1 \leq A_2$ if $\mu_{A_1} \leq \mu_{A_2}$ and $\nu_{A_1} \geq \nu_{A_2}$;
(iv) $A^c = (\nu_A, \mu_A)$, where A^c indicates the inverse of A.

where $\vee$ and $\wedge$ represents the *max* and *min*, respectively.

Definition 11.5 Consider a scenario where there exist n alternative outcomes, each having a chance of occurrence of p. Suppose $p = \{p_1, p_2, ..., p_3\}$ be the probability distribution, such that $p_i \geq 0$, for all i, $\sum_i p_i = 1$ for i in 1 and n. The entropy measure for this distribution is defined by $E = -k \sum_i p_i \ln p_i$, for i between 1 and n, and where k is an arbitrary positive constant.

Definition 11.6 (Liu and Wang, 2018) Let $A = (\mu_A, \nu_A)_q$ be a q-ROFN, then the score function S of A is given by $S(A) = \mu_A^q - \nu_A^q$, again the accuracy function P of A is defined by $P(A) = \mu_A^q + \nu_A^q$, $q \geq 1$.

Definition 11.7 (Liu and Wang, 2018) Let $A_1 = (\mu_{A_1}, \nu_{A_1})_q$ and $A_2 = (\mu_{A_2}, \nu_{A_2})_q$ be two q-ROFNs, the score functions of A_1 and A_2 are denoted by $S(A_1)$ and $S(A_2)$, respectively. Again, the accuracy function of A_1 and A_2 are denoted by $P(A_1)$ and $P(A_2)$, respectively. Then

(i) If $S(A_1) > S(A_2)$, then $A_1 > A_2$;
(ii) If $S(A_1) = S(A_2)$, then
  (a) if $P(A_1) > P(A_2)$, then $A_1 > A_2$;
  (b) if $P(A_1) = P(A_2)$, then $A_1 = A_2$.

11.3 MADM METHOD BASED ON ENTROPY AND TOPSIS

In the context of q-ROF, this section proposes a methodology for calculating attribute weights and ranking order for each alternative.

11.3.1 Problem description

Assume that $D = \{D_1, D_2, ..., D_m\}$ is an alternative set made up of m non-inferior alternatives and that $C = \{C_1, C_2, ..., C_n\}$ is the collection of attributes. If there are no other alternatives that can result in an improvement in one quality without generating a decrease in another, then the option is not inferior. Let $B = \{b_{ij}\}$ be the decision matrix, where $\{b_{ij}\}$ are the function of membership and non-membership of q-ROFN of every alternative D_i on attribute C_j.

11.3.2 Entropy measure

Zhang and Yu (2012) illustrated the importance of entropy measure in MADM problem. Entropy measure calculates the weight of the attributes C_j corresponding to the alternatives D_i. The steps are as follows:

Step 1: Design a decision matrix $B = \{b_{ij}\}$, where b_{ij} is the jth, $(j = 1, 2, \ldots, n)$ attribute value of ith, ith (i=1, 2, …, m) alternative.

Step 2: We normalize the attribute values b_{ij} to b'_{ij} with the given formulae:

$$b'_{ij} = \frac{b_{ij}}{\sum_i b_{ij}}, \ \forall\ i, j. \tag{11.1}$$

Step 3: Evaluate the information entropy $F(j)$ using the given equation:

$$F(j) = -\frac{\sum_i b'_{ij} \ln b'_{ij}}{\ln m}, \forall j. \tag{11.2}$$

In case, $b'_{ij} = 0$, we assume $ln\ b'_{ij} = 0$.

Step 4: Determine the weight of the attributes using the given formulae:

$$W_j = \frac{1 - F(j)}{\sum_j (1 - F(j))}, \forall j. \tag{11.3}$$

Here W_j is the jth, $(j = 1, 2, \ldots, n)$ attribute weight.

11.3.3 TOPSIS method

The TOPSIS method for solving MADM problems, developed by Hwang and Yoon (1981), aims to select the option that is closest to the positive ideal solution (PIS) and the furthest away from the negative ideal solution (NIS) (Rahim et al., 2023). The following steps are as follows:

Step 5: Calculate weighted normalized matrix $X = x_{ij}$ using the given equation:

$$x_{ij} = W_j b'_{ij}, \forall\ i, j. \tag{11.4}$$

Step 6: Evaluate the positive ideal solution (PIS) R^+ and negative ideal solution (NIS) R^-:

$$\begin{aligned} R+ &= \{(\max_i x_{ij} \mid j \in J)(\min_i x_{ij} \mid j \in J') \mid i = 1, 2, \ldots, m\} \\ &= \{x_1^+, x_2^+, \ldots, x_n^+\}, \end{aligned} \tag{11.5}$$

$$R- = \{(\min_i x_{ij} \mid j \in J)(\max_i x_{ij} \mid j \in J')\mid i = 1,2,\ldots,m\} = \left\{x_1^-, x_2^-, \ldots, x_n^-\right\}. \quad (11.6)$$

where J and J' are the set of benefit attributes and cost attributes, respectively.

Step 7: Determine Euclidean distance of alternatives from PIS and NIS using the following formulae:

$$R_i^+ = \sqrt{\sum_j \left(x_{ij} - x_j^+\right)^2}, \forall\, i \quad (11.7)$$

$$R_i^- = \sqrt{\sum_j \left(x_{ij} - x_j^-\right)^2}, \forall\, i \quad (11.8)$$

Step 8: Calculate the closeness coefficient (CC) of the alternatives by the given equation:

$$C_i = \frac{R_i^-}{R_i^+ + R_i^-}, \forall\, I \quad (11.9)$$

The closeness coefficient indicates the superiority of alternatives. The maximum value of C_i reflects that ith alternatives are better, otherwise, the minimum value of C_i indicates that ith alternatives are poorer.

11.4 ILLUSTRATIVE EXAMPLE AND RESULT

In this section, two examples are set for applicability and validating the proposed study.

Example 11.1 The study focuses on sustainable supplier selection which contains economical, environmental, and social issues. Here, economical issues are taken with cost (C_1) and product quality attributes (C_2), environmental issues include pollution production (G_1), waste management (G_2), and environment management (G_3), also social issues are made with safety (S_1), health of staff (S_2), and job creation (S_3). Utilizing the attributes, we have to select a better sustainable supplier. We are comfortable to provide the value of each attribute corresponding to six alternatives (D_1, D_2, D_3, D_4, D_5, D_6) as a q-ROFN. According to given information, we have to rank the sustainable supplier using entropy measure-based TOPSIS method. The following steps are as follows:

Step 1: We establish a decision matrix (see Table 11.1) which contains six alternatives with eight attributes. For simplicity, we take the value of q is 3.

Table 11.1 Decision matrix of attributes corresponding to alternatives (Example 11.1)

	C_1	C_2	G_1	G_2	G_3	S_1	S_2	S_3
D_1	$(0.7,0.3)_3$	$(0.9,0.4)_3$	$(0.8,0.6)_3$	$(0.6,0.1)_3$	$(0.7,0.3)_3$	$(0.7,0.2)_3$	$(0.6,0.5)_3$	$(0.8,0.7)_3$
D_2	$(0.5,0.4)_3$	$(0.6,0.5)_3$	$(0.5,0.3)_3$	$(0.9,0.6)_3$	$(0.9,0.1)_3$	$(0.6,0.3)_3$	$(0.7,0.5)_3$	$(0.9,0.6)_3$
D_3	$(0.7,0.3)_3$	$(0.6,0.4)_3$	$(0.7,0.6)_3$	$(0.8,0.4)_3$	$(0.6,0.3)_3$	$(0.9,0.5)_3$	$(0.8,0.4)_3$	$(0.5,0.2)_3$
D_4	$(0.6,0.2)_3$	$(0.8,0.4)_3$	$(0.9,0.6)_3$	$(0.8,0.3)_3$	$(0.7,0.1)_3$	$(0.8,0.6)_3$	$(0.7,0.4)_3$	$(0.7,0.3)_3$
D_5	$(0.9,0.4)_3$	$(0.9,0.6)_3$	$(0.8,0.6)_3$	$(0.7,0.5)_3$	$(0.9,0.7)_3$	$(0.9,0.6)_3$	$(0.8,0.6)_3$	$(0.9,0.5)_3$
D_6	$(0.7,0.2)_3$	$(0.8,0.6)_3$	$(0.7,0.5)_3$	$(0.8,0.3)_3$	$(0.6,0.4)_3$	$(0.7,0.3)_3$	$(0.9,0.4)_3$	$(0.7,0.6)_3$

Step 2: Calculate the score value of each q-ROFN by Definition 11.5 and normalized form of each attribute by utilizing Eq. (11.1), which are shown in Tables 11.2 and 11.3, respectively.

Step 3: Determine the information entropy and depicted in Table 11.4 by using Eq. (11.2).

Step 4: Using Eq. (11.3), the weights of the attributes are calculated as 0.158, 0.159, 0.128, 0.053, 0.121, 0.064, 0.140, and 0.177.

Step 5: After evaluating the weight of the attributes, the weighted normalized matrix is established by using Eq. (11.4), and it is depicted in Table 11.5.

Step 6: The PIS and NIS are calculated by utilizing Eqs. (11.5, 11.6) and the result is shown in Table 11.6.

Table 11.2 Score value of each q-ROFN (Example 11.1)

	C_1	C_2	G_1	G_2	G_3	S_1	S_2	S_3
D_1	0.316	0.665	0.296	0.215	0.316	0.335	0.091	0.169
D_2	0.061	0.091	0.098	0.513	0.728	0.189	0.218	0.513
D_3	0.316	0.152	0.127	0.448	0.189	0.604	0.448	0.117
D_4	0.208	0.448	0.513	0.485	0.342	0.296	0.279	0.316
D_5	0.665	0.513	0.296	0.218	0.386	0.513	0.296	0.604
D_6	0.335	0.296	0.218	0.485	0.152	0.316	0.665	0.127

Table 11.3 Normalized value of the attributes (Example 11.1)

	C_1	C_2	G_1	G_2	G_3	S_1	S_2	S_3
D_1	0.166	0.307	0.191	0.091	0.150	0.149	0.046	0.092
D_2	0.032	0.042	0.063	0.217	0.345	0.084	0.109	0.278
D_3	0.166	0.070	0.082	0.190	0.089	0.268	0.224	0.063
D_4	0.109	0.207	0.331	0.205	0.162	0.131	0.140	0.171
D_5	0.350	0.237	0.191	0.092	0.183	0.228	0.148	0.327
D_6	0.176	0.137	0.141	0.205	0.072	0.140	0.333	0.069

Table 11.4 Information entropy of the attributes (Example 11.1)

	C_1	C_2	G_1	G_2	G_3	S_1	S_2	S_3
D_1	−0.298	−0.363	−0.316	−0.218	−0.284	−0.283	−0.141	−0.219
D_2	−0.110	−0.133	−0.175	−0.332	−0.367	−0.208	−0.242	−0.356
D_3	−0.298	−0.186	−0.205	−0.315	−0.216	−0.353	−0.335	−0.175
D_4	−0.242	−0.326	−0.366	−0.325	−0.295	−0.267	−0.275	−0.302
D_5	−0.367	−0.341	−0.316	−0.220	−0.311	−0.337	−0.283	−0.366
D_6	−0.306	−0.272	−0.276	−0.325	−0.189	−0.276	−0.366	−0.184

Table 11.5 Weighted normalized matrix of each attribute (Example 11.1)

	C_1	C_2	G_1	G_2	G_3	S_1	S_2	S_3
D_1	0.050	0.106	0.038	0.011	0.038	0.021	0.013	0.030
D_2	0.010	0.014	0.013	0.027	0.088	0.012	0.030	0.091
D_3	0.050	0.024	0.016	0.024	0.023	0.039	0.063	0.021
D_4	0.033	0.071	0.066	0.026	0.041	0.019	0.039	0.056
D_5	0.105	0.081	0.038	0.012	0.047	0.033	0.041	0.107
D_6	0.053	0.047	0.028	0.026	0.018	0.020	0.093	0.023

Table 11.6 The outcomes of the PIS and NIS (Example 11.1)

	C_1	C_2	G_1	G_2	G_3	S_1	S_2	S_3
X^+	0.010	0.106	0.013	0.027	0.088	0.039	0.093	0.107
X^-	0.105	0.014	0.066	0.011	0.018	0.012	0.013	0.021

Table 11.7 Euclidean distance, CC, and rank of the alternatives (Example 11.1)

	R_i^+	R_i^-	CC	Rank
D_1	0.133	0.113	0.458	4
D_2	0.115	0.149	0.565	1
D_3	0.145	0.095	0.395	6
D_4	0.112	0.106	0.484	3
D_5	0.122	0.122	0.498	2
D_6	0.134	0.109	0.449	5

Step 7: Euclidean distance of each alternative from PIS and NIS is determined by applying Eqs. (11.7, 11.8). The outcomes are depicted in Table 11.7.

Step 8: Lastly, we calculate the CC of each alternative by utilizing Eq. (11.9), and the results are shown in Table 11.7. Thereafter, we rank the sustainable supplier. From Table 11.7, we see that the alternatives D_4 as a better supplier than others. Similarly, D_2 be the worst supplier compared to others.

Example 11.2 Here, we take same attributes as in Example 11.1 for sustainable supplier selection. The value of each attribute for each of the seven options (P_1, P_2, P_3, P_4, P_5, P_6, P_7) is comfortable to be provided as a q-ROFN. Using the above data, we use the TOPSIS method, which is entropy measure-based to rank the sustainable supplier. The steps are as follows:

Step 1: We build a decision matrix (see Table 11.8) which contains seven alternatives with eight attributes. We assume q has a value of 3 to keep things simple.

Step 2: Evaluate the score value of each q-ROFN by Definition 11.5 and the normalized form of each attribute by utilizing Eq. (11.1), which are shown in Tables 11.9 and 11.10, respectively.

Step 3: Calculate the information entropy and it is depicted in Table 11.11 by using Eq. (11.2).

Step 4: Using Eq. (11.3), the weight of the attributes is calculated as 0.125, 0.159, 0.116, 0.061, 0.192, 0.044, 0.027, and 0.275.

Step 5: After determining the weight of the attributes, the weighted normalized matrix is established by using Eq. (11.4), and it is shown in Table 11.12.

Table 11.8 Decision matrix of attributes corresponding to alternatives (Example 11.2)

	C_1	C_2	G_1	G_2	G_3	S_1	S_2	S_3
P_1	$(0.6,0.4)_3$	$(0.8,0.5)_3$	$(0.7,0.6)_3$	$(0.9,0.5)_3$	$(0.6,0.4)_3$	$(0.8,0.2)_3$	$(0.9,0.7)_3$	$(0.8,0.4)_3$
P_2	$(0.8,0.3)_3$	$(0.6,0.5)_3$	$(0.8,0.3)_3$	$(0.9,0.6)_3$	$(0.7,0.4)_3$	$(0.8,0.3)_3$	$(0.8,0.3)_3$	$(0.7,0.6)_3$
P_3	$(0.7,0.3)_3$	$(0.6,0.4)_3$	$(0.7,0.6)_3$	$(0.8,0.4)_3$	$(0.6,0.3)_3$	$(0.9,0.5)_3$	$(0.8,0.4)_3$	$(0.5,0.2)_3$
P_4	$(0.6,0.3)_3$	$(0.9,0.5)_3$	$(0.8,0.6)_3$	$(0.7,0.3)_3$	$(0.9,0.7)_3$	$(0.9,0.4)_3$	$(0.8,0.6)_3$	$(0.9,0.3)_3$
P_5	$(0.9,0.4)_3$	$(0.9,0.6)_3$	$(0.8,0.6)_3$	$(0.7,0.3)_3$	$(0.9,0.7)_3$	$(0.9,0.4)_3$	$(0.8,0.6)_3$	$(0.9,0.3)_3$
P_6	$(0.7,0.2)_3$	$(0.8,0.4)_3$	$(0.7,0.5)_3$	$(0.8,0.2)_3$	$(0.6,0.4)_3$	$(0.7,0.3)_3$	$(0.9,0.6)_3$	$(0.7,0.6)_3$
P_7	$(0.7,0.5)_3$	$(0.8,0.7)_3$	$(0.9,0.6)_3$	$(0.9,0.7)_3$	$(0.8,0.6)_3$	$(0.7,0.5)_3$	$(0.6,0.4)_3$	$(0.8,0.7)_3$

Table 11.9 Score value of each *q*-ROFN (Example 11.2)

	C_1	C_2	G_1	G_2	G_3	S_1	S_2	S_3
P_1	0.152	0.387	0.127	0.604	0.152	0.504	0.386	0.448
P_2	0.485	0.091	0.485	0.513	0.279	0.485	0.485	0.127
P_3	0.316	0.152	0.127	0.448	0.189	0.604	0.448	0.117
P_4	0.189	0.604	0.296	0.189	0.728	0.296	0.279	0.702
P_5	0.665	0.513	0.296	0.316	0.386	0.665	0.296	0.702
P_6	0.335	0.448	0.218	0.504	0.152	0.316	0.513	0.127
P_7	0.218	0.169	0.513	0.386	0.296	0.218	0.152	0.169

Step 6: The PIS and NIS are calculated by utilizing Eqs. (11.5, 11.6) and the result is depicted in Table 11.13.

Step 7: Euclidean distance of each alternative from PIS and NIS is determined by applying Eqs. (11.7, 11.8). The outcomes are shown in Table 11.14.

Step 8: Lastly, we evaluate the CC of each alternative by using Eq. (11.9), and the results are depicted in Table 11.14. Thereafter, we rank the sustainable supplier. From Table 11.14, we see the alternatives P_2 as a better supplier than others. Similarly, P_3 is the worst supplier compared to others.

11.5 COMPARATIVE STUDY WITH OTHER METHODS

In this section, to determine the effectiveness and viability of the suggested alternatives, we provide comparisons and discussions. Here, the proposed methods are compared to other existing methods in a comparison analysis. For this purpose, the weights of the attributes are obtained from other two methods, such as AHP and ITARA. Thereafter, TOPSIS method is integrated with AHP and ITARA for determining the rank of the alternatives. The outcomes of the AHP-TOPSIS and ITARA-TOPSIS methods are depicted in Figures 11.2 and 11.3 for Examples 11.1 and 11.2, respectively.

Table 11.10 Normalized value of the attributes (Example 2).

	C_1	C_2	G_1	G_2	G_3	S_1	S_2	S_3
P_1	0.071	0.176	0.082	0.235	0.081	0.176	0.160	0.202
P_2	0.226	0.041	0.313	0.199	0.148	0.169	0.201	0.057
P_3	0.148	0.069	0.082	0.174	0.100	0.210	0.186	0.053
P_4	0.088	0.275	0.191	0.073	0.386	0.103	0.116	0.316
P_5	0.310	0.234	0.191	0.123	0.205	0.232	0.123	0.316
P_6	0.156	0.204	0.141	0.196	0.081	0.110	0.213	0.057
P_7	0.102	0.079	0.239	0.180	0.138	0.102	0.071	0.079

Table 11.11 Information entropy of the attributes (Example 2).

	C_1	C_2	G_1	G_2	G_3	S_1	S_2	S_3
P_1	−0.188	−0.306	−0.205	−0.340	−0.203	−0.305	−0.294	−0.323
P_2	−0.336	−0.132	−0.364	−0.321	−0.283	−0.300	−0.323	−0.164
P_3	−0.282	−0.185	−0.205	−0.304	−0.231	−0.328	−0.313	−0.155
P_4	−0.214	−0.355	−0.316	−0.192	−0.367	−0.234	−0.250	−0.364
P_5	−0.363	−0.340	−0.316	−0.257	−0.325	−0.339	−0.258	−0.364
P_6	−0.290	−0.324	−0.276	−0.319	−0.203	−0.243	−0.329	−0.164
P_7	−0.233	−0.200	−0.342	−0.309	−0.273	−0.233	−0.188	−0.200

Table 11.12 Weighted normalized matrix of each attribute (Example 11.2)

	C_1	C_2	G_1	G_2	G_3	S_1	S_2	S_3
P_1	0.019	0.062	0.015	0.037	0.029	0.022	0.010	0.123
P_2	0.061	0.014	0.056	0.031	0.054	0.022	0.013	0.035
P_3	0.040	0.024	0.015	0.027	0.036	0.027	0.012	0.032
P_4	0.024	0.096	0.034	0.012	0.140	0.013	0.008	0.193
P_5	0.083	0.082	0.034	0.019	0.074	0.030	0.008	0.193
P_6	0.042	0.071	0.025	0.031	0.029	0.014	0.014	0.035
P_7	0.027	0.021	0.064	0.048	0.037	0.027	0.019	0.021

Table 11.13 The outcomes of the PIS and NIS (Example 11.2)

	C_1	C_2	G_1	G_2	G_3	S_1	S_2	S_3
X^+	0.019	0.096	0.015	0.037	0.140	0.030	0.014	0.193
X^-	0.083	0.014	0.056	0.012	0.029	0.013	0.008	0.032

Table 11.14 Euclidean distance, CC, and rank of the alternatives (Example 11.2)

	R_i^+	R_i^-	CC	Rank
P_1	0.135	0.131	0.491	4
P_2	0.207	0.040	0.162	7
P_3	0.206	0.065	0.241	6
P_4	0.037	0.221	0.857	1
P_5	0.097	0.182	0.653	2
P_6	0.197	0.079	0.288	5
P_7	0.071	0.103	0.590	3

Furthermore, other two ranking methods such as COPRAS and MOORA are utilized to rank the alternatives' rank. Therefore, we utilize entropy method with COPRAS and MOORA methods for ranking the alternatives, and the results are shown in Figures 11.2 and 11.3 for Examples 11.1 and 11.2, respectively. From Figure 11.2, according to the ranking outcomes for each alternative using different approaches, D_2 and D_3 are in the first and last positions, respectively. From Figure 11.3, according to the ranking outcomes for each alternative using different approaches, P_4 and P_2 are in the first and last positions, respectively. With the exception of ITARA-TOPSIS in Example 11.1, for which there are a few variations in the ranking outcome as D_4 and D_5 are in the second and third positions, respectively, the remaining alternatives' ranking orders using the existing methods are equivalent to the proposed method. This leads to the conclusion that the ranking outcomes of the alternatives generated by the proposed entropy-TOPSIS method are acceptable and reliable.

11.6 ADVANTAGES OF THE STUDY

Here we address some advantages of the proposed study.

1. In contrast with existing methods, the suggested score function provides a suitable defuzzified value of q-ROFN that is easier to apply to decision-making issues.
2. It considers the numerical values of q-ROFNs as well as membership and nonmembership values.

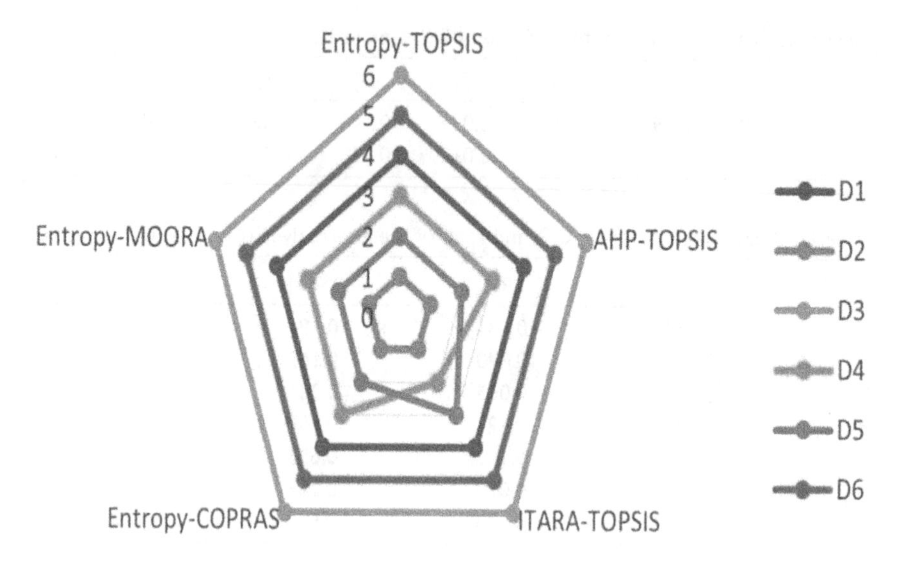

Figure 11.2 Ranking order of the alternatives obtained by different methods from Example 11.1.

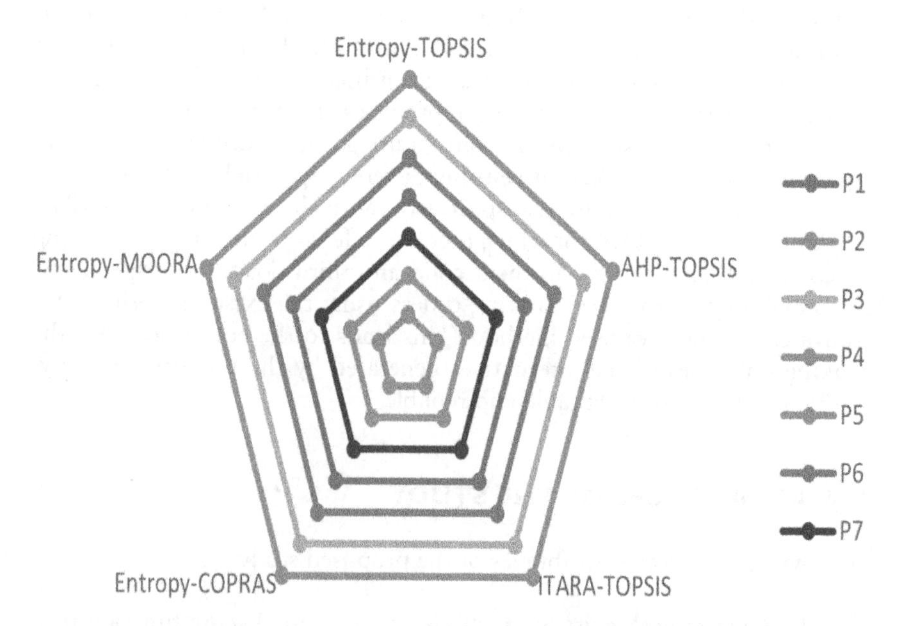

Figure 11.3 Ranking order of the alternatives obtained by different methods from Example 11.2.

3. The developed methodology for evaluating an MADM problem can be used to help organizations for implementing an integrating framework to select the best sustainable supplier selection under different criteria.
4. This approach can be successfully handled with imprecision and vagueness of input parameters and changing significant weight of criteria in MADM problem.
5. Integrating entropy measure with TOPSIS method determines the issue that includes interaction among the criteria. Thereafter, the proposed method can produce an attribute set of criteria and the weights of the supplier, and select suppliers in a logical and credible way through a straight computational process for suppliers.

11.7 CONCLUSION

In MADM problems, attribute weights play a crucial role because various attribute weights may result in different ranking orders for the alternatives. In order to create a suitable MADM solution, the first contribution combines the q-ROF set with the TOPSIS method. Due to shortage of data and insufficient information, DMs depend on the q-ROF set. The attribute weight value has been calculated by the q-ROF entropy as a replacement for the assumed weight that decision-makers specify directly in the TOPSIS method is the second contribution. In other words, weight of the q-ROF entropy is calculated rather than assumed, which lessens any bias that may result from the subjective assessment.

The classically weighted TOPSIS method cannot fully reflect reality, it can only determine a distance relative to the ideal solution. In order to pick sustainable suppliers, this research suggests a new MADM model that combines with the q-ROF entropy-based TOPSIS method. The paper uses a sustainable supplier selection as an example to demonstrate the model's viability and efficacy by using two numerical examples. As a result, entropy measure-based TOPSIS method under q-ROF environment is better than other mentioned approaches. Furthermore, we observe that D_4 is a better supplier to the alternatives. Comparatively speaking, D_2 is the worst supplier for Example 11.1. In comparison to other options, we can see that P_2 is a better supplier. Likewise, P_3 is the worst supplier when compared to the others in Example 11.2. Due to the fact that it assumes a comparatively objective, naive, and hazy assessment of the decision-makers' opinions, the entropy-based TOPSIS method merged with q-ROF set offers significant potential for success in MADM. In future, the novel multi-criteria assessment model can be applied to deal with fuzzy and uncertainties in MADM themes including planning options, construction options, site selection, and management decision concerns in many other fields (Ding et al., 2023; Wang et al., 2023).

REFERENCES

Amoozad Mahdiraji, H., Arzaghi, S., Stauskis, G., & Zavadskas, E. K. (2018). A hybrid fuzzy BWM-COPRAS method for analyzing key factors of sustainable architecture. *Sustainability*, 10(5), 1626.

Atanassov, K. (1986). Intuitionistic fuzzy sets. *Fuzzy Sets and Systems*, 20(1), 87–96.

Debnath, K., & Roy, S. K. (2023). Power partitioned neutral aggregation operators for T-spherical fuzzy sets: An application to H2 refuelling site selection. *Expert Systems with Applications*, 216, 119470.

Din¸cer, H., Yu¨ksel, S., & Mart´ınez, L. (2019). Interval type 2-based hybrid fuzzy evaluation of financial services in E7 economies with DEMATEL-ANP and MOORA methods. *Applied Soft Computing*, 79, 186–202.

Ding, S., Li, R., & Guo, J. (2023). An entropy-based TOPSIS and optimized grey prediction model for spatiotemporal analysis in strategic emerging industry. *Expert Systems with Applications*, 213, 119169.

Ghosh, S., Roy, S. K., & Weber, G. W. (2023). Interactive strategy of carbon cap-and-trade policy on sustainable multi-objective solid transportation problem with twofold uncertain waste management. *Annals of Operations Research*, 1–41. https://doi.org/10.1007/s10479-023-05347-w.

Giri, B. K., & Roy, S. K. (2022). Neutrosophic multi-objective green four-dimensional fixed-charge transportation problem. *International Journal of Machine Learning and Cybernetics*, 13(10), 3089–3112.

Giri, B. K., Roy, S. K., & Deveci, M. (2023). Fuzzy robust flexible programming with *Me* measure for electric sustainable supply chain. *Applied Soft Computing*, 145, 110614.

Giri, B. K., Roy, S. K., & Deveci, M. (2023). Projection based regret theory on three-way decision model in probabilistic interval-valued q-rung orthopair hesitant fuzzy set and its application to medicine company. *Artificial Intelligence Review*, 56, 3617–3649.

Hwang, C. L., & Yoon, K. (1981). Methods for multiple attribute decision making. In *Multiple Attribute Decision Making: Methods and Applications a State-of-the-Art Survey*, Springer, 186, 58–191.

Jana, J., & Roy, S. K. (2023). Linguistic Pythagorean hesitant fuzzy matrix game and its application in multi-criteria decision making. *Applied Intelligence*, 53(1), 1–22.

Liu, P., & Wang, P. (2018). Some q-rung orthopair fuzzy aggregation operators and their applications to multiple-attribute decision making. *International Journal of Intelligent Systems*, 33(2), 259–280.

Giri, B. K., & Roy, S. K. (2024). Fuzzy-random robust flexible programming on sustainable closed-loop renewable energy supply chain. *Applied Energy*, 363, 123044.

Mondal, A., Giri, B. K., & Roy, S. K. (2023). An integrated sustainable bio-fuel and bio-energy supply chain: A novel approach based on DEMATEL and fuzzy-random robust flexible programming with Me measure. *Applied Energy*, 343, 121225.

Mondal, A., & Roy, S. K. (2021). Multi-objective sustainable opened-and closed-loop supply chain under mixed uncertainty during COVID-19 pandemic situation. *Computers & Industrial Engineering*, 159, 107453.

Rahim, M., Garg, H., Amin, F., Perez-Dominguez, L., & Alkhayyat, A. (2023). Improved cosine similarity and distance measures-based TOPSIS method for cubic Fermatean fuzzy sets. *Alexandria Engineering Journal*, 73, 309–319.

Wang, X., Liu, C., Su, M., Li, F., & Dong, M. (2023). Machine learning-based AI approaches for personalized smart education systems using entropy and TOPSIS approach. *Soft Computing*, 1–17. https://doi.org/10.1007/s00500-023-08392-6.

Yager, R. R. (2016). Generalized orthopair fuzzy sets. *IEEE Transactions on Fuzzy Systems*, 25(5), 1222–1230.

Yager, R. R., & Abbasov, A. M. (2013). Pythagorean membership grades, complex numbers, and decision making. *International Journal of Intelligent Systems*, 28(5), 436452.

Zadeh, L. (1965). Fuzzy sets. *Information Control*, 8, 338–353.

Zhang, H., & Yu, L. (2012). MADM method based on cross-entropy and extended TOPSIS with interval-valued intuitionistic fuzzy sets. *Knowledge-Based Systems*, 30, 115–120.

Chapter 12

A novel similarity measure for advanced fuzzy set with applications to multiple attribute decision-making

Priya Yadav and Gagandeep Kaur

12.1 INTRODUCTION

Multi-Attribute Decision-Making (MCDM) is a complex problem-solving process that considers various factors such as salary, work location, promotion opportunity, and colleagues. This decision-making process can be rational or irrational, and can be influenced by physiological, biological, cultural, social, and other factors. The complexity level of a decision-making (DM) process can be affected by authority and risk levels. Mathematical equations, statistics, mathematics, economic theories, and computer devices can help solve complex decision-making problems automatically. Multi-Criteria Decision Making (MCDM) or Multi-Criteria Decision Analysis (MCDA) is one of the most accurate methods of decision-making and is considered a revolution in this field as stated by Martin Aruldoss (2013). Benjamin Franklin developed MCDM in his research on moral algebra. Several empirical and theoretical scientists (Hajduk, 2021) have worked on MCDM methods since the 1950s to examine their mathematical modelling capability. An MADM problem can be expressed in matrix format as:

$$D = \begin{bmatrix} x_{11} & x_{12} & \cdots & x_{1n} \\ x_{21} & x_{22} & \cdots & x_{2n} \\ \vdots & \vdots & \ddots & \vdots \\ x_{m1} & x_{m2} & \cdots & x_{mn} \end{bmatrix}$$

where A_i, i=1, 2,..., m are the alternatives; X_j, j=1, 2, ..., n are attributes with which alternative performances are measured; x_{ij} is the performance of alternative A_i with respect to attribute X_j.

Fuzzy set (FS) theory, introduced by Zadeh (1965), is a successful method for representing uncertainty in data. In FS theory, the membership of an element to a fuzzy set is a single value between 0 and 1. But in reality, the degree of non-membership (DoNM) in a fuzzy set may not always be certain and is just equal to 1 minus the degree of membership (DoM), and this uncertainty can be represented by aggregation operators (AOs). Atanassov

DOI: 10.1201/9781003497219-12

(2016) extended the theory to an intuitionistic fuzzy sets, which is characterized by membership and non-membership functions. Atanassov and Gargov (1999) extended the IFS to interval-valued intuitionistic fuzzy sets (IVIFSs), which contain degrees of agreeness and disagreeness as interval values. Experts have chosen appropriate AOs to address the nature of the DM problem. In recent decades, researchers have applied IFSs and IVIFSs to various fields, such as aggregating various interval-valued intuitionistic fuzzy numbers (IVIFNs). To handle it, Torra (2010) came up with the idea of hesitant fuzzy sets (HFSs). Bin Zhu (2012) enhanced the idea of dual hesitant fuzzy sets (DHFSs) by assigning equal importance to possible non-membership values as membership values in HFSs. Xu (2011) established different operators to aggregate values, and Garg and Arora (2018) presented some AOs under the dual hesitant fuzzy soft set environment and applied them to solve MCDM problems.

Despite efficient uncertainties capture, these approaches often fail to model situations where expert refusal dominates. To address this, Xu (2018) proposed probabilistic hesitant fuzzy sets (PHFSs) and introduced AOs on interval-valued PHFSs (IVPHFSs). They also explored preference relations based on IVPHFSs, Hao (2017) introduced probabilistic dual hesitant fuzzy sets (PDHFSs).

Distance and similarity measures are crucial in various scientific fields like decision-making, pattern recognition, machine learning, and market prediction. They have gained attention in recent decades due to their application across various areas. Despite the development of distance and similarity measures for FSs, IFSs, HFSs, DHFSs, and PDHFSs, there is limited research on PDHFSs, necessitating the development of similarity measures under PDHFSs.

In this chapter, we will talk about the distance and similarity measures of different FS. Firstly, we define all the definitions which we will consider in our approach and then we introduce our new similarity measure based on PDHFSs and their properties. We will consider one simple example and then apply our approach to real-life problems. After this, we compare some previous results with our approach.

12.2 PRELIMINARIES

In this section, we introduce basic terms and definitions related to HFSs, DHFSs, PHFSs, PDHFSs, and a cosine similarity measure for different FSs, which will be needed further.

Definition 12.1 On the universal set X, Atanassov (2016) defined an IFS A as:

$$A = \{ x,\ \mu_A(x),\ \nu_A(x) > | \, x \in X\},$$

where the functions:

$$\mu_A : x \to [0,1]$$

and

$$v_A : x \to [0,1]$$

define DoM and DoNM of $x \in X$, respectively, and for every $x \in X$:

$$0 \leq \mu_A(x) + v_A(x) \leq 1$$

Definition 12.2 Let X be a fixed set, an HFS on X was expressed by Xu, (2011) mathematically as:

$$A = \{< x, h_A(x) > | x \in X\}$$

where $h_A(x)$ is a set of some values in [0,1] denoting DoMs of $x \in X$ to A.

Definition 12.3 Let X be a fixed set, then DHFS D on X is given by Bin Zhu (2012) as:

$$D = \{< x, h(x), g(x) > | x \in X\},$$

in which $h(x)$ and $g(x)$ are two sets of some values in [0,1], denoting the possible DoMs and DoNMs of the element $x \in X$ to the set D, respectively, with the conditions:

$$0 \leq \gamma, \eta \leq 1, 0 \leq \gamma^+, \eta^+ \leq 1$$

where $\gamma \in h(x)$, $\eta \in g(x)$, $\gamma^+ \in h^+(x) = \cup_{\gamma \in h(x)} max\{\gamma\}$ and $\eta^+ \in g^+(x) = \cup_{\eta \in g(x)} max\{\eta\}$ for all $x \in X$.

Definition 12.4 Let X be a fixed set, and a PDHFS on X is derived by Hao (2017) and is described as:

$$P = \{<x, h(x) \mid p(x), g(x) \mid q(x)> \mid x \in X\}$$

Here $h(x) \mid p(x)$ and $g(x) \mid q(x)$ are sets of possible elements where $h(x)$ and $g(x)$ represent the HFS DoMs and DoNMs to the set X, respectively. $p(x)$ and $q(x)$ are the corresponding probabilities. Also, there is $0 \leq \gamma, \eta \leq 1$, $0 \leq \gamma^+, \eta^+ \leq 1$ and $p_i \in [0,1], q_j \in [0,1], \sum_{i+1}^{h} p_i = 1, \sum_{i+1}^{g} q_i = 1$

where $\gamma \in h(x)$, $\eta \in g(x)$, $\gamma^{+} \in h^{+}(x) = \cup_{\gamma \in h(x)} \max\{\gamma\}$ and $\eta^{+} \in g^{+}(x) = \cup\, \eta \in g(x)$ $\max\{\eta\}$, $p_i \in p(x)$, and $q_i \in q(x)$. h and g are total numbers of elements in $h(x)$ | $p(x)$, $g(x)$ | $q(x)$, respectively.

12.3 NEW MEASURE AND THEIR PROPERTIES

In this section, a similarity measure and a weighted similarity measure between PDHFSs will be proposed over a universal set $X = \{x_1, x_2, \ldots, x_n\}$. Here, we use notations which will be given below:

Notations	*Meaning*
N	elements in X
h_A	hesitant DoMs of A
g_A	hesitant DoNMs of A
M_A	elements in h_A
N_A	elements in g_A
p_A	probability for hesitant DoMs of A
q_A	probability for hesitant DoNMs of A
W	weight vector

Let $A=\{(x, h_{A_j}(x)|p_{A_j}(x), g_{A_k}(x)|q_{A_k}(x))|x\in X\}$ where j=1, 2, …, M_A; k=1, 2, …, N_A and $B=\{(x, h_{B_{j'}}(x)|p_{B_{j'}}(x), g_{B_{k'}}(x)|q_{B_{k'}}(x))|x\in X\}$ where j'=1, 2, …, M_B; k'=1, 2, …, N_B be two PDHFSs. Suppose $M = max\{M_A, M_B\}$ and $N=max\{N_A, N_B\}$ be two real numbers, and we define similarity between A and B as:

$$CPDHFS(A,B)=\frac{1}{n}\sum_{i=1}^{n}\frac{\left(\frac{1}{M_A}\sum_{j=1}^{M_A}\gamma_{A_j}(x_i)p_{\gamma_{A_j}}(x_i)\right)\left(\frac{1}{M_B}\sum_{j'=1}^{M_B}\gamma_{B_{j'}}(x_i)p_{\gamma_{B_{j'}}}(x_i)\right)+\left(\frac{1}{N_A}\sum_{k=1}^{N_A}\eta_{A_k}(x_i)q_{\eta_{A_k}}(x_i)\right)\left(\frac{1}{N_B}\sum_{k'=1}^{N_B}\eta_{B_{k'}}(x_i)q_{\eta_{B_{k'}}}(x_i)\right)}{\sqrt{\frac{1}{M_A}\sum_{j=1}^{M_A}\gamma_{A_j}^2(x_i)p_{\gamma_{A_j}}^2(x_i)+\frac{1}{N_A}\sum_{k=1}^{N_A}\eta_{A_k}^2(x_i)q_{\eta_{A_k}}^2(x_i)}\sqrt{\frac{1}{M_B}\sum_{j'=1}^{M_B}\gamma_{B_{j'}}^2(x_i)p_{\gamma_{B_{j'}}}^2(x_i)+\frac{1}{N_B}\sum_{k'=1}^{N_B}\eta_{B_{k'}}^2(x_i)q_{\eta_{B_{k'}}}^2(x_i)}}$$

Here $\gamma_{A_j} \in h_{A_j}, \gamma_{B_j} \in h_{B_{j'}}, \eta_{A_j} \in g_{A_j}, \eta_{B_j} \in g_{B_{j'}}$.

Taking $n = 1$, similarity between PDHFSs A and B gets converted to correlation coefficient subjected to IFS, i.e., $CPDHFS\ (A, B) = k\ (A, B)$.

Therefore, the proposed measure between PDHFSs A and B also satisfies the following properties:

(P1): $0 \le CPDHFS\,(A, B) \le 1$;
(P2): $CPDHFS\,(A, B) = CPDHFS\,(B, A)$
(P3): $CPDHFS\,(A, B) = 1$ if $A = B$, i.e. $\gamma_{A_j}(x_i) = \gamma_{B_{j'}}(x_i)$ and $\eta_{A_k}(x_i) = \eta_{B_{k'}}(x_i)$ for $i=1,2, \ldots, n$

Proof:

(P1): For membership, since

$$0 \le \gamma_{A_j}(x_i) \le 1$$

and

$$0 \le p_{\gamma_{A_j}}(x_i) \le 1$$

Multiplying Eqs. (2) and (3), we get

$$0 \le \gamma_{A_j}(x_i)\, p_{\gamma_{A_j}}(x_i) \le 1$$

Taking summation from j=1 to M_A, we get

$$0 \le \sum_{j=1}^{M_A} \gamma_{A_j}(x_i)\, p_{\gamma_{A_j}}(x_i) \le M_A$$

Dividing by M_A, we get

$$0 \le \frac{1}{M_A} \sum_{j=1}^{M_A} \gamma_{A_j}(x_i)\, p_{\gamma_{A_j}}(x_i) \le 1$$

Similarly, it follows for PDHFS 'B' and for non-membership, as below

$$0 \le \frac{1}{M_B} \sum_{j=1}^{M_B} \gamma_{B_{j'}}(x_i)\, p_{\gamma_{B_{j'}}}(x_i) \le 1,$$

$$0 \le \frac{1}{N_A} \sum_{k=1}^{N_A} \eta_{A_k}(x_i)\, q_{\eta_{A_k}}(x_i) \le 1,$$

and

$$0 \le \frac{1}{N_B} \sum_{k'=1}^{N_B} \eta_{B_{k'}}(x_i)\, q_{\eta_{B_{k'}}}(x_i) \le 1$$

Now squaring both Eqs. (12.2) and (12.3) and then multiplying, we get

$$0 \leq \gamma_{A_j}^2(x_i)p_{\gamma A_j}^2(x_i) \leq 1$$

Taking summation from j=1 to M_A, we get

$$0 \leq \sum_{j=1}^{M_A} \gamma_{A_j}^2(x_i)p_{\gamma A_j}^2(x_i) \leq M_A$$

Dividing by M_A, we get

$$0 \leq \frac{1}{M_A}\sum_{j=1}^{M_A} \gamma_{A_j}^2(x_i)p_{\gamma A_j}^2(x_i) \leq 1$$

Similarly, it follows for PDHFS 'B' and for non-membership, as below

$$0 \leq \frac{1}{M_B}\sum_{j=1}^{M_B} \gamma_{B_{j'}}^2(x_i)p_{\gamma B_{j'}}^2(x_i) \leq 1,$$

$$0 \leq \frac{1}{N_A}\sum_{k=1}^{N_A} \eta_{A_k}^2(x_i)q_{\eta A_k}^2(x_i) \leq 1$$

and

$$0 \leq \frac{1}{N_B}\sum_{k'=1}^{N_B} \eta_{B_{k'}}^2(x_i)q_{\eta B_{k'}}^2(x_i) \leq 1$$

From the above notion, we conclude that

$$0 \leq \frac{1}{n}\sum_{i=1}^{n} \frac{\left(\frac{1}{M_A}\sum_{j=1}^{M_A}\gamma_{A_j}(x_i)p_{\gamma A_j}(x_i)\right)\left(\frac{1}{M_B}\sum_{j'=1}^{M_B}\gamma_{B_{j'}}(x_i)p_{\gamma B_{j'}}(x_i)\right) + \left(\frac{1}{N_A}\sum_{k=1}^{N_A}\eta_{A_k}(x_i)q_{\eta A_k}(x_i)\right)\left(\frac{1}{N_B}\sum_{k'=1}^{N_B}\eta_{B_{k'}}(x_i)q_{\eta B_{k'}}(x_i)\right)}{\sqrt{\frac{1}{M_A}\sum_{j=1}^{M_A}\gamma_{A_j}^2(x_i)p_{\gamma A_j}^2(x_i) + \frac{1}{N_A}\sum_{k=1}^{N_A}\eta_{A_k}^2(x_i)q_{\eta A_k}^2(x_i)}\sqrt{\frac{1}{M_B}\sum_{j'=1}^{M_B}\gamma_{B_{j'}}^2(x_i)p_{\gamma B_{j'}}^2(x_i) + \frac{1}{N_B}\sum_{k'=1}^{N_B}\eta_{B_{k'}}^2(x_i)q_{\eta B_{k'}}^2(x_i)}} \leq 1$$

which indicate $0 \leq CPDHFS\,(A, B) \leq 1$

(P2): As

$$CPDHFS(A,B)=\frac{1}{n}\sum_{i=1}^{n}\frac{\left(\frac{1}{M_A}\sum_{j=1}^{M_A}\gamma_{A_j}(x_i)p_{\gamma_{A_j}}(x_i)\right)\left(\frac{1}{M_B}\sum_{j'=1}^{M_B}\gamma_{B_{j'}}(x_i)p_{\gamma_{B_{j'}}}(x_i)\right)+\left(\frac{1}{N_A}\sum_{k=1}^{N_A}\eta_{A_k}(x_i)q_{\eta_{A_k}}(x_i)\right)\left(\frac{1}{N_B}\sum_{k'=1}^{N_B}\eta_{B_{k'}}(x_i)q_{\eta_{B_{k'}}}(x_i)\right)}{\sqrt{\frac{1}{M_A}\sum_{j=1}^{MA}\gamma_{A_j}^2(x_i)p_{\gamma_{A_j}}^2(x_i)+\frac{1}{N_A}\sum_{k=1}^{N_A}\eta_{A_k}^2(x_i)q_{\eta_{A_k}}^2(x_i)}\sqrt{\frac{1}{M_B}\sum_{j'=1}^{M_B}\gamma_{B_{j'}}^2(x_i)p_{\gamma_{B_{j'}}}^2(x_i)+\frac{1}{N_B}\sum_{k'=1}^{N_B}\eta_{B_{k'}}^2(x_i)q_{\eta_{B_{k'}}}^2(x_i)}}$$

and

$$CPDHFS(B,A)=\frac{1}{n}\sum_{i=1}^{n}\frac{\left(\frac{1}{M_B}\sum_{j'=1}^{M_B}\gamma_{B_{j'}}(x_i)p_{\gamma_{B_{j'}}}(x_i)\right)\left(\frac{1}{M_A}\sum_{j=1}^{M_A}\gamma_{A_j}(x_i)p_{\gamma_{A_j}}(x_i)\right)+\left(\frac{1}{N_B}\sum_{k'=1}^{N_B}\eta_{B_{k'}}(x_i)q_{\eta_{B_{k'}}}(x_i)\right)\left(\frac{1}{N_A}\sum_{k=1}^{N_A}\eta_{A_k}(x_i)q_{\eta_{A_k}}(x_i)\right)}{\sqrt{\frac{1}{N_A}\sum_{k=1}^{N_A}\eta_{A_k}^2(x_i)q_{\eta_{A_k}}^2(x_i)+\frac{1}{M_A}\sum_{j=1}^{M_A}\gamma_{A_j}^2(x_i)p_{\gamma_{A_j}}^2(x_i)}\sqrt{\frac{1}{N_B}\sum_{k'=1}^{N_B}\eta_{B_{k'}}^2(x_i)q_{\eta_{B_{k'}}}^2(x_i)+\frac{1}{M_B}\sum_{j'=1}^{M_B}\gamma_{B_{j'}}^2(x_i)p_{\gamma_{B_{j'}}}^2(x_i)}}$$

Re-writing the above equation, we get

$$CPDHFS(B,A)=\frac{1}{n}\sum_{i=1}^{n}\frac{\left(\frac{1}{M_A}\sum_{j=1}^{M_A}\gamma_{A_j}(x_i)p_{\gamma_{A_j}}(x_i)\right)\left(\frac{1}{M_B}\sum_{j'=1}^{M_B}\gamma_{B_{j'}}(x_i)p_{\gamma_{B_{j'}}}(x_i)\right)+\left(\frac{1}{N_A}\sum_{k=1}^{N_A}\eta_{A_k}(x_i)q_{\eta_{A_k}}(x_i)\right)\left(\frac{1}{N_B}\sum_{k'=1}^{N_B}\eta_{B_{k'}}(x_i)q_{\eta_{B_{k'}}}(x_i)\right)}{\sqrt{\frac{1}{M_A}\sum_{j=}^{M_A}\gamma_{A_j}^2(x_i)p_{\gamma_{A_j}}^2(x_i)+\frac{1}{N_A}\sum_{k=1}^{N_A}\eta_{A_k}^2(x_i)q_{\eta_{A_k}}^2(x_i)}\sqrt{\frac{1}{M_B}\sum_{j'=1}^{M_B}\gamma_{B_{j'}}^2(x_i)p_{\gamma_{B_{j'}}}^2(x_i)+\frac{1}{N_B}\sum_{k'=1}^{N_B}\eta_{B_{k'}}^2(x_i)q_{\eta_{B_{k'}}}^2(x_i)}}$$

$$\Longrightarrow CPDHFS\,(A,\,B) = CPDHFS\,(B,\,A).$$

Hence, it follows the symmetric property.

(P3): If $A=B$, then

$$CPDHFS(A,A)=\frac{1}{n}\sum_{i=1}^{n}\frac{\left(\frac{1}{M_A}\sum_{j=1}^{M_A}\gamma_{A_j}(x_i)p_{\gamma_{A_j}}(x_i)\right)^2+\left(\frac{1}{N_A}\sum_{k=1}^{N_A}\eta_{A_k}(x_i)q_{\eta_{A_k}}(x_i)\right)^2}{\sqrt{\frac{1}{M_A}\sum_{j=1}^{M_A}\gamma_{A_j}^2(x_i)p_{\gamma_{A_j}}^2(x_i)+\frac{1}{N_A}\sum_{k=1}^{N_A}\eta_{A_k}^2(x_i)q_{\eta_{A_k}}^2(x_i)}^{\,2}}$$

$$CPDHFS(A,A)=\frac{1}{n}\sum_{i=1}^{n}\frac{\frac{1}{M_A}\sum_{j=1}^{M_A}\gamma_{A_j}^2(x_i)p_{\gamma_{A_j}}^2(x_i)+\frac{1}{N_A}\sum_{k=1}^{N_A}\eta_{A_k}^2(x_i)q_{\eta_{A_k}}^2(x_i)}{\frac{1}{M_A}\sum_{j=1}^{M_A}\gamma_{A_j}^2(x_i)p_{\gamma_{A_j}}^2(x_i)+\frac{1}{N_A}\sum_{k=1}^{N_A}\eta_{A_k}^2(x_i)q_{\eta_{A_k}}^2(x_i)}$$

$$CPDHFS(A,A)=\frac{1}{n}\sum_{i=1}^{n}1$$

$$CPDHFS(A,A)=\frac{1}{n}\times n$$

$CPDHFS$ $(A, A)=1$ for $i=1, 2, \ldots, n$.

Considering the weights of x_i, a weighted similarity measure is given as:

$$WPDHFS(A,B)=\frac{1}{n}\sum_{i=1}^{n}w_i\frac{\left(\frac{1}{M_A}\sum_{j=1}^{M_A}\gamma_{A_j}(x_i)p_{\gamma_{A_j}}(x_i)\right)\left(\frac{1}{M_B}\sum_{j'=1}^{M_B}\gamma_{B_{j'}}(x_i)p_{\gamma_{B_{j'}}}(x_i)\right)+\left(\frac{1}{N_A}\sum_{k=1}^{N_A}\eta_{A_k}(x_i)q_{\eta_{A_k}}(x_i)\right)\left(\frac{1}{N_B}\sum_{k'=1}^{N_B}\eta_{B_{k'}}(x_i)q_{\eta_{B_{k'}}}(x_i)\right)}{\sqrt{\frac{1}{M_A}\sum_{j=1}^{M_A}\gamma_{A_j}^2(x_i)p_{\gamma_{A_j}}^2(x_i)+\frac{1}{N_A}\sum_{k=1}^{N_A}\eta_{A_k}^2(x_i)q_{\eta_{A_k}}^2(x_i)}\sqrt{\frac{1}{M_B}\sum_{j'=1}^{M_B}\gamma_{B_{j'}}^2(x_i)p_{\gamma_{B_{j'}}}^2(x_i)+\frac{1}{N_B}\sum_{k'=1}^{N_B}\eta_{B_{k'}}^2(x_i)q_{\eta_{B_{k'}}}^2(x_i)}}$$

where $w_i \in [0,1]$, $i = 1, 2, \ldots, n$, and $\sum_{i=1}^{n} w_i = 1$.

Obviously, the weighted similarity measure also satisfies the following properties:

(P1): $0 \leq WCPDHFS\,(A, B) \leq 1$
(P2): $WCPDHFS\,(A, B) = WCPDHFS\,(B, A)$
(P3): $WCPDHFS\,(A, B) = 1$ if $A = B$ i.e. $\gamma_{A_j}(x_i) = \gamma_{B_{j'}}(x_i)$ *and* $\eta_{A_k}(x_i) = \eta_{B_{k'}}(x_i)$ *for* $i = 1, 2, \ldots, n$

Example 12.1 Let A_1 and A_2 are given by
A_1=({0.2 | 0.4, 0.3 | 0.6, 0.6 | 0.1} {0.2 | 0.3, 0.4 | 0.7}) and A_2 = ({0.8 | 0.4, 0.6 | 0.6} {0.3 | 1 })
Then calculate the similarity for A_1 and A_2, i.e., find $CPDHFS\,(A_1, A_2)$.

Proof: $CPDHFS\,(A_1, A_2)$ is given by

$$CPDHFS(A_1, A_2) = \frac{1}{n}\sum_{i=1}^{n} \frac{\left(\frac{1}{M_A}\sum_{j=1}^{M_A}\gamma_{A_j}(x_i)p_{\gamma_{A_j}}(x_i)\right)\left(\frac{1}{M_B}\sum_{j'=1}^{M_B}\gamma_{B_{j'}}(x_i)p_{\gamma_{B_{j'}}}(x_i)\right) + \left(\frac{1}{N_A}\sum_{k=1}^{N_A}\eta_{A_k}(x_i)q_{\eta_{A_k}}(x_i)\right)\left(\frac{1}{N_B}\sum_{k'=1}^{N_B}\eta_{B_{k'}}(x_i)q_{\eta_{B_{k'}}}(x_i)\right)}{\sqrt{\frac{1}{M_A}\sum_{j=1}^{M_A}\gamma_{A_j}^2(x_i)p_{\gamma_{A_j}}^2(x_i) + \frac{1}{N_A}\sum_{k=1}^{N_A}\eta_{A_k}^2(x_i)q_{\eta_{A_k}}^2(x_i)}\ \sqrt{\frac{1}{M_B}\sum_{j'=1}^{M_B}\gamma_{B_{j'}}^2(x_i)p_{\gamma_{B_{j'}}}^2(x_i) + \frac{1}{N_B}\sum_{k'=1}^{N_B}\eta_{B_{k'}}^2(x_i)q_{\eta_{B_{k'}}}^2(x_i)}}$$

Here, $n=1, M_{A1}=3, M_{A2}=3, N_{A1}=2, N_{A2}=2$

$$CPDHFS(A_1, A_2) = \frac{1}{1}\sum_{i=1}^{1} \frac{\left(\frac{1}{3}\sum_{j=1}^{3}\gamma_{A_j}(x_i)p_{\gamma_{A_j}}(x_i)\right)\left(\frac{1}{3}\sum_{j'=1}^{3}\gamma_{B_{j'}}(x_i)p_{\gamma_{B_{j'}}}(x_i)\right) + \left(\frac{1}{2}\sum_{k=1}^{2}\eta_{A_k}(x_i)q_{\eta_{A_k}}(x_i)\right)\left(\frac{1}{2}\sum_{k'=1}^{2}\eta_{B_{k'}}(x_i)q_{\eta_{B_{k'}}}(x_i)\right)}{\sqrt{\frac{1}{3}\sum_{j=1}^{3}\gamma_{A_j}^2(x_i)p_{\gamma_{A_j}}^2(x_i) + \frac{1}{2}\sum_{k=1}^{2}\eta_{A_k}^2(x_i)q_{\eta_{A_k}}^2(x_i)}\ \sqrt{\frac{1}{3}\sum_{j'=1}^{3}\gamma_{B_{j'}}^2(x_i)p_{\gamma_{B_{j'}}}^2(x_i) + \frac{1}{2}\sum_{k'=1}^{2}\eta_{B_{k'}}^2(x_i)q_{\eta_{B_{k'}}}^2(x_i)}}$$

$$CPDHFS(A_1,A_2)=\frac{\left[\frac{1}{3}(0.2\times 0.4+0.3\times 0.6+0.6\times 1)\right]\left[\frac{1}{3}(0.8\times 0.4+0.6\times 0.6+0.8\times 0.4)\right]+\left[\frac{1}{2}(0.2\times 0.3+0.4\times 0.7)\right]\left[\frac{1}{2}(0.3\times 1+0.3\times 1)\right]}{\sqrt{\frac{1}{3}\left(0.2^2\times 0.4^2+0.3^2\times 0.6^2+0.6^2\times 1^2\right)+\frac{1}{2}\left(0.2^2\times 0.3^2+0.4^2\times 0.7^2\right)}\times\sqrt{\frac{1}{3}\left(0.8^2\times 0.4^2+0.6^2\times 0.6^2+0.8^2\times 0.4^2\right)+\frac{1}{2}\left(0.3^2\times 1^2+0.3^2\times 1^2\right)}}$$

$$CPDHFS(A_1,A_2)=\frac{\frac{1}{3}((0.08\times +0.18+0.6))\left(\frac{1}{3}(0.32+0.36+0.36)\right)+\left(\frac{1}{2}(0.06+0.28)\right)\left(\frac{1}{2}(0.3+0.3)\right)}{\sqrt{\frac{1}{3}(0.0064+0.0324+0.36)+\frac{1}{2}(0.0036+0.0784)}\times\sqrt{\frac{1}{3}(0.1024+0.1296+0.1296)+\frac{1}{2}(0.09+0.09)}}$$

$$CPDHFS(A_1,A_2)=\frac{(0.2866667\times 0.3466667)+(0.17\times 0.3)}{\sqrt{0.132933+0.041}\ \sqrt{0.120533+0.09}}$$

$$CPDHFS(A_1,A_2)=\frac{0.150377}{0.191358}$$

$$CPDHFS\ (A_1,A_2)=0.7858412.$$

12.4 DM ON THE BASIS OF NEW SIMILARITY MEASURE OF PDHFSS

This section presented the DM method using the similarity measure of PDHFSs described above, followed by a real-life illustrative example to demonstrate the approach.

12.4.1 DM approach

Accordingly, it is assumed that there are m alternatives $(A_1, A_2, \ldots, A_m)$ that the expert analyzed based on the given n criteria $(C_1, C_2, \ldots, C_n)$ and gave his opinion in the form of the PDHFSs. Let w_q $(q=1, 2, \ldots, n)$ be the weight of the criteria C_q such that $w_q > 0$ and $\sum_{q=1}^{n} w_q = 1$. Thus, the ranking for each alternative in PDHFSs is given by:

$$A_r = \{(C_1, \alpha r_1), (C_2, \alpha r_2), \ldots, (C_v, \alpha r_v)\},$$

where $\alpha_{rv} = (h_{rv} \mid p_{rv}, g_{rv} \mid p_{rv})$ where $r=1, 2, \ldots, m$; $v= 1, 2, \ldots, n$. Then the steps are followed to solve the DM problem using proposed measures:

Step 1: Construct the matrix R for 'd' number of decision-makers or experts in the form of PDHFSs as:

$$R^{(d)} = \begin{pmatrix} \left(h_{11}^{(d)} \middle| p_{11}^{(d)}, g_{11}^{(d)} \middle| q_{11}^{(d)}\right) & \left(h_{12}^{(d)} \middle| p_{12}^{(d)}, g_{12}^{(d)} \middle| q_{12}^{(d)}\right) & \cdots & \left(h_{1n}^{(d)} \middle| p_{1n}^{(d)}, g_{1n}^{(d)} \middle| q_{1n}^{(d)}\right) \\ \left(h_{21}^{(d)} \middle| p_{21}^{(d)}, g_{21}^{(d)} \middle| q_{21}^{(d)}\right) & \left(h_{22}^{(d)} \middle| p_{22}^{(d)}, g_{22}^{(d)} \middle| q_{22}^{(d)}\right) & \cdots & \left(h_{2n}^{(d)} \middle| p_{2n}^{(d)}, g_{2n}^{(d)} \middle| q_{2n}^{(d)}\right) \\ \vdots & \vdots & \ddots & \vdots \\ \left(h_{m1}^{(d)} \middle| p_{m1}^{(d)}, g_{m1}^{(d)} \middle| q_{m1}^{(d)}\right) & \left(h_{m2}^{(d)} \middle| p_{m2}^{(d)}, g_{m2}^{(d)} \middle| q_{m2}^{(d)}\right) & \cdots & \left(h_{mn}^{(d)} \middle| p_{mn}^{(d)}, g_{mn}^{(d)} \middle| q_{mn}^{(d)}\right) \end{pmatrix}$$

where $\left(h_{rv}^{(d)} \middle| p_{rv}^{(d)}, g_{rv}^{(d)} \middle| q_{rv}^{(d)}\right) = (\{\gamma_{rv}^{(d)} | p_{rv}^{(d)}\}, \{\eta_{rv}^{(d)} | q_{rv}^{(d)}\})$, here $r=1, 2, \ldots, m$ and $v=1, 2, \ldots, n$.

Step 2: If $d = 1$, then $\left(h_{rv}^{(d)} \middle| p_{rv}^{(d)}, g_{rv}^{(d)} \middle| q_{rv}^{(d)}\right) = \left(h_{rv} \middle| p_{rv}, g_{rv} \middle| q_{rv}\right) = \left(\{\gamma_{rv} | p_{rv}\}, \{\eta_{rv} | q_{rv}\}\right)$ and then move to the next step, otherwise if $d \geq 2$, then a matrix is formed is given as:

$$R = \begin{pmatrix} \left(h_{11} \middle| p_{11}, g_{11} \middle| q_{11}\right) & \left(h_{12} \middle| p_{12}, g_{12} \middle| q_{12}\right) & \cdots & \left(h_{1n} \middle| p_{1n}, g_{1n} \middle| q_{1n}\right) \\ \left(h_{21} \middle| p_{21}, g_{21} \middle| q_{21}\right) & \left(h_{22} \middle| p_{22}, g_{22} \middle| q_{22}\right) & \cdots & \left(h_{r2n} \middle| p_{2n}, g_{2n} \middle| q_{2n}\right) \\ \vdots & \vdots & \ddots & \vdots \\ \left(h_{m1} \middle| p_{m1}, g_{m1} \middle| q_{m1}\right) & \left(h_{m2} \middle| p_{m2}, g_{m2} \middle| q_{m2}\right) & \cdots & \left(h_{mn} \middle| p_{mn}, g_{mn} \middle| q_{mn}\right) \end{pmatrix}$$

where $\left(h_{rv}\middle|p_{rv}, g_{rv}\middle|q_{rv}\right) = \left(\left\{\gamma_{rv}\middle|p_{rv}\right\}, \eta_{rv}\middle|q_{rv}\right)$ here $r = 1, 2, \ldots, m$ and $v = 1, 2, \ldots, n$.

Step 3: Calculate the overall measures *CPDHFS* of alternatives using the equation given below:

$$CPDHFS(A,B) = \frac{1}{n}\sum_{i=1}^{n} \frac{\left(\frac{1}{M_A}\sum_{j=1}^{M_A}\gamma_{A_j}(x_i)p_{\gamma A_j}(x_i)\right)\left(\frac{1}{M_B}\sum_{j'=1}^{M_B}\gamma_{B_{j'}}(x_i)p_{\gamma B_{j'}}(x_i)\right) + \left(\frac{1}{N_A}\sum_{k=1}^{N_A}\eta_{A_k}(x_i)q_{\eta A_k}(x_i)\right)\left(\frac{1}{N_B}\sum_{k'=1}^{N_B}\eta_{B_{k'}}(x_i)q_{\eta B_{k'}}(x_i)\right)}{\sqrt{\frac{1}{M_A}\sum_{j=1}^{M_A}\gamma_{A_j}^2(x_i)p_{\gamma A_j}^2(x_i) + \frac{1}{N_A}\sum_{k=1}^{N_A}\eta_{A_k}^2(x_i)q_{\eta A_k}^2(x_i)}\sqrt{\frac{1}{M_B}\sum_{j'=1}^{M_B}\gamma_{B_{j'}}^2(x_i)p_{\gamma B_{j'}}^2(x_i) + \frac{1}{N_B}\sum_{k'=1}^{N_B}\eta_{B_{k'}}^2(x_i)q_{\eta B_{k'}}^2(x_i)}}$$

12.4.2 Numerical example

We are considering the medical diagnosis problem to indicate the approach.

Example 12.2 Consider a medical diagnosis problem (pattern recognition problem). A set of diseases are listed as $A^*=(A_1, A_2, \ldots, A_m)$ and the values of symptoms are listed as $C=(C_1, C_2, \ldots, C_n)$.

A*	A_1	A_2	A_3	A_4
	Typhoid	Malaria	Viral fever	Stomach problem

C	C_1	C_2	C_3	C_4
	Temperature	Headache	Stomach-ache	Cough

	C_1	C_2	C_3	C_4
A_1	{0.4|0.2, 0.6|0.8} {0.2|1}	{0.2|0.3, 0.3|0.7} {0.4|0.6, 0.45|0.4}	{0.3|1} {0.6|1}	{0.2|0.8, 0.3|0.2} {0.2|0.3, 0.4|0.7}
A_2	{0.23|0.3, 0.32|0.7} {0.6|0.2, 0.66|0.8}	{0.32|1} {0.66|1}	{0.72|0.2, 0.75|0.8} {0.1|0.5, 0.15|0.5}	{0.60|1} {0.2|1}
A_3	{0.4|1} {0.2|0.6,0.28|0.4}	{0.2|0.3, 0.25|0.7} {0.4|0.6,0.5|0.4}	{0.5|1} {0.4|1}	{0.2|0.8,0.25|0.2} {0.4|1}
A_4	{0.6|0.3, 0.25|0.7} {0.4|0.6, 0.6|0.4}	{0.3|1} {0.23|0.7,0.12|0.3}	{0.32|0.1, 0.8|0.9} {0.4|1}	{0.8|0.4, 0.2|0.6} {0.2|0.7, 0.4|0.3}

Suppose a patient A has been check out by an expert to find that which symptoms have maximum adherence with that of diseases A_1, A_2, A_3, A_4. For this, they have noted the alternatives of patient A with all the symptoms indicating by the following set:

$$A = \langle x_1,\{0.2 \mid 1\}\{0.6 \mid 0.4, 0.7 \mid 0.6\}\rangle, \langle x_2,\{0.3 \mid 1\}\{0.2 \mid 1\}\rangle, \langle x_3,\{0.25 \mid 0.4, 0.28 \mid 0.6\}\{0.3 \mid 1\}\rangle, \langle x_4,\{0.2 \mid 1\}\{0.7 \mid 1\}\rangle$$

Now, the goal of this problem is to point out the disease of patient A. For it uses designed unweighted measures to compute measurement values of A_k (k=1, 2, 3, 4) from A. The outcome agreed with these measured values is given as:

CPDHFS(A_1, A) = 0.0268, CPDHFS(A_2, A) = 0.0272, CPDHFS(A_3, A) = 0.0294, CPDHFS(A_4, A) = 0.0278.

Hence, we conclude that A_3 is the disease that patient A suffers from.

12.5 COMPARISON

In this section, we have taken a few distances and the similarity concept depends on IFSs, HFSs, DHFSs, and PDHFSs. In IFSs only membership and non-membership values will be taken, but in HFSs hesitant value will be taken. Similarly in case of DHFSs, we have taken the hesitant DoM and DoNM and in PDHFSs we added their probabilities with them. Distance and similarities are opposite concept to each other. From the below table, we conclude that distance measures between two IFSs, HFSs and DHFSs, will give us the same result, while PDHFSs will give us a better result which is different from the others because in this approach we will use their probabilities with them. As we look for the similarity concept, the HFSs and DHFSs will give the same result. Similarity measure of PDHFSs will give us better results than we consider in our approach because we counted their probabilities of occurrence with their membership and non-membership values. On comparing all concepts of FS which are considered in our table, our approach will give much better result than all the others.

12.6 CONCLUSION

In this chapter, we have studied similarity measures and weighted similarity measures. Firstly, we look at some previous distance and similarity measures of HFSs, DHFSs and PHFSs. In the current chapter, we enlarge the theory of DHFSs and PHFSs to PDHFSs. We have invented the new similarity measure for PDHFSs which is useful in many real-world problems one of them already discussed above in Example 11.2 including medical diagnosis problems. We have used the DM approach and the distance and

Authors	*Environment*	*Formula*	A_1	A_2	A_3	A_4	*Ranking*
Ye (2011)	IFS	Cosine	0.4456	0.4941	0.4843	0.3706	$A_2 \succ A_3 \succ A_1 \succ A_4$
Torra (2010)	HFS	Cosine	0.0462	0.0450	0.0352	0.1008	$A_4 \succ A_1 \succ A_2 \succ A_3$
Zhang (2019)	DHFS	Cosine	0.0443	0.0444	0.0349	0.0754	$A_4 \succ A_2 \succ A_1 \succ A_3$
Wang & Xin (2005)	IFS	Distance	0.1094	0.0206	0.1269	0.2000	$A_2 \succ A_1 \succ A_3 \succ A_4$
Wenyi Zeng (2016)	HFS	Distance	0.0562	0	0.0738	0.0750	$A_2 \succ A_1 \succ A_3 \succ A_4$
Zhang (2020)	DHFS	Distance	0.5938	0.5138	0.5988	0.6750	$A_2 \succ A_1 \succ A_3 \succ A_4$
Garg & Kaur (2018)	PDHFS	Distance	0.5419	0.5156	0.5900	0.5850	$A_1 \succ A_2 \succ A_4 \succ A_3$
Our approach	PDHFS	Cosine	0.0268	0.0272	0.0294	0.0278	$A_3 \succ A_4 \succ A_2 \succ A_1$

similarity measures of DHFSs and PHFSs. Thus, it is clear that with the proposed method, we achieve the desired results and also additional information about the real scenarios.

REFERENCES

Atanassov, K. (2016). Intuitionistic fuzzy sets. *International Journal Bioautomation, 20*, 1–33.

Atanassov, K. T., & Gargov, G. (1999). Interval valued intuitionistic fuzzy sets. In Krassimir T. Atanassov*Intuitionistic Fuzzy Sets: Theory and Applications* (pp. 139–177). Physica-Verlag HD

Bin Zhu, Z. X. (2012). Dual hesitant fuzzy sets. *Journal of Applied Mathematics, 2012*l, 1–13, https://doi.org/10.1155/2012/879629

Garg, H., & Kaur, G. (2018). Algorithm for probabilistic dual hesitant fuzzy multi-criteria decision-making based on aggregation operators with new distance measures. *Mathematics*,10, 280.

Garg, H., & Arora, R. (2018). Dual hesitant fuzzy soft aggregation operators and their application in decision-making. *Cognitive Computation, 10*, 769–789.

Hajduk, S. (2021). Multi-criteria analysis in the decision-making approach for the linear ordering of urban transport based on TOPSIS technique. *Energies, 15*(1), 274.

Hao, Z. (2017). Probabilistic dual hesitant fuzzy set and its application in risk evaluation. *Knowledge-Based Systems, 127*, 16–28.

Martin Aruldoss, T. M. (2013). A survey on multi criteria decision making methods and its applications. *American Journal of Information Systems, 1(1)*,31–43.

Torra, V. (2010). Hesitant fuzzy sets. *International Journal of Intelligent Systems, 26*(6), 529–539.

Wang, W., & Xin, X. (2005). Distance measure between intuitionistic fuzzy sets. *Pattern Recognition Letters, 26*(13), 2063–2069.

Wenyi Zeng, D. L. (2016). Distance and similarity measures between hesitant fuzzy sets and their application in pattern recognition. *Pattern Recognition Letters, 84*, 267–271.

Xu, B. Z. (2018). Probability-hesitant fuzzy sets and the representation of preference relations. *Technological and Economic Development of Economy, 24*(3), 1029–1040.

Xu, M. X. (2011). Hesitant fuzzy information aggregation in decision making. *International Journal of Approximate Reasoning, 52*(3), 395–407.

Ye, J. (2011). Cosine similarity measures for intuitionistic fuzzy sets and their applications. *Mathematical and Computer Modelling, 53*(1), 91–97.

Zadeh, L. (1965). Fuzzy sets. *Inform Control, 8*, 338–353.

Zhang, H. (2020). Distance and entropy measures for dual hesitant fuzzy sets. *Computational and Applied Mathematics, 39*, 1–16.

Zhang, Y. (2019). A new concept of Cosine similarity measures based on dual hesitant fuzzy sets and its possible applications. *Cluster Computing, 22*(6), 15483–15492.

Chapter 13

Fermatean fuzzy sets in multiset framework with application in diagnostic process based on composite relations

Doonen Zuakwagh, Paul Augustine Ejegwa, and Abolape Deborah Akwu

13.1 INTRODUCTION: BACKGROUND AND DRIVING FORCES

Medical diagnosis is the procedure that helps in deciding which disease or situation describes a patient's signs and symptoms. It can also be expressed as the process of scrutinizing the symptoms and causes of some ailments by either laboratory tests or oral examination. Imprecision and vagueness are often witnessed in the process of medical diagnosis, which need to be resolved to guarantee a reliable patient's medical report. Zadeh (1965) proposed fuzzy set (FS) to resolve vague and imprecise information in everyday life experiences, and to serve as a reliable way to handle decision-making problems (DMP). On the other hand, FS considers only membership degree. By adopting the idea of bags (Knuth, 1981), Yager (1986) generalized FS to fuzzy multiset (FMS) by allowing membership degree to repeat to enhance the model of cases involving duplicity. Some theoretical properties of FMS using grade sequence were discussed and applied to language query, fuzzy data retrieval, and automatic classification (Kim and Miyamoto, 1996; Miyamoto, 2004).

Atanassov (1986) developed the term intuitionistic fuzzy sets (IFSs) which characterizes imprecise information by incorporating the degrees of membership and non-membership to enhance precise models of vague data. IFS has been used in different areas of application like pattern recognition, medical diagnosis, etc. (Nwokoro *et al.*, 2023; Ejegwa and Onasanya, 2019; Li and Chen, 2002; Ejegwa and Ahemen, 2023; Ejegwa *et al.*, 2023a; Wang and Xin, 2005; Ejegwa and Agbetayo, 2023; Ejegwa *et al.*, 2022; Liu *et al.*, 2017; Ejegwa and Onyeke, 2020). Shinoj and Sunil (2013) hybridized the ideas of FMS and IFS to form a new concept called intuitionistic fuzzy multiset (IFMS) which is resourceful in addressing imprecision in cases of duplicity. Applications of IFMS have been discussed by Shinoj and Sunil (2013). However important the ideas of IFSs and IFMSs, in a case where the addition of membership degree and the non-membership degree is larger than 1, IFS and IFMS are no longer useful.

DOI: 10.1201/9781003497219-13

Yager (2013) developed Pythagorean fuzzy sets (PFSs) to resolve the limitation of IFSs. PFS can model situations where the sum of its defining parameters, membership degree, and non-membership degree exceeds 1, but the sum of squares of the parameters is at most 1. PFS has been widely applied to different real-world problems (Ejegwa *et al.*, 2023b; Yager and Abbasov, 2013; Shahzadi *et al.*, 2020; Ejegwa 2019; Reformat *et al.*, 2014; Ejegwa and Onyeke, 2021; Zhang, 2016 Ejegwa *et al.*, 2020; Zeng *et al.*, 2018; Ejegwa *et al.*, 2021a; Ejegwa and Jana, 2021; Wu *et al.*, 2022; Akram *et al.*, 2019; Ejegwa *et al.*, 2021b). By allowing the repetition of the parameters of PFSs, the idea of Pythagorean fuzzy multisets (PFMSs) was introduced and applied in sundry areas (Riaz *et al.*, 2020; Ejegwa, 2020; Porchelvi and Jayapriya, 2020; Ejegwa *et al.*, 2020).

PFS and PFMS are helpful in resolving fuzzy decision problems, but with apparent flaws, especially in certain cases where the sum of squares of the defining parameters exceeds 1. To resolve this obvious limitation, Senapati and Yager (2019a) developed Fermatean fuzzy set (FFS), in which the sum of the third power of the defining parameters must be less than 1. An FFS can describe indeterminate information easily by increasing the spatial scope of the defining parameters in comparison to IFS and PFS. Numerous applications of FFS have been discussed. Wang *et al.* (2019) introduced a hesitant Fermatean fuzzy multicriteria decision-making technique via Archimedean Bonferroni mean operators, Senapati and Yager (2019b) introduced some Fermatean fuzzy information weighted aggregation operators, and Liu *et al.* (2019a) discussed a Fermatean fuzzy linguistic set with real-life applications. The notion of Fermatean fuzzy composite relation (FFCR) was developed (Ejegwa *et al.*, 2021b; Ejegwa and Zuakwagh, 2022) to aid the applications of FFSs. Other applications of FFSs were discussed, such as in COVID-19 testing facilities (Garg *et al.*, 2020), supplier evaluation (Keshavarz *et al.*, 2020), disease diagnosis (Ejegwa *et al.*, 2022; Ejegwa and Sarkar, 2023), decision-making (Liu *et al.*, 2019b), capital budgeting (Sergi and Sari, 2020), and career allocation (Onyeke and Ejegwa, 2023). Nonetheless, the idea of FFSs cannot model situations where repetitions are allowed. This setback of FFS is the motivation for this present study. Although the concepts of IFMS and PFMS facilitate the resolution of fuzzy decision problems in the presence of duplicity, some shortcomings are observed with the concepts, especially in some decision-making scenarios. On account of the deficiencies of IFMS and PFMS, the Fermatean fuzzy multiset (FFMS) and its composite relations need to be examined with application in decision-making.

In this chapter, we adopt the method in (Shinoj and Sunil, 2012; Riaz *et al.*, 2020) to develop a new soft computing tool called Fermatean fuzzy multiset (FFMS) and discuss its application in disease diagnosis based on composite relations. To achieve this, some specific objectives are considered which are as follows:

i. Propose FFMS and discuss its properties;
ii. Investigate *max–min–max* composite relation on FFMSs;
iii. Modify *max–min–max* composite relation on FFMSs to avoid error of exclusion;
iv. Establish a decision application framework under Fermatean fuzzy multisets in medical diagnosis based on modified composite relation.

The rest of the chapter is structured as follows: Section 13.2 discusses FFSs; Section 13.3 introduces the idea of FFMSs and discusses its properties; Section 13.4 introduces FFCRs and presents their properties; Section 13.5 discusses disease diagnosis based on composite relations using Fermatean fuzzy information in multiset framework; and Section 13.6 concludes the chapter and gives some recommendations for further study.

13.2 PRELIMINARIES

Under this section, a number of existing definitions are reiterated to be used in the research. Let X be a nonempty set throughout the chapter.

Definition 13.1 (Senapati and Yager, 2019a). An FFS represented by F in X is described by

$$F = \left\{ \left\langle x, \alpha_F(x), \beta_F(x) \right\rangle : x \in X \right\}, \tag{13.1}$$

where $\alpha_F : X \to [0, 1]$ and $\beta_F : X \to [0, 1]$, such that

$$0 \le \alpha_F^3(x) + \beta_F^3(x) \le 1, \tag{13.2}$$

for all $x \in X$. The numbers $\alpha_F(x)$ and $\beta_F(x)$ are the degrees of membership and non-membership of the element x in F. For any FFS F and $x \in X$,

$$\pi_A(x) = \sqrt[3]{1 - \left(\alpha_F^3(x) - \beta_F^3(x)\right)}$$

is the degree of indeterminacy of x to F. For easiness, Senapati and Yager (2019a) called the number $(\alpha_F(x), \beta_F(x))$ as Fermatean fuzzy number (FFN) designated by $F=(\alpha_F, \beta_F)$.

Definition 13.2 (Senapati and Yager, 2019a). Let $F=(\alpha_F,\beta_F)$, $F_1 = (\alpha_{F_1}, \beta_{F_1})$, and $F_2 = (\alpha_{F_2}, \beta_{F_2})$ be IFNs, then

i. $F_1 \cap F_2 = \left(min\{\alpha_{F_1}, \alpha_{F_2}\}, max\{\beta_{F_1}, \beta_{F_2}\}\right)$
ii. $F_1 \cup F_2 = \left(max\{\alpha_{F_1}, \alpha_{F_2}\}, min\{\beta_{F_1}, \beta_{F_2}\}\right)$
iii. $F^c = (\beta_F, \alpha_F)$.

13.3 FERMATEAN FUZZY MULTISETS

The section introduces the idea of FFMS as an extension of FFS.

Definition 13.3 An FFMS $\mathcal{A}$ drawn from X is of the form:

$$\mathcal{A} = \{\langle x, CM_{\mathcal{A}}(x), CN_{\mathcal{A}}(x)\rangle : x \in X\}, \tag{13.3}$$

where

$$CM_{\mathcal{A}}(x) = \mathcal{M}_{\mathcal{A}}^{i}(x), \ \ldots, \mathcal{M}_{\mathcal{A}}^{P}(x)$$

and

$$CN_{\mathcal{A}}(x) = \aleph_{\mathcal{A}}^{i}(x), \ldots, \aleph_{\mathcal{A}}^{P}(x)$$

are the count membership and count non-membership degrees defined by $CM_{\mathcal{A}} : X \rightarrow N^{[0,1]}$ and $CN_{\mathcal{A}} : X \rightarrow N^{[0,1]}$ such that

$$0 \leq CM_{\mathcal{A}}^{3}(x) + CN_{\mathcal{A}}^{3}(x) \leq 1. \tag{13.4}$$

The count hesitation margin of x in A is $CH_{\mathcal{A}}(x) = \sqrt[3]{1 - [CM_{\mathcal{A}}(x)]^3 - [CN_{\mathcal{A}}(x)]^3}$, where

$$CH_{\mathcal{A}}(x) = \pi_{\mathcal{A}}^{1}(x), \ldots, \pi_{\mathcal{A}}^{P}(x). \tag{13.5}$$

In fact, $CH_{\mathcal{A}}(x)$ is the degree of determinacy of $x \in X$ to A and $CH_{\mathcal{A}}(x) \in [0,1]$. We assume the set of FFMSs in X to be denoted by FFMSs (X).

Definition 13.4. For any two FFMSs $\mathcal{A}_1$, and $\mathcal{A}_2$ in X defined by

$\mathcal{A}_1 = \{\langle x, (\mathcal{M}_{\mathcal{A}_1}^{1}(x), \mathcal{M}_{\mathcal{A}_1}^{2}(x), \ldots, \mathcal{M}_{\mathcal{A}_1}^{p}(x)), (\aleph_{\mathcal{A}_1}^{1}(x), \aleph_{1}^{2}(x), \ldots, \aleph_{\mathcal{A}_1}^{p}(x))\rangle x \in X\}$ and

$\mathcal{A}_2 = \{\langle x, (\mathcal{M}_{\mathcal{A}_2}^{1}(x), \mathcal{M}_{\mathcal{A}_2}^{2}(x), \ldots, \mathcal{M}_{\mathcal{A}_2}^{p}(x)), (\aleph_{\mathcal{A}_2}^{1}(x), \aleph_{\mathcal{A}_2}^{2}(x), \ldots, \aleph_{\mathcal{A}_2}^{p}(x))\rangle x \in X\}$,

The following operations and relations hold:

i. $\mathcal{A}_1 \subset \mathcal{A}2 \Leftrightarrow \mathcal{M}'^{j}_{\mathcal{A}_1}(x) \leq \mathcal{M}'^{j}_{\mathcal{A}_1}(x)$ and $\aleph'^{j}_{\mathcal{A}_1}(x) \leq \aleph'^{j}_{\mathcal{A}_1}(x)$ for j=1, 2, ..., $P \ \forall x \in X$, and $\mathcal{A}_1 = \mathcal{A}_2 \Leftrightarrow \mathcal{A}_1 \subset \mathcal{A}_2$ and $\mathcal{A}_2 \subset \mathcal{A}_1$.

ii. $\mathcal{A}^c = \{\langle x, (\aleph'^{1}_{\mathcal{A}_1}(x), \ldots, \aleph'^{p}_{\mathcal{A}_1}(x)), (\mathcal{M}'^{1}_{\mathcal{A}_1}(x), \ldots, \mathcal{M}'^{p}_{\mathcal{A}_1}(x))\rangle : x \in X\}$.

iii. Union of FFMSs $\mathcal{A}_1$ and $\mathcal{A}_2$ denoted as $\mathcal{A}_1 \cup \mathcal{A}_2$ is defined by the count membership and count non-membership values as;

$$\left.\begin{array}{l}\mathcal{M}^{j}_{\mathcal{A}_1 \cup \mathcal{A}_2}(x) = max\left\{\mathcal{M}^{j}_{\mathcal{A}_1}(x), \mathcal{M}^{j}_{\mathcal{A}_2}(x)\right\} \\ \aleph^{j}_{\mathcal{A}_1 \cup \mathcal{A}_2}(x) = min\left\{\aleph^{j}_{\mathcal{A}_1}(x), \aleph^{j}_{\mathcal{A}_2}(x)\right\}\end{array}\right\}, \tag{13.6}$$

for $j = 1, 2, \ldots, P \; \forall \, x \in X$.

iv. Intersection of FFMSs $\mathcal{A}_1$ and $\mathcal{A}_2$ denoted as $\mathcal{A}_1 \cap \mathcal{A}_2$ is defined by the count membership and count non-membership values as;

$$\left.\begin{array}{l}\mathcal{M}^{j}_{\mathcal{A}_1 \cap \mathcal{A}_2}(x) = min\left\{\mathcal{M}^{j}_{\mathcal{A}_1}(x), \mathcal{M}^{j}_{\mathcal{A}_2}(x)\right\} \\ \aleph^{j}_{\mathcal{A}_1 \cap \mathcal{A}_2}(x) = max\left\{\aleph^{j}_{\mathcal{A}_1}(x), \aleph^{j}_{\mathcal{A}_2}(x)\right\}\end{array}\right\}, \tag{13.7}$$

for $j=1, 2, \ldots, P \; \forall \, x \in X$.

v. Addition of FFMSs $\mathcal{A}_1$ and $\mathcal{A}_2$ denoted as $\mathcal{A}_1 \oplus \mathcal{A}_2$ is defined by the count membership and count non-membership values as;

$$\left.\begin{array}{c}\mathcal{M}^{j}_{\mathcal{A}_1 \oplus \mathcal{A}_2}(x) = \mathcal{M}^{j}_{\mathcal{A}_1}(x) + \mathcal{M}^{j}_{\mathcal{A}_2}(x) - \mathcal{M}^{j}_{\mathcal{A}_1}(x)\mathcal{M}^{j}_{\mathcal{A}_2}(x) \\ \aleph^{j}_{\mathcal{A}_1 \oplus \mathcal{A}_2}(x) = \aleph^{j}_{\mathcal{A}_1}(x)\aleph^{j}_{\mathcal{A}_2}(x)\end{array}\right\}, \tag{13.8}$$

for $j=1, 2, \ldots, P \; \forall \, x \in X$.

vi. Multiplication of FFMSs $\mathcal{A}_1$ and $\mathcal{A}_2$ denoted as $\mathcal{A}_1 \otimes \mathcal{A}_2$ is defined by the count membership and count non-membership values as;

$$\left.\begin{array}{c}\mathcal{M}^{j}_{\mathcal{A}_1 \otimes \mathcal{A}_2}(x) = \mathcal{M}^{j}_{\mathcal{A}_1}(x)\mathcal{M}^{j}_{\mathcal{A}_2}(x) \\ \aleph^{j}_{\mathcal{A}_1 \otimes \mathcal{A}_2}(x) = \aleph^{j}_{\mathcal{A}_1}(x) + \aleph^{j}_{\mathcal{A}_2}(x) - \aleph^{j}_{\mathcal{A}_1}(x)\aleph^{j}_{\mathcal{A}_2}(x)\end{array}\right\}, \tag{13.9}$$

for $j=1, 2, \ldots, P \; \forall \, x \in X$.

Definition 13.5 Given an FFMS $\mathcal{A}$ in X as

$$\mathcal{A} = \left\{x, \left\langle\left(\mathcal{M}^1, \aleph^1\right), \left(\mathcal{M}^2, \aleph^2\right), \ldots, \left(\mathcal{M}^P, \aleph^P\right)\right\rangle : x \in X\right\}. \tag{13.10}$$

We observe that the membership and non-membership degrees repeated nth times. Now, we can transform $\mathcal{A}$ into an FFS F as follows:

$$F = \left\{\left\langle x, \frac{\mathcal{M}^1 + \mathcal{M}^2 + \ldots + \mathcal{M}^P}{n}, \frac{\aleph^1 + \aleph^2 + \ldots + \aleph^P}{n}\right\rangle : x \in X\right\}. \tag{13.11}$$

For an example, given an FFMS $\mathcal{A} = \{x_1,(0.5,0.4, 0.7,0.2, x_2 0.5,0.1,0.8,0.2)\}$ in $X = \{x_1, x_2\}$, then the equivalent FFS incorporating the properties of FFMS is

$$F= \{\langle x_1, 0.6,0.3\rangle, \langle x_2, 0.65,0.15\rangle\}.$$

13.4 FERMATEAN FUZZY COMPOSITE RELATIONS IN MULTISET FRAMEWORK

In this section, the idea of composite relations on FFMSs is considered with application in medical diagnosis by extending the works in Sanchez (1976) De *et al.* (2001), and Ejegwa (2020b).

13.13 *Max–min–max* composite relation based on FFMSs

Here, we explicate the Fermatean fuzzy *max–min–max* composite relation (FFMMMCR).

Definition 13.6 Let X and Y be nonempty sets. Then the Fermatean fuzzy multi-relation (FFMR) denoted by $\boldsymbol{\phi}$ from X to Y is an FFMS in X×Y consisting of count membership degree (CMD), $\mathcal{M}_{\phi}^{j}$ and count non-membership degree (CNMD), $\aleph_{\phi}^{j}$ which is represented by $\boldsymbol{\phi}$ (X→Y).

Definition 13.7 Suppose $\mathcal{A}_1$ is an FFMS of X, then a *max–min–max* composite relation (MMMCR), $\boldsymbol{\phi}$ (X→Y), with $\mathcal{A}_1$ as an FFMS $\boldsymbol{B}$ in Y represented by $\boldsymbol{B} = \phi o \mathcal{A}_1$, where its CMD and CNMD are:

$$\left.\begin{array}{c}\mathcal{M}_{B}^{j}(y)=\max\left\{\min\left\{\ \mathcal{M'}_{A_1}^{j}(x),\mathcal{M}_{\phi}^{j}(x,y)\right\}\right\}\\ \aleph_{B}^{j}(y)=\min\left\{\max\left\{\ \aleph'^{j}_{A_1}(x),\aleph_{\phi}^{j}(x,y)\right\}\right\}\end{array}\right\}, \tag{13.12}$$

for all $x \in X$ and $y \in Y$.

In addition, the FFMMMCR on FFMSs denoted by $\boldsymbol{\phi}$ o μ, where μ(X →Y) and ϕ (Y→Z) are FFMRs, is an FFMR from X to Z such that its CMD and CNMD are:

$$\left.\begin{array}{c}\mathcal{M}_{\phi o \mu}(x,z)=\max\left\{\min\left\{\mathcal{M}_{\phi}^{j}(x,y),\mathcal{M}_{\mu}^{j}(y,z)\right\}\right\}\\ \aleph_{\phi o \mu}(x,z)=\min\left\{\max\left\{\ \aleph_{\phi}^{j}(x,y),\aleph_{\mu}^{j}(y,z)\right\}\right\}\end{array}\right\}. \tag{13.13}$$

Combining Eqs. (13.12) and (13.13), the FFMMMCR $\boldsymbol{B}$ and $\boldsymbol{\phi}$ o μ can be computed by:

$$\left.\begin{array}{c}\boldsymbol{B} = M_{B}^{j}(y)-\aleph_{B}^{j}(y)\pi_{B}^{j}(y),\ for\ all\ y \in Y\\ \boldsymbol{\phi}\ o\ \mu = M_{\phi o \mu}^{j}(x,z)-\aleph_{\phi o \mu}^{j}(x,z)\pi_{\phi o \mu}^{j}(x,z),\ for\ all\,(x,z) \in X \times Z\end{array}\right\}. \tag{13.14}$$

13.4.2 Modified FFMMMCR based on multiset framework

The idea of FFMMMCR uses a maximum of the minimum of CMD and a minimum of the maximum CNMD. Here, we modify FFMMMCR based on the maximum average method to solve the issue of exclusion.

Definition 13.8 If ϕ and μ are FFMRs of $X\times Y$ and $Y\times Z$ denoted by $\phi(X\times Y)$ and $\mu(Y\times Z)$, respectively, then the modified Fermatean fuzzy multi-composite relation denoted by $\eta=\phi \ o\ \mu$ is an FFMS in $X\times Z$ of the form:

$$\eta = \left\{ \left\langle (x,z), \mathcal{M}_{\eta}^{j}(x,z), \aleph_{\eta}^{j}(x,z) \right\rangle : (x,z) \in X \times Z \right\}, \tag{13.15}$$

where

$$\left. \begin{aligned} \mathcal{M}_{\eta}(x,z) &= \max\left\{ \text{average}\left(\mathcal{M}_{\phi}^{j}(x,y), \mathcal{M}_{\mu}^{j}(y,z) \right) \right\} \\ \aleph_{\eta}(x,z) &= \min\left\{ \text{average}\left(\aleph_{\phi}^{j}(x,y), \aleph_{\mu}^{j}(y,z) \right) \right\} \end{aligned} \right\}, \tag{13.16}$$

for all $(x,y) \in X\times Y$ and $(y,z) \in Y\times Z$ and $(x,z) \in X\times Z$, respectively. Similarly, suppose $\mathcal{A}_1$ is an FFMS of X, then the modified FFMMMCR denoted by ϕ $(X\to Y)$ with $\mathcal{A}_1$ is an FFMS $\hat{B}$ in Y represented by $\hat{B} = \phi o \mathcal{A}_1$, where its CMD and CNMD are:

$$\left. \begin{aligned} \mathcal{M}_{\hat{B}}^{j}(y) &= \max\left\{ \text{average}\left(\mathcal{M}_{\mathcal{A}_1}^{j}(x), \mathcal{M}_{\phi}^{j}(x,y) \right) \right\} \\ \aleph_{\hat{B}}^{j}(y) &= \min\left\{ \text{average}\left(\aleph_{\mathcal{A}_1}^{j}(x), \aleph_{\phi}^{j}(x,y) \right) \right\} \end{aligned} \right\}, \tag{13.17}$$

for all $x \in X$ and $y \in Y$. By combining Eqs. (13.16) and (13.17), the modified FFMMMCR η and $\hat{B}$ can be computed by:

$$\left. \begin{aligned} \eta &= M_{\eta}^{j}(x,z) - \aleph_{\eta}^{j}(x,z)\pi_{\eta}^{j}(x,z), \textit{ for all } (x,z) \in X \times Z, \\ \hat{B} &= M_{\hat{B}}^{j}(y) - \aleph_{\hat{B}}^{j}(y)\pi_{\hat{B}}^{j}(y), \textit{ for all } y \in Y \end{aligned} \right\}. \tag{13.18}$$

13.5 MEDICAL DIAGNOSIS BASED ON FERMATEAN FUZZY MULTI-COMPOSITE RELATIONS

In this section, the application of FFMSs using composite relations in medical diagnosis is discussed. The data for this work are taken from the work by Shinoj and Sunil (2012).

13.5.1 Numerical example of medical diagnosis

An example of a medical diagnosis involving four patients is discussed. Assume four patients represented by $P = \{P_1, P_2, P_3, P_4\}$ visit a medical facility for medical diagnosis on the set of diseases represented by $D = \{VF, TB, TF, TD\}$, where VF is viral fever, TB is tuberculosis, TF is typhoid fever, and TD is throat disease, respectively. The patients are showing symptoms, S given by $S = \{t, c, th, h, b\}$, where t is temperature, c is cough, th is throat pain, h is headache, and b is body pain, respectively.

One of the query that may arise is whether by taking only one-time assessment we can arrive at a medical decision that a particular patient has a disease or not. Every now and then, patients may show symptoms of different diseases, so giving a proper diagnosis is challenging. Our aim is to make a proper diagnosis for each of the patients. In Table 13.1, each of the symptoms, S_i, is described by count membership ($\mathcal{M}^i$) and count non-membership ($\aleph^i$), respectively.

The samples were taken from the patients at three different timings in a day; 7.00 am, 12.00 pm, and 7.00 pm, respectively. The first set of values in Table 13.2 represents the count membership values at 7.00 am, 12.00 pm,

Table 13.1 Symptoms and diseases

Symptoms	*VF*	*TB*	*TF*	*TD*
Temperature	(0.8,0.1)	(0.2,0.7)	(0.5,0.3)	(0.1,0.7)
Cough	(0.2,0.7)	(0.9,0.0)	(0.3,0.5)	(0.3,0.6)
Throat pain	(0.3,0.5)	(0.7,0.2)	(0.2,0.7)	(0.8,0.1)
Headache	(0.5,0.3)	(0.6,0.3)	(0.2,0.6)	(0.1,0.8)
Body pain	(0.5,0.4)	(0.7,0.2)	(0.4,0.4)	(0.1,0.8)

Table 13.2 Patients and symptoms

Patients	*Temperature*	*Cough*	*Throat pain*	*Headache*	*Body pain*
P_1	$\begin{pmatrix}0.6,0.7,0.5\\0.2,0.1,0.4\end{pmatrix}$	$\begin{pmatrix}0.4,0.3,0.4\\0.3,0.6,0.4\end{pmatrix}$	$\begin{pmatrix}0.1,0.2,0.0\\0.7,0.7,0.8\end{pmatrix}$	$\begin{pmatrix}0.5,0.6,0.7\\0.4,0.3,0.2\end{pmatrix}$	$\begin{pmatrix}0.2,0.3,0.4\\0.6,0.4,0.4\end{pmatrix}$
P_2	$\begin{pmatrix}0.4,0.3,0.5\\0.5,0.4,0.4\end{pmatrix}$	$\begin{pmatrix}0.7,0.6,0.8\\0.2,0.2,0.1\end{pmatrix}$	$\begin{pmatrix}0.6,0.5,0.4\\0.3,0.3,0.4\end{pmatrix}$	$\begin{pmatrix}0.3,0.6,0.2\\0.7,0.3,0.7\end{pmatrix}$	$\begin{pmatrix}0.8,0.7,0.5\\0.1,0.2,0.3\end{pmatrix}$
P_3	$\begin{pmatrix}0.1,0.2,0.1\\0.7,0.6,0.9\end{pmatrix}$	$\begin{pmatrix}0.3,0.2,0.1\\0.6,0.0,0.7\end{pmatrix}$	$\begin{pmatrix}0.8,0.7,0.8\\0.0,0.1,0.1\end{pmatrix}$	$\begin{pmatrix}0.3,0.2,0.2\\0.6,0.7,0.6\end{pmatrix}$	$\begin{pmatrix}0.4,0.3,0.2\\0.4,0.7,0.7\end{pmatrix}$
P_4	$\begin{pmatrix}0.5,0.4,0.5\\0.4,0.4,0.3\end{pmatrix}$	$\begin{pmatrix}0.4,0.3,0.4\\0.5,0.3,0.5\end{pmatrix}$	$\begin{pmatrix}0.2,0.1,0.0\\0.7,0.6,0.7\end{pmatrix}$	$\begin{pmatrix}0.5,0.6,0.3\\0.4,0.3,0.6\end{pmatrix}$	$\begin{pmatrix}0.4,0.5,0.3\\0.6,0.4,0.3\end{pmatrix}$

and 7.00 pm, respectively, while the second set of values represents the corresponding count non-membership values.

The data in Table 13.2 are separated into Tables 13.4, 13.7, and 13.10, respectively.

13.5.1.1 Medical analysis at 7.00 am based on FFSs

Using Eqs. (13.12)–(13.14) and Eqs. (13.16)–(13.18) on Tables 13.1 and 13.2, we get Table 13.3.

Table 13.3 Analysis of patients and symptoms at 7.00 am using FFMMMCR and modified FFMMMCR

Patients	*VF*	*TB*	*TF*	*TD*
P_1	0.4162 0.5698	0.2160 0.5154	0.2160 0.2258	−0.2468 0.0719
P_2	0.2169 0.4269	0.5268 0.6667	0.0179 0.3710	0.3266 0.5268
P_3	0.0179 0.3162	0.5268 0.6667	0.0179 0.1708	0.7213 0.7606
P_4	0.1269 0.4269	0.1269 0.4269	0.1269 0.1708	−0.2468 0.0851

In Table 13.3, the first result is for FFMMMCR and the second result is for modified FFMMMCR, respectively. From the results using FFMMMCR, it is revealed that P_1 has mild viral fever, P_2 has tuberculosis, P_3 has throat disease and tuberculosis, while P_4 has no significant symptoms for any of the diseases. The results using modified FFMMMCR reveal that P_1 has viral fever with mild symptoms of tuberculosis, P_2 has tuberculosis and throat disease, P_3 has throat disease and tuberculosis, while P_4 has a raising case of tuberculosis and viral fever which cannot be ignored.

13.5.1.2 Medical analysis at 12.00 noon based on FFSs

Using Eqs. (13.12)–(13.14) and Eqs. (13.16)–(13.18) on Tables 13.1 and 13.2, we get Table 13.4.

From Table 13.4 using FFMMMCR, we discover that P_1 has a raising case of viral fever, P_2 suffers from tuberculosis, P_3 suffers from tuberculosis and throat diseases, while P_4 has no significant symptoms. Using modified FFMMMCR, the results reveal that P_1 has symptoms of viral fever, P_2 suffers from tuberculosis, P_3 suffers from tuberculosis and throat disease, while P_4 has a raising case of tuberculosis.

Table 13.4 Analysis of patients and symptoms at 12.00 noon using FFMMMCR and modified FFMMMCR

Patients	*VF*	*TB*	*TF*	*TD*
P_1	0.4044 0.4559	0.3266 0.3266	0.0550 0.3093	−0.2468 0.1164
P_2	0.2160 0.4162	0.5254 0.6667	0.0179 0.2707	0.2160 0.4709
P_3	−0.1733 0.2160	0.7000 0.7000	−0.1733 0.2092	0.6131 0.6667
P_4	0.2160 0.3710	0.3266 0.4619	0.0179 0.1164	−0.2468 0.1164

Table 13.5 Analysis of patients and symptoms at 7.00 pm using FFMMMCR and modified FFMMMCR

Patients	*VF*	*TB*	*TF*	*TD*
P_1	0.2160 0.4269	0.3266 0.4709	0.1269 0.1708	−0.1378 −0.0254
P_2	0.1269 0.4269	0.7213 0.8136	0.4995 0.2707	−0.2468 0.2833
P_3	−0.1733 0.2707	0.6131 0.7411	−0.3514 0.1269	0.1419 0.4116
P_4	0.2160 0.4709	0.1094 0.4269	0.2160 0.2160	−0.3001 0.0718

13.5.1.3 Medical analysis at 7.00pm based on FFSs

Using Eqs. (13.12)–(13.14) and Eqs. (13.16)–(13.18) on Tables 13.1 and 13.2, we obtain Table 13.5.

From Table 13.5 using FFMMMCR, the result show that P_1 has no significant symptoms to be considered, P_2 suffers from tuberculosis with mid symptoms of typhoid fever, P_3 also suffers from tuberculosis, while P_4 has no significant symptoms. The results from the modified FFMMMCR reveal that P_1 suffers from tuberculosis and a raising case of viral fever, P_2 suffers from a high rate of tuberculosis and mild viral fever, P_3 also has a high rate of tuberculosis, and P_4 has a raising case of viral fever and tuberculosis.

13.5.1.4 Combined medical analysis based on FFMSs

By transforming the data in Table 13.2 into FFSs, we get Table 13.6.

Using Eqs. (13.12)–(13.14) and Eqs. (13.16)–(13.18) on the data in Tables 13.1 and 13.6, we get Table 13.7.

Table 13.6 Combined medical analysis of patients

Patients	*Temperature*	*Cough*	*Throat pain*	*Headache*	*Body pain*
P_1	(0.6000,0.2333)	(0.3667,0.4333)	(0.1000,0.7333)	(0.6000,0.3000)	(0.3000,0.4667)
P_2	(0.4000,0.4333)	(0.7000,0.1667)	(0.5000,0.3333)	(0.3667,0.5667)	(0.6667,0.1667)
P_3	(0.1333,0.7333)	(0.2000,0.4333)	(0.7667,0.0667)	(0.2333,0.6333)	(0.3000,0.6000)
P_4	(0.4667,0.3667)	(0.3667,0.4333)	(0.1000,0.6667)	(0.4667,0.4333)	(0.4000,0.4333)

Table 13.7 Composite relation between patients and diseases using both FFMMMCR and modified FFMMMCR

Patients	*VF*	*TB*	*TF*	*TD*
P_1	0.3860 0.5554	0.3266 0.4011	0.1260 0.3561	−0.2468 0.0574
P_2	0.1269 0.3561	0.5554 0.6562	0.0178 0.2675	0.1856 0.4562
P_3	−0.1733 0.2674	0.5268 0.5509	−0.1730 0.1230	0.6848 0.4087
P_4	0.1195 0.4229	0.0616 0.4377	0.1995 0.1678	−0.2468 0.0864

From Table 13.7, the highest scores give the proper diagnosis. Using FFMMMCR, P_1 has mild symptoms of viral fever, P_2 has tuberculosis, P_3 has throat disease and tuberculosis, and P_4 shows no symptoms of any disease. However, using modified FFMMMCR, P_1 suffers from viral fever with some considerable symptoms of tuberculosis and typhoid fever, P_2 suffers from tuberculosis with some considerable symptoms of throat disease and viral fever, P_3 suffers from tuberculosis with some considerable symptoms of throat disease, and P_4 has mild tuberculosis and viral fever. By comparison, the modified FFMMMCR is more informative because it gives a clear direction of drug administration and treatment for a reliable recovery.

13.6 CONCLUSION

We have introduced FFMSs as generalization of IFMSs and PFMSs in such a way that the count membership degrees, non-membership degrees, and hesitation margins are allowed to repeat by incorporating the properties of FFSs. Also, we studied FFMSs and described some of their properties. The application of FFMSs in medical diagnosis was shown using FFMMMCR and modified FFMMMCR, respectively. In addition, a comparison analysis was conducted to ascertain the most precise, trusted, and reliable composite relation. The distinctive feature of this new technique is that it makes use of all the parameters of FFMSs, and due to its reliability in soft computing, we recommend it to be applied in discussing multiple-attribute decision-making problems.

REFERENCES

Akram, M., Dudek, W. A. and Dar, J. M. (2019). Pythagorean Dombi fuzzy aggregation operators with application in multicriteria decision-making, *International Journal of Intelligent Systems*, 34(11): 3000–3019.

Atanassov, K. T. (1986). Intuitionistic fuzzy sets, *Fuzzy Sets and Systems*, 20(1): 87–96.

De, S. K., Biswas, R. and Roy, A. R. (2001). An application of intuitionistic fuzzy sets in medical diagnosis, *Fuzzy Sets Systems*, 117(2): 209–213.

Ejegwa, P. A. (2019). Pythagorean fuzz sets and its application in career placement based on academic performance using max-min-max composition, *Complex and Intelligence Systems*, 5: 165–175.

Ejegwa, P. A. (2020a). Pythagorean fuzzy multiset and its application to course placements, *Open Journal of Discrete Applied Mathematics*, 3(1): 55–74.

Ejegwa, P. A. (2020b). Improved composite relation for Pythagorean fuzzy sets and its application to medical diagnosis, *Granular Computing*, 5(2): 277–286.

Ejegwa, P. A. and Agbetayo, J. M. (2023). Similarity-distance decision-making technique and its applications via intuitionistic fuzzy pairs, *Journal of Computational and Cognitive Engineering*, 2(1): 68–74.

Ejegwa, P. A. and Ahemen, S. (2023). Enhanced intuitionistic fuzzy similarity operator with applications in emergency management and pattern recognition, *Granular Computing*, 8: 361–372.

Ejegwa, P. A., Feng, Y. and Zhang, W. (2020). Pattern recognition based on an improved Szmidt and Kacprzyk's correlation coefficient in Pythagorean fuzzy environment, In: Han, M. *et al.* (Eds.); *Advances in Neural Networks – 17th International Symposium on Neural Networks (ISNN 2020)*, Lecture Notes in Computer Science (LNCS) 12557, pp. 190–206, Springer.

Ejegwa, P. A. and Jana, C. (2021). Some new weighted correlation coefficients between Pythagorean fuzzy sets and their applications, In: Garg, H. (Eds.); *Pythagorean Fuzzy Sets*, pp. 39–64, Springer.

Ejegwa, P. A., Jana, C. and Pal, M. (2022). Medical diagnostic process based on modified composite relation on Pythagorean fuzzy multisets, *Granular Computing*, 7: 15–23.

Ejegwa, P. A., Muhiuddin, G., Algehyne, E. A., Agbetayo, J. M. and Al-Kadi, D. (2022). An enhanced Fermatean fuzzy composition relation based on a maximum-average approach and its application in diagnostic analysis, *Journal of Mathematics*, Article ID 1786221, 12 pages.

Ejegwa, P. A., Nwankwo, K. N., Ahmed, M., Ghazal, T. M. and Khan, M. A. (2021). Composite relation under Fermatean fuzzy context and its application in diseases diagnosis, *Informatica*, 32(10): 87–101.

Ejegwa, P. A. and Onasanya, B. O. (2019). Improved intuitionistic fuzzy composite relation and its application to medical diagnosis process, *Note intuitionistic Fuzzy Sets*, 25(1): 43–58.

Ejegwa, P. A. and Onyeke, I. C. (2020). Medical diagnostic analysis on some selected patients based on modified Thao et al.'s correlation coefficient of intuitionistic fuzzy sets via an algorithmic approach, *Journal of Fuzzy Extension and Applications*, 1(2): 130–141.

Ejegwa, P. A. and Onyeke, I. C. (2021). A robust weighted distance measure and its applications in decision making via Pythagorean fuzzy information, *Journal of the Institute of Electronics and Computer*, 3: 87–97.

Ejegwa, P. A., Onyeke, I. C., Kausar, N. and Kattel, P. (2023a). A new partial correlation coefficient technique based on intuitionistic fuzzy information and its pattern recognition application, *International Journal of Intelligent Systems*, Article ID 5540085, 14 pages.

Ejegwa, P. A., Onyeke, I. C., Terhemen, B. T., Onoja, M. P., Ogiji, A. and Opeh, C. U. (2022). Modified Szmidt and Kacprzyk's intuitionistic fuzzy distances and their applications in decision-making, *Journal of the Nigerian Society of Physical Sciences*, 4: 175–182.

Ejegwa, P. A. and Sarkar, A. (2023). Fermatean fuzzy approach of diseases diagnosis based on new correlation coefficient operators. In: Garg, H., Chatterjee, J. M. (Eds.); *Deep Learning in Personalized Healthcare and Decision Support*, pp. 23–38, Academic Press.

Ejegwa, P. A., Sarkar, A. and Onyeke, I. C. (2023b). New methods of computing correlation coefficient based on pythagorean fuzzy information and their applications in disaster control and diagnostic analysis. In: Jana, C., Pal, M., Muhiuddin, G., Liu, P. (Eds.); *Fuzzy Optimization, Decision-Making and Operations Research*. Springer, Cham.

Ejegwa, P. A., Wen, S., Feng, Y. and Zhang, W. (2021a). Determination of pattern recognition problems based on a Pythagorean fuzzy correlation measure from statistical viewpoint. In: *Proceedings of the 13th International Conference of Advanced Computational Intelligence*, pp. 132–139, Wanzhou, China.

Ejegwa, P. A., Wen, S., Feng, Y., Zhang, W. and Chen, J. (2021b). Some new Pythagorean fuzzy correlation techniques via statistical viewpoint with applications to decision-making problems, Journal of Intelligent and Fuzzy Systems, 40(5): 9873–9886.

Ejegwa, P. A. and Zuakwagh, D. (2022). Fermatean fuzzy modified composite relation and its application in pattern recognition, *Journal of Fuzzy Extension and Applications*, 3(2): 140–151.

Garg, H., Shahzadi, G. and Akram, M. (2020). Decision-making analysis base on Fermatean fuzzy Yager aggregation operator with application in COVID-19 testing facility, *Mathematical Problems in Engineering*, Article ID 7279027.

Keshavarz, M., Amiri, M. and Tabatabaei, H. (2020). A new decision-making approach based on Fermatean fuzzy sets and WASPAS for green construction supplier evaluation, *Mathematics*, 8(12): 22202.

Kim, K. S. and Miyamoto, S. (1996). Application of fuzzy multisets to fuzzy database systems, Soft Computing in Intelligence Systems and Information Processing. In: *Proceedings of the 1996 Asian Fuzzy Systems Symposium*, Kenting, Taiwan, pp. 115–120.

Knuth, D. (1981). The art of computer programming, *Semi Numerical Algorithms, Second Edition*, Vol. 2, Addison-Wesley, Reading.

Li, D. and Chen, C. (2002). New similarity measures of intuitionistic fuzzy sets and application to pattern recognitions, *Pattern Recognition Letters*, 23(1): 221–225.

Liu, Y., Bi, J. W. and Fan, Z. P. (2017). Ranking products through online reviews: A method based on sentiment analysis technique and intuitionistic fuzzy set theory, *Information Fusion*, 36: 149–161.

Liu, D., Liu, Y. and Chen, X. (2019a). Fermatean fuzzy linguistic set and its application in multi-criteria decision making, *International Journal Intelligent Systems*, 34: 878–894.

Liu, D., Liu, Y. and Wang, L. (2019b). Distance measure for Fermatean fuzzy linguistic term sets based on linguistic scale function: An illustration of the TODIM and TOPSIS methods, *International Journal of Intelligent Systems*, 34(37): 2807–2834.

Miyamoto, S. (2004). Multisets and fuzzy multisets as a framework of information systems. In: Torra, V., Narukawa, Y. (Eds.); *Modeling Decisions for Artificial Intelligence*, MDAI 2004, Lecture Notes in Computer Science, Vol. 3131, Springer, Berlin, Heidelberg.

Nwokoro, C. O., Inyang, U. G., Eyoh, I. J. and Ejegwa, P. A. (2023). Intuitionistic fuzzy approach for predicting maternal outcomes. In: Jana, C., Pal, M., Muhiuddin, G., Liu, P. (Eds.); *Fuzzy Optimization, Decision-making and Operations Research*. Springer, Cham. https://doi.org/10.1007/978-3-031-35668-1_18.

Onyeke, I. C. and Ejegwa, P. A. (2023). Modified Senapati and Yager's Fermatean fuzzy distance and its application in students' course placement in tertiary institution. In: Sahoo, L., Senapati, T., Yager, R. R. (Eds.); *Real Life Applications of Multiple Criteria Decision Making Techniques in Fuzzy Domain; Studies in Fuzziness and Soft Computing*, vol. 420, pp. 237–253, Springer.

Porchelvi, R. S. and Jayapriya, V. (2020). Pythagorean fuzzy multiset and its applications in fish feed for Indian major carp, *Advances in Mathematics*, 9(11): 9803–9811.

Reformat, M. Z. and Yager, R. R. (2014). Suggesting recommendations using Pythagorean fuzzy sets illustrated using netflix movie data. In: Laurent, A., Strauss, O., Bouchon-Meunier, B., Yager, R. R. (Eds.). Information Processing and Management of Uncertainty in Knowledge-Based Systems. IPMU 2014. *Communications in Computer and Information Science*, 442:546–556.

Riaz, M., Naeem, K., Peng, X. and Deeba Afzal, D. (2020). Pythagorean fuzzy multisets and their applications to therapeutic analysis and pattern recognition, *Punjab University Journal of Mathematics*, 52(4): 15–40.

Sanchez, E. (1976). Resolution of composition fuzzy relation equations, *Information Control*, 30: 38–48.

Senapati, T. and Yager, R. R. (2019a). Some new operations over Fermatean fuzzy number and application of Fermatean fuzzy WPM in multi-criteria decision making. *Informatica*, 30: 391–412.

Senapati, T. and Yager, R. R. (2019b). Fermateran fuzzy weighted averaging geometric operators and its application in multi-criteria decision making methods, *Engineering Application of Artificial Intelligence*, 85: 112–121.

Sergi, D. and Sari, I. U. (2020). Fuzzy capital budgeting using Fermatean fuzzy sets. In: *Proceeding of International Conference on Intelligence and Fuzzy Systems*, pp. 448–456.

Shahzadi, G., Akram, M. and Al-Kenani, A. N. (2020). Decision making approach under Pythagorean fuzzy Yager weighted operators. *Mathematics*, 8(1): 70.

Shinoj, T. K. and Sunil, J. J. (2012). Intuitionistic fuzzy multisets and its application in medical diagnosis, *International Journal of Mathematical and Computational Science*, 6: 34–38.

Wang, H., Wang, X. and Wang, L. (2019). Multi-criteria decision-making based on Archimedean Bonferroni mean operators of Hesitant Fermatean 2-tuple linguistic terms, *Complexity*, Article ID 5705907.

Wang, W. and Xin, X. (2005). Distance measure between intuitionistics fuzzy sets, *Pattern Recognition Letters*, 26: 2063–2069.

Wu, K., Ejegwa, P. A., Feng, Y., Onyeke, I. C., Johnny, S. E. and Ahemen, S. (2022). Some enhanced distance measuring approaches based on Pythagorean fuzzy information with applications in decision making. *Symmetry*, 14: 2669.

Yager, R. R. (1986). On the theory of bags, *International Journal of General Systems*, 13: 23–37.

Yager, R. R. (2013). Pythagorean fuzzy subsets, In: *Proceedings Joint IFSA World Congress and NAFIPS Annual Meeting*, Edmonton, pp. 57–61.

Yager, R. R. and Abbasov, A. M. (2013). Pythagorean membership grades, complex numbers and decision making. *International Journal of Intelligent Systems*, 28: 436–452.

Zadeh, L. A. (1965). Fuzzy sets, *Information Control*, 8(3): 338–353.

Zhang, X. (2016). A novel approach based on similarity measure for Pythagorean fuzzy multiple criteria group decision making, *International Journal of Intelligent Systems*, 31: 593–611.

Zeng, S., Mu, Z. and Balezentis, T. (2018). A novel aggregation method for Pythagorean fuzzy multiple attributes group decision making, *International Journal of Intelligent Systems*, 33(3): 573–585.

Chapter 14

Advanced possibility degree measure for linguistic *q*-rung orthopair fuzzy set and its application in multi-attribute decision-making

Neelam, Bhuvneshvar Kumar, and Reeta Bhardwaj

14.1 INTRODUCTION

In the real world, people must deal with a lot of problems every day and must make the best choices for a better life. But one of the most important things for decision-makers (DMks) to do in a decision-making (DM) situation is to pick the right setting for giving performance ratings for the attributes of the options. Decision-makers who are unable to make exact numerical judgements face a more difficult and ambiguous problem. To eliminate these kinds of ambiguities, Zadeh (1965) initially proposed the idea of a "fuzzy set" (FS). This concept can be used to characterize fuzzy data in the form of membership degree. Subsequently, Atanassov (1986) proposed the intuitionistic fuzzy set (IFS), an extension of FS, which can describe fuzzy information more precisely than FS in the form of membership degree (MD) and non-membership degree (NMD). Researchers have developed several adaptations in the last few years to address uncertainty. These include the interval-valued Pythagorean fuzzy set (IVPFS) (Garg, 2016; Garg, 2017); Ohlan 2022; Yager 2013; and Atanassov 1999). When faced with multi-attribute decision-making (MADM) challenges, the DMks' primary concern is determining the best setting to provide performance ratings for attribute alternatives. Advanced possibility degree measures were introduced by Dhankhar and Kumar (2023) to tackle MADM problems in an environment with intuitionistic fuzzy numbers (IFNs). A Pythagorean fuzzy power Muirhead mean operator was defined by Li et al. (2018), along with an MADM method. An enhanced possibility degree measure and MADM approach for the IFNs environment were proposed by Liu et al. (2020).

The q-rung orthopair fuzzy set (q-ROFS), which is an extension of the fuzzy set with $q \geq 1$, was recently presented by Yager (2016). The flexibility and generality of the q-ROFSs can be increased by adjusting the values of parameter "q." As specific examples of q-ROFSs, IFS and PFS have values of 1 and 2, respectively, for the parameter "q." The interval-valued q-rung orthopair fuzzy set (IVq-ROFS) is presented by Joshi et al. (2018). In the

DOI: 10.1201/9781003497219-14

IVq-ROFS context, Garg (2021) introduced a fresh possibility degree metric. Prioritized aggregation operators were suggested by Riaz et al. (2020) and their applications to green supplier chain management in a q-ROFS context are provided. Power averaging operators was presented by Yager (2001). All three IFSs, PFSs, and q-ROFSs provide a quantitative decision. They occasionally are unable to resolve the qualitative form issue. For instance, qualitative terms like "bad," "average," "good," and "excellent" are used in place of quantitative terms to rate the quality of the meals at the restaurant. Decision-makers employ language variables in these situations. To address the qualitative issues, Zadeh (1975) initially developed the idea of linguistic variable (LV). After that, several fuzzy set extensions in the form of LV are shown. An MADM technique is presented by Garg (2018) under the linguistic Pythagorean fuzzy set (LPFS). The correlation coefficient, entropy measure, and TOPSIS approach were defined by Lin et al. (2019) in an LPFS context. To overcome MAGDM difficulties, Liu and Liu (2019a) presented power Bonferroni operators of linguistic q-rung orthopair fuzzy numbers (Lq-ROFNs). A multi-criteria group decision-making (MCGDM) approach was defined by Zhao (2022) in a linguistic q-rung orthopair fuzzy set (Lq-ROFS) context. Prioritized aggregation operators based on the Hamacher t-norm and t-conorm are presented by Deb et al. (2022) along with an MCGDM technique for Lq-ROFNs. A MAGDM approach proposed by Liu and Liu (2019b) on the basis of power Muirhead mean operators under Lq-ROFS environment.

We use rankings in our daily lives. Numerous ranking tools are available for number comparison, such as the possibility degree measure (PDM), score function, and accuracy function. PDM is a more useful tool when comparing numbers and objects. In this chapter, we present a novel advanced possibility degree measure (APDM) for comparing numbers in the Lq-ROFS environment. We also discuss the different aspects of the proposed APDM for Lq-ROFNs. Then, we present a novel MADM method for Lq-ROFNs, based on the proposed APDM.

The remainder of this chapter is organized as follows: An overview of the key ideas involved in this investigation is provided in Section 14.2. An improved possibility degree measure (APDM) for Lq-ROFNs was introduced in Section 14.3. We introduce a novel MADM strategy based on the proposed APDM in Section 14.4. Numerical examples are provided in Section 14.5 to demonstrate the proposed MADM technique. The chapter is finally concluded in Section 14.6.

14.2 PRELIMINARIES

Definition 14.1 (Herrera and Martinez, 2001) A finite linguistic term set (LTS) $S = \{s_0, s_1, \ldots, s_h\}$ of odd cardinality, where s_t denotes an appropriate value for a linguistic variable. For example, while evaluating a restaurant's "location," we can consider five linguistic terms as s_0 = "bad," s_1 = "normal," s_2 = "good," s_3 = "very good," and s_4 = "excellent."

The LTS S satisfies the following criteria: (Herrera and Martinez, 2001).

i. $s_k \le s_t \Leftrightarrow k \le t$;
ii. $Neg(s_k) = s_{h-k}$;
iii. $\max(s_k, s_t) = s_k \Leftrightarrow s_k \ge s_t$;
iv. $\min(s_k, s_t) = s_k \Leftrightarrow s_k \ge s_t$.

Later, extended from discrete LTS S to continuous LTS (CLTS) by Xu (2004) as:

$$S_{[0,h]} = \{s_z \mid s_0 \le s_z \le s_h\}. \quad (14.1)$$

Definition 14.2 (Liu and Liu, 2019) A linguistic q-rung orthopair fuzzy set (Lq-ROFS) Ψ in finite universal set χ is defined as follows:

$$\Psi = \{\langle x, s_{\upsilon(x)}, s_{\omega(x)}\rangle \mid x \in \chi\} \quad (14.2)$$

where $s_{\upsilon(x)}$ and $s_{\omega(x)}$ denote the linguistic membership degree (LMD) and linguistic non-membership degree (LNMD) of x to Ψ, respectively, $s_{\upsilon(x)} \in S_{[0,h]}, s_{\omega(x)} \in S_{[0,h]}$, $0 \le \upsilon(x)^q + \omega(x)^q \le h^q$ and $q \ge 1$. The hesitance degree of x to Ψ is defined as $s_{\pi(x)} = s_{\left(h^q - \upsilon(x)^q - \omega(x)^q\right)^{1/q}}$.

Usually, the pair $\langle \upsilon, \omega\rangle$ is termed as linguistic q-rung orthopair fuzzy number (Lq-ROFN) in the Lq-ROFS Ψ.

Let $S_{[0,h]}$ be a collection of Lq-ROFNs in the CLTS $S_{[0,h]}$.

Definition 14.3 (Liu and Liu, 2019) For aggregating the Lq-ROFNs $\Psi_1 = \langle s_{\upsilon_1}, s_{\omega_1}\rangle$, $\Psi_2 = \langle s_{\upsilon_2}, s_{\omega_2}\rangle$, ..., $\Psi_k = \langle s_{\upsilon_k}, s_{\omega_k}\rangle$ $(k = 1, 2, ..., n)$, the linguistic q-rung orthopair fuzzy weighted power average (Lq-ROFWPA) operator is defined as

$$\text{Lq-ROFPA}(\Psi_1, \Psi_2, ..., \Psi_n) = \left\langle s_{h\left(1-\prod_{k=1}^{n}\left(1-\left(\upsilon_k^q/h^q\right)\right)^{r_k}\right)^{1/q}}, s_{h\left(\prod_{k=1}^{n}(\omega_k/h)\right)^{r_k}}\right\rangle \quad (14.3)$$

where r_k is defined as: $r_k = \dfrac{\omega_k\left(1+T(\Psi_k)\right)}{\sum_{k=1}^{n}\omega_k\left(1+T(\Psi_k)\right)}$, where ω_k is the weight of Ψ_k such that $\omega_k > 0$ and $\sum_{k=1}^{n}\omega_k = 1$ and $T(\Psi_k) = \sum_{\substack{k=1 \\ k\ne g}}^{n} Sup(\Psi_k, \Psi_g)$.

14.3 ADVANCED POSSIBILITY DEGREE MEASURE FOR LINGUISTIC Q-RUNG ORTHOPAIR FUZZY NUMBERS

In this section, we propose a novel possibility degree measure for Lq-ROFNs.

Definition 14.4 Consider $\Psi_1 = \langle s_{\upsilon_1}, s_{\omega_1} \rangle$ and $\Psi_2 = \langle s_{\upsilon_2}, s_{\omega_2} \rangle$ be any two Lq-ROFNs, then the advanced possibility degree measure (APDM) $p(\Psi_1 \geq \Psi_2)$ of $\Psi_1 \geq \Psi_2$ is defined as:

(1) If either $\Psi_1 \neq 0$ or $\Psi_2 \neq 0$, then

$$p(\Psi_1 \geq \Psi_2) = \min\left(\max\left(\frac{h^q + \upsilon_1^q - 2\upsilon_2^q - \omega_1^q + 2\upsilon_2^q \omega_2^q}{\pi_1^q + \pi_2^q + 2\upsilon_1^q \omega_1^q + 2\upsilon_2^q \omega_2^q}, 0\right), 1\right); \tag{14.4}$$

(2) If $\pi_1 = \pi_2 = 0$, then

$$p(\Psi_1 \geq \Psi_2) = \begin{cases} 1 & : \upsilon_1 > \upsilon_2 \\ 0 & : \upsilon_1 < \upsilon_2 . \\ 0.5 & : \upsilon_1 = \upsilon_2 \end{cases} \tag{14.5}$$

Example 14.1 Consider $\Psi_1 = \langle s_3, s_2 \rangle$ and $\Psi_2 = \langle s_4, s_3 \rangle$ be two Lq-ROFNs, where $\Psi_1, \Psi_2 \in S_{[0,8]}$. We calculate the proposed APDM between Ψ_1 and Ψ_2 using Eq. (14.4) for $q = 3$ as follows:

$$p(\Psi_1 \geq \Psi_2) = \min\left(\max\left(\frac{h^q + \upsilon_1^q - 2\upsilon_2^q - \omega_1^q + 2\upsilon_2^q \omega_2^q}{\pi_1^q + \pi_2^q + 2\upsilon_1^q \omega_1^q + 2\upsilon_2^q \omega_2^q}, 0\right), 1\right)$$

$$= \min\left(\max\left(\frac{8^3 + 3^3 - 2 \times 4^3 - 2^3 + 2 \times 4^3 \times 3^3}{477 + 421 + 2 \times 3^3 \times 2^3 + 2 \times 4^3 \times 3^3}, 0\right), 1\right)$$

$$= 0.8063.$$

Theorem 14.1 Consider any two Lq-ROFNs for Ψ_1 and Ψ_2, then the proposed advanced possibility degree measure $p(\Psi_1 \geq \Psi_2)$ satisfies the following properties:

(a) $0 \leq p(\Psi_1 \geq \Psi_2) \leq 1$.
(b) $p(\Psi_1 \geq \Psi_2) = 0.5$ if $\Psi_1 = \Psi_2$.
(c) $p(\Psi_1 \geq \Psi_2) + p(\Psi_2 \geq \Psi_1) = 1$.

Proof. Consider two Lq-ROFNs $\Psi_1 = \langle s_{\upsilon_1}, s_{\omega_1} \rangle$ and $\Psi_2 = \langle s_{\upsilon_2}, s_{\omega_2} \rangle$, where $\Psi_1, \Psi_2 \in S_{[0,h]}$, then we have

(a) Since $p(\Psi_1 \geq \Psi_2) \geq 0$ is trivial, now we will prove $p(\Psi_1 \geq \Psi_2) \leq 1$. For this, let consider

$$a = \left(\frac{h^q + \upsilon_1^q - 2\upsilon_2^q - \omega_1^q + 2\upsilon_2^q\omega_2^q}{\pi_1^q + \pi_2^q + 2\upsilon_1^q\omega_1^q + 2\upsilon_2^q\omega_2^q} \right).$$

Currently, the following three circumstances arise:

(1) If $a \geq 1$, then
$$p(\Psi_1 \geq \Psi_2) = min\ (max\ (a, 0), 1) = min\ (a, 1) = 1.$$
(2) If $0 < a < 1$, then
$$p(\Psi_1 \geq \Psi_2) = min\ (max\ (a, 0), 1) = \min\ (a, 1) = a.$$
(3) If $a \leq 0$, then
$$p(\Psi_1 \geq \Psi_2) = min\ (max\ (a, 0), 1) = min\ (0, 1) = 0.$$

From all the three cases, we can conclude that $0 \leq p(\Psi_1 \geq \Psi_2) \leq 1$.

(b) Consider $\Psi_1 = \langle s_{\upsilon_1}, s_{\omega_1} \rangle$ and $\Psi_2 = \langle s_{\upsilon_2}, s_{\omega_2} \rangle$ be any two Lq-ROFNs. If $\Psi_1 = \Psi_2$ which implies that $\upsilon_1 = \upsilon_2$ and $\omega_1 = \omega_2$. By utilizing Eq. (14.4), we have

$$p(\Psi_1 \geq \Psi_2) = \min\left(\max\left(\frac{h^q + \upsilon_1^q - 2\upsilon_2^q - \omega_1^q + 2\upsilon_2^q\omega_2^q}{\pi_1^q + \pi_2^q + 2\upsilon_1^q\omega_1^q + 2\upsilon_2^q\omega_2^q}, 0 \right), 1 \right)$$

$$= \min\left(\max\left(\frac{h^q - \upsilon_1^q - \omega_1^q + 2\upsilon_1^q\omega_1^q}{2\pi_1^q + 4\upsilon_1^q\omega_1^q}, 0 \right), 1 \right)$$

$$= \min\left(\max\left(\frac{\pi_1^q + 2\upsilon_1^q\omega_1^q}{2\pi_1^q + 4\upsilon_1^q\omega_1^q}, 0 \right), 1 \right) = 0.5$$

(c) Consider $\Psi_1 = \langle s_{\upsilon_1}, s_{\omega_1} \rangle$ and $\Psi_2 = \langle s_{\upsilon_2}, s_{\omega_2} \rangle$ be any two Lq-ROFNs. Suppose

$$u = \left(\frac{h^q + \upsilon_1^q - 2\upsilon_2^q - \omega_1^q + 2\upsilon_2^q\omega_2^q}{\pi_1^q + \pi_2^q + 2\upsilon_1^q\omega_1^q + 2\upsilon_2^q\omega_2^q} \right),$$

$$v = \left(\frac{h^q + \upsilon_2^q - 2\upsilon_1^q - \omega_2^q + 2\upsilon_1^q\omega_1^q}{\pi_1^q + \pi_2^q + 2\upsilon_1^q\omega_1^q + 2\upsilon_2^q\omega_2^q} \right).$$

Then, the following cases arise:

(1) If $u \leq 0, v \geq 1$, then

$$p(\Psi_1 \geq \Psi_2) + p(\Psi_2 \geq \Psi_1) = \min(\max(u, 0), 1) + \min(\max(v, 0), 1) = \min(0, 1) + \min(v, 1) = 1.$$

(2) If $u > 0, v < 1$, then

$$p(\Psi_1 \geq \Psi_2) + p(\Psi_2 \geq \Psi_1) = \min(\max(u, 0), 1) + \min(\max(v, 0), 1) = \min(v, 1) + \min(v, 1) = 1.$$

(3) If $u \geq 1, v \leq 0$, then

$$p(\Psi_1 \geq \Psi_2) + p(\Psi_1 \geq \Psi_2) = \min(\max(u, 0), 1) + \min(\max(v, 0), 1) = \min(u, 1) + \min(0, 1) = 1.$$

Theorem 14.2 Consider any two Lq-ROFNs for $\Psi_1 = \langle s_{\upsilon_1}, s_{\omega_1} \rangle$ and $\Psi_2 = \langle s_{\upsilon_2}, s_{\omega_2} \rangle$, then the proposed advanced possibility degree measure $p(\Psi_1 \geq \Psi_2)$ satisfies the following properties:

(a) $p(\Psi_1 \geq \Psi_2) = 1$ if $\upsilon_1^q - \upsilon_2^q \geq \frac{\pi_2^q}{2} + \upsilon_1^q \omega_1^q$;

(b) $p(\Psi_1 \geq \Psi_2) = 0$ if $\upsilon_2^q - \upsilon_1^q \leq \frac{\pi_1^q}{2} + \upsilon_2^q \omega_2^q$.

Proof. For any two Lq-ROFNs $\Psi_1 = \langle s_{\upsilon_1}, s_{\omega_1} \rangle$ and $\Psi_2 = \langle s_{\upsilon_2}, s_{\omega_2} \rangle$, we have

(a) Consider $\upsilon_1^q - \upsilon_2^q \geq \frac{\pi_2^q}{2} + \upsilon_1^q \omega_1^q$, we have

$$\left(\frac{h^q + \upsilon_1^q - 2\upsilon_2^q - \omega_1^q + 2\upsilon_2^q \omega_2^q}{\pi_1^q + \pi_2^q + 2\upsilon_1^q \omega_1^q + 2\upsilon_2^q \omega_2^q} \right) = \left(\frac{h^q + 2\upsilon_1^q - \upsilon_1^q - 2\upsilon_2^q - \omega_1^q + 2\upsilon_2^q \omega_2^q}{\pi_1^q + \pi_2^q + 2\upsilon_1^q \omega_1^q + 2\upsilon_2^q \omega_2^q} \right)$$

$$\geq \left(\frac{\pi_1^q + \pi_2^q + 2\upsilon_1^q \omega_1^q + 2\upsilon_2^q \omega_2^q}{\pi_1^q + \pi_2^q + 2\upsilon_1^q \omega_1^q + 2\upsilon_2^q \omega_2^q} \right) = 1$$

Therefore,

$$\min\left(\max\left(\frac{h^q + \upsilon_1^q - 2\upsilon_2^q - \omega_1^q + 2\upsilon_2^q \omega_2^q}{\pi_1^q + \pi_2^q + 2\upsilon_1^q \omega_1^q + 2\upsilon_2^q \omega_2^q}, 0 \right), 1 \right) = 1. \text{ Hence } p(\Psi_1 \geq \Psi_2) = 1.$$

(b) Consider $\upsilon_2^q - \upsilon_1^q \leq \frac{\pi_1^q}{2} + \upsilon_2^q \omega_2^q$, we have

$$\left(\frac{h^q + \upsilon_1^q - 2\upsilon_2^q - \omega_1^q + 2\upsilon_2^q \omega_2^q}{\pi_1^q + \pi_2^q + 2\upsilon_1^q \omega_1^q + 2\upsilon_2^q \omega_2^q} \right) = \left(\frac{h^q + 2\upsilon_1^q - \upsilon_1^q - 2\upsilon_2^q - \omega_1^q + 2\upsilon_2^q \omega_2^q}{\pi_1^q + \pi_2^q + 2\upsilon_1^q \omega_1^q + 2\upsilon_2^q \omega_2^q} \right)$$

$$= \left(\frac{h^q - \upsilon_1^q - \omega_1^q - 2(\upsilon_2^q - \upsilon_1^q) + 2\upsilon_2^q \omega_2^q}{\pi_1^q + \pi_2^q + 2\upsilon_1^q \omega_1^q + 2\upsilon_2^q \omega_2^q} \right)$$

$$\leq \left(\frac{\pi_1^q - \pi_1^q - 2\upsilon_2^q \omega_2^q + 2\upsilon_2^q \omega_2^q}{\pi_1^q + \pi_2^q + 2\upsilon_1^q \omega_1^q + 2\upsilon_2^q \omega_2^q} \right) = 0$$

Therefore, $\min\left(\max\left(\frac{h^q+\upsilon_1^q-2\upsilon_2^q-\omega_1^q+2\upsilon_2^q\omega_2^q}{\pi_1^q+\pi_2^q+2\upsilon_1^q\omega_1^q+2\upsilon_2^q\omega_2^q},0\right),1\right)=0.$ Hence $p(\Psi_1 \geq \Psi_2) = 0$.

However, we develop the PDM $P = \left[p_{ij}\right]_{n\times n} = \left[p(\Psi_1 \geq \Psi_2)\right]_{n\times n}$ where i, $j = 1, 2, ..., n$ in order to rank n Lq-ROFNs $\Psi_1, \Psi_2, \cdots, \Psi_n$, by utilizing Eq. (14.4) as follows:

$$P_{mx} = \begin{bmatrix} p_{11} & p_{12} & \cdots & p_{1n} \\ p_{21} & p_{22} & \cdots & p_{2n} \\ \vdots & \vdots & \ddots & \vdots \\ p_{n1} & p_{n2} & \cdots & p_{nn} \end{bmatrix}. \tag{14.6}$$

Now, for Lq-ROFNs Ψ_i, we conclude the ranking value Φ_i as

$$\Phi_i = \frac{1}{n(n-1)}\left(\sum_{j=1}^{n} p_{ij} + \frac{n}{2} - 1\right). \tag{14.7}$$

As a result, the rank of Lq-ROFNs is based on the descending order of Φ_i, $i = 1, 2, ..., n$.

Example 14.2 Consider $\Psi_1 = \langle s_5, s_1\rangle$ and $\Psi_2 = \langle s_2, s_4\rangle$ be two Lq-ROFNs, where $\Psi_1, \Psi_2 \in S_{[0,8]}$. We calculate the proposed APDM between Ψ_1 and Ψ_2 using Eq. (14.4) for $q = 3$ as follows:

$$\begin{aligned} p(\Psi_1 \geq \Psi_2) &= \min\left(\max\left(\frac{h^q+\upsilon_1^q-2\upsilon_2^q-\omega_1^q+2\upsilon_2^q\omega_2^q}{\pi_1^q+\pi_2^q+2\upsilon_1^q\omega_1^q+2\upsilon_2^q\omega_2^q},0\right),1\right) \\ &= \min\left(\max\left(\frac{8^3+5^3-2\times 2^3-1^3+2\times 2^3\times 4^3}{386+440+2\times 5^3\times 1^3+2\times 2^3\times 4^3},0\right),1\right) \\ &= 0.7829. \end{aligned}$$

And $p(\Psi_2 \geq \Psi_1) = 0.2171$.

By using Eq. (14.6), we construct the possibility degree matrix as

$$\mathrm{P} = \begin{bmatrix} 0.5000 & 0.7829 \\ 0.2171 & 0.5000 \end{bmatrix}.$$

By using Eq. (14.7), the ranking values Φ_1 and Φ_2 of the Lq-ROFNs Ψ_1 and Ψ_2, where $\Phi_1 = 0.6414$ and $\Phi_2 = 0.3586$. Since $\Phi_1 > \Phi_2$, therefore $\Psi_1 > \Psi_2$.

14.4 PROPOSED MADM APPROACH BASED ON PROPOSED APDM

In this section, we propose a novel MADM approach for Lq-ROFNs based on the proposed advanced possibility degree measure.

Consider the alternatives $O_1, O_2, \ldots, O_m$ and the attributes $C_1, C_2, \ldots, C_n$ with weights $w_1, w_2, \ldots, w_n$ such that $w_t > 0$ and $\sum_{t=1}^{n} w_t = 1$. Decision-maker evaluates the alternatives O_z towards the attributes C_t by utilizing the Lq-ROFNs $\widetilde{\Psi}_{zt} = \langle s_{\widetilde{\upsilon_{zt}}}, s_{\widetilde{\omega_{zt}}} \rangle$, to construct the decision matrix $\widetilde{D} = \left(\widetilde{\Psi}_{zt}\right)_{m \times n}$ where $z = 1, 2, \ldots, m$ and $t = 1, 2, \ldots, n$. The proposed MADM approach consists of the following steps:

Step 1: Collect the decision-maker assessment in the form of the decision matrix $\widetilde{D} = \left(\widetilde{\Psi}_{zt}\right)_{m \times n}$ as

$$\widetilde{D} = \begin{array}{c} \\ O_1 \\ O_2 \\ \vdots \\ O_m \end{array} \begin{array}{c} \begin{array}{cccc} C_1 & C_2 & \cdots & C_n \end{array} \\ \begin{pmatrix} \widetilde{\Psi}_{11} & \widetilde{\Psi}_{12} & \cdots & \widetilde{\Psi}_{1n} \\ \widetilde{\Psi}_{21} & \widetilde{\Psi}_{22} & \cdots & \widetilde{\Psi}_{2n} \\ \vdots & \vdots & \ddots & \vdots \\ \widetilde{\Psi}_{m1} & \widetilde{\Psi}_{m2} & \cdots & \widetilde{\Psi}_{mn} \end{pmatrix} \end{array}$$

Step 2: Convert the decision matrix $\widetilde{D} = \left(\widetilde{\Psi}_{zt}\right)_{m \times n}$ into the normalized decision matrix $D = \left(\Psi_{zt}\right)_{m \times n} = \left(\langle s_{\widetilde{\upsilon_{zt}}}, s_{\widetilde{\omega_{zt}}} \rangle\right)_{m \times n}$ as

$$\Psi_{zt} = \begin{cases} \langle s_{\widetilde{\upsilon_{zt}}}, s_{\widetilde{\omega_{zt}}} \rangle; & \text{for benefit-type attributes} \\ \langle s_{\widetilde{\omega_{zt}}}, s_{\widetilde{\upsilon_{zt}}} \rangle; & \text{for cost-type attributes} \end{cases} \tag{14.8}$$

Step 3: By utilizing Eq. (14.3), calculate the aggregated Lq-ROFNs $\Psi_z = \langle \upsilon_z, \omega_z \rangle$ of the alternative O_z, $z = 1, 2, \ldots, m$ as

$$\Psi_z = \left\langle s_{h\left(1-\prod_{t=1}^{n}\left(1-\left(\upsilon_{zt}^{q}/h^{q}\right)\right)^{rk}\right)^{1/q}}, s_{h\left(\prod_{t=1}^{n}\left(\omega_{zt}/h\right)^{rk}\right)} \right\rangle. \tag{14.9}$$

Step 4: Calculate the possibility degree matrix $P = \left[p_{zr}\right]_{m\times m}$, $k,r = 1, 2, \ldots, m$ as

$$P = \begin{bmatrix} p_{11} & p_{12} & \cdots & p_{1m} \\ p_{21} & p_{22} & \cdots & p_{2m} \\ \vdots & \vdots & \ddots & \vdots \\ p_{m1} & p_{m2} & \cdots & p_{mm} \end{bmatrix}, \tag{14.10}$$

where p ($p_{zr} = p(\Psi_z \geq \Psi_r)$) is the APDM given in Definition 14.4.

Step 5: By utilizing Eq. (14.7), conclude the ranking value $\Phi_1, \Phi_2, \ldots, \Phi_m$ of the alternative's $O_1, O_2, \ldots, O_m$ as

$$\Phi_z = \frac{1}{m(m-1)}\left(\sum_{r=1}^{m} p_{kr} + \frac{m}{2} - 1\right). \tag{14.11}$$

Step 6: Compute the ranking order of the alternatives $O_1, O_2, \ldots, O_m$ corresponding to the decreasing order of $\Phi_1, \Phi_2, \ldots, \Phi_m$.

14.5 ILLUSTRATIVE EXAMPLE

We provide an example of the proposed MADM approach in this section.

Example 14.3 A car manufacturer company wants to launch a company in the market. There are four different countries O_1, O_2, O_3 and O_4 selected as alternatives. For the business management sets, the following four attributes C_1 (sales), C_2 (marketing), C_3 (research), and C_4 (development) with weights $w_1 = 0.20$, $w_2 = 0.35$, $w_3 = 0.25$, and $w_4 = 0.20$. The choices are better when the values for these four criteria are higher. All the attributes are of benefit-type attributes. The decision-makers evaluate each alternative corresponding to each attribute using Lq-ROFNs for $q = 3$, whose MD and NMD are from the following linguistic term set (LTS):

$S = \{s_0$ = extremely poor, s_1 = very poor, s_2 = poor, s_3 = slightly poor, s_4 = fair, s_5 = slightly good, s_6 = good, s_7 = very good, s_8 = extremely good$\}$.

For the analysis, the alternatives obtain the decision matrix $\widetilde{D} = \left(\widetilde{\Psi}_{zt}\right)_{4\times 4}$ as follows:

$$D = \begin{matrix} & C_1 & C_2 & C_3 & C_4 \\ O_1 \\ O_2 \\ O_3 \\ O_4 \end{matrix} \begin{pmatrix} \langle s_6, s_2 \rangle & \langle s_5, s_3 \rangle & \langle s_6, s_1 \rangle & \langle s_4, s_3 \rangle \\ \langle s_5, s_2 \rangle & \langle s_6, s_2 \rangle & \langle s_4, s_3 \rangle & \langle s_5, s_3 \rangle \\ \langle s_4, s_3 \rangle & \langle s_5, s_1 \rangle & \langle s_6, s_1 \rangle & \langle s_4, s_4 \rangle \\ \langle s_4, s_3 \rangle & \langle s_2, s_5 \rangle & \langle s_7, s_1 \rangle & \langle s_4, s_3 \rangle \end{pmatrix}$$

Step 1: The decision matrix $\widetilde{D} = \left(\widetilde{\Psi}_{zt}\right)_{4\times 4}$ is obtained by evaluating the alternatives in terms of the attributes using the Lq-ROFNs.

Step 2: All the attributes are of benefit-type attributes, so no need to normalize the decision matrix.

Step 3: Aggregate all the Lq-ROFNs of the alternatives O_z, $z = 1, 2, 3, 4$ by utilizing Eq. (14.9) as follows:

$\Psi_1 = \langle s_{0.9382}, s_{2.0933} \rangle$, $\Psi_2 = \langle s_{0.8413}, s_{2.4066} \rangle$, $\Psi_3 = \langle s_{0.7203}, s_{1.6545} \rangle$, $\Psi_4 = \langle s_{0.7145}, s_{2.9087} \rangle$.

Step 4: Using Eq. (14.10), compute the possibility degree matrix $P = [p_{kr}]_{4\times 4}$ as

$$P = \begin{bmatrix} 0.5000 & 0.5033 & 0.4927 & 0.5096 \\ 0.4967 & 0.5000 & 0.4893 & 0.5062 \\ 0.5073 & 0.5107 & 0.5000 & 0.5171 \\ 0.4904 & 0.4938 & 0.4829 & 0.5000 \end{bmatrix}.$$

Step 5: By utilizing Eq. (14.11), calculate the ranking value $\Phi_1 = 0.2505$, $\Phi_2 = 0.2493$, $\Phi_3 = 0.2529$ and $\Phi_4 = 0.2473$ of the alternative's $O_1, O_2, \ldots, O_m$, respectively.

Step 6: Since $\Phi_3 > \Phi_1 > \Phi_2 > \Phi_4$, therefore $O_3 > O_1 > O_2 > O_4$. Hence, O_3 is the best alternative.

14.6 CONCLUSION

In this chapter, we have proposed the multi-attribute decision-making (MADM) approach under the linguistic q-rung orthopair fuzzy sets (Lq-ROFS) environment. For comparing the numbers in Lq-ROFS, we have developed an advanced possibility degree measure (APDM). We provided a description of the characteristics of the advanced possibility degree measure (APDM) for the linguistic q-rung orthopair fuzzy numbers (Lq-ROFNs). Some examples are illustrated based on the proposed APDM. According to the obtained results, we have concluded that the proposed APDM gives more satisfactory results. Furthermore, we provide a novel multi-attribute decision-making (MADM) approach based on the proposed APDM under the Lq-ROFS environment. To validate the effectiveness of the proposed MADM approach, we solved the numerical example for multi-attribute decision-making problems.

REFERENCES

Atanassov, K. T. (1986). Intuitionistic fuzzy sets. *Fuzzy Sets and Systems*, *20*(1), 87–96.

Atanassov, K. T. (1999). Interval valued intuitionistic fuzzy sets. In: *Intuitionistic fuzzy sets. Studies in fuzziness and soft computing*, vol 35. Physica, Heidelberg. https://doi.org/10.1007/978-3-7908-1870-3_2

Deb, N., Sarkar, A., & Biswas, A. (2022). Linguistic q-rung orthopair fuzzy prioritized aggregation operators based on Hamacher t-norm and t-conorm and their applications to multicriteria group decision making. *Archives of Control Sciences*, *32* (2) 451–484.

Dhankhar, C., & Kumar, K. (2023). Multi-attribute decision-making based on the advanced possibility degree measure of intuitionistic fuzzy numbers. *Granular Computing*, *8*(3), 467–478.

Garg, H. (2016). A novel accuracy function under interval-valued Pythagorean fuzzy environment for solving multicriteria decision making problem. *Journal of Intelligent & Fuzzy Systems*, *31*(1), 529–540.

Garg, H. (2017). A novel improved accuracy function for interval valued Pythagorean fuzzy sets and its applications in the decision-making process. *International Journal of Intelligent Systems*, *32*(12), 1247–1260.

Garg, H. (2018). Linguistic Pythagorean fuzzy sets and its applications in multiattribute decision-making process. *International Journal of Intelligent Systems*, *33*(6), 1234–1263.

Garg, H. (2021). A new possibility degree measure for interval-valued q-rung orthopair fuzzy sets in decision-making. *International Journal of Intelligent Systems*, *36*(1), 526–557.

Herrera, F., & Martínez, L. (2001). A model based on linguistic 2-tuples for dealing with multigranular hierarchical linguistic contexts in multi-expert decision-making. *IEEE Transactions on Systems, Man, and Cybernetics, Part B (Cybernetics)*, *31*(2), 227–234.

Joshi, B. P., Singh, A., Bhatt, P. K., & Vaisla, K. S. (2018). Interval valued q-rung orthopair fuzzy sets and their properties. *Journal of Intelligent & Fuzzy Systems*, *35*(5), 5225–5230.

Li, L., Zhang, R., Wang, J., Zhu, X., & Xing, Y. (2018). Pythagorean fuzzy power Muirhead means operators with their application to multi-attribute decision making. *Journal of Intelligent & Fuzzy Systems*, *35*(2), 2035–2050.

Lin, M., Huang, C., & Xu, Z. (2019). TOPSIS method based on correlation coefficient and entropy measure for linguistic Pythagorean fuzzy sets and its application to multiple attribute decision making. *Complexity*, *2019*, 1–16.

Liu, H., Tu, J., & Sun, C. (2020). Improved possibility degree method for intuitionistic fuzzy multi-attribute decision making and application in aircraft cockpit display ergonomic evaluation. *IEEE Access*, *8*, 202540–202554.

Liu, P., & Liu, W. (2019a). Multiple-attribute group decision-making based on power Bonferroni operators of linguistic q-rung orthopair fuzzy numbers. *International Journal of Intelligent Systems*, *34*(4), 652–689.

Liu, P., & Liu, W. (2019b). Multiple-attribute group decision-making method of linguistic q-rung orthopair fuzzy power Muirhead mean operators based on entropy weight. *International Journal of Intelligent Systems*, *34*(8), 1755–1794.

Ohlan, A. (2022). Novel entropy and distance measures for interval-valued intuitionistic fuzzy sets with application in multi-criteria group decision-making. *International Journal of General Systems*, *51*(4), 413–440.

Riaz, M., Pamucar, D., Athar Farid, H. M., & Hashmi, M. R. (2020). q-Rung orthopair fuzzy prioritized aggregation operators and their application towards green supplier chain management. *Symmetry*, *12*(6), 976.

Xu, Z. (2004). A method based on linguistic aggregation operators for group decision making with linguistic preference relations. *Information Sciences*, *166*(1–4), 19–30.

Yager, R. R. (2001). The power average operator. *IEEE Transactions on Systems, Man, and Cybernetics-Part A: Systems and Humans*, *31*(6), 724–731.

Yager, R. R. (2013). Pythagorean membership grades in multicriteria decision making. *IEEE Transactions on Fuzzy Systems*, *22*(4), 958–965.

Yager, R. R. (2016). Generalized orthopair fuzzy sets. *IEEE Transactions on Fuzzy Systems*, *25*, 1222–1230.

Zadeh, L. A. (1975). The concept of a linguistic variable and its application to approximate reasoning—I. *Information Sciences*, *8*(3), 199–249.

Zadeh, L. A. (1965). Fuzzy sets. *Information and Control*, *8*(3), 338–353.

Zhao, S. (2022). Selection of wind turbines with multi-criteria group decision making approach in linguistic Q-Rung orthopair fuzzy environment. *Advances in Computer, Signals and Systems*, *6*(1), 52–66.

Chapter 15

Future implications and scope

K. Kumar, G. Kaur, and R. Arora

15.1 FUTURE IMPLICATIONS AND SCOPE

This chapter sheds light on potential outcomes and future development paths. Fuzzy set and its generalizations, i.e., intuitionistic fuzzy set, interval-valued intuitionistic fuzzy set, Type-2 intuitionistic fuzzy set, linguistic intuitionistic fuzzy set, *q*-rung orthopair fuzzy set, linguistic *q*-rung orthopair fuzzy set, and neutrosophic cubic set, have greatly expanded and become effective methods for dealing with ambiguity. In the last few decades, research on these sets has taken numerous paths. These have also found practical use in a variety of real-world challenges. In the context of uncertainty and hesitancy, decision-making (DM) has applications in many real-world challenges such as supplier selection, supply chain management, resource allocation, human resource development, and so on. In the last few decades, research on these sets has gone in numerous directions. These have also found practical use in a variety of real-world challenges. To develop the DM methods under the fuzzy set, intuitionistic fuzzy set, interval-valued intuitionistic fuzzy set, Type-2 intuitionistic fuzzy set, linguistic intuitionistic fuzzy set, *q*-rung orthopair fuzzy set, linguistic *q*-rung orthopair fuzzy set, the neutrosophic cubic set environment is the main attraction of this book *Strategic Fuzzy Extensions and Decision-making Techniques*.

The book mainly focuses on the development of some computational methods, score functions, distance measures, similarity measures, correlation coefficients, PDMs, aggregation operators (AOs), and TOPSIS method for DM problems in the context of fuzzy set and its extensions. In this book, the authors have developed an enhanced Possibility Degree Measure (EPDM) to effectively rank IFNs. Besides that, based on the proposed EPDM, the authors have proposed the multi-attribute decision-making (MADM) method to solve the MADM problems. In the future, a multi-attribute group decision-making (MAGDM) approach can be developed based on the EPDM for solving the MAGDM issues under the intuitionistic fuzzy numbers (IFNs) environment. In the future, the authors can also develop EPDM for the other extensions of fuzzy sets like interval-valued intuitionistic fuzzy set, Type-2 intuitionistic fuzzy set, linguistic intuitionistic fuzzy

DOI: 10.1201/9781003497219-15

set, q-rung orthopair fuzzy set, linguistic q-rung orthopair fuzzy set, and neutrosophic cubic set environment.

This book also proposed generalized intuitionistic fuzzy Hamy mean (GIFHM) and generalized intuitionistic fuzzy weighted harmonic mean (GIFWHM) to aggregate experts' preferences over many aspects. The most notable feature of these operators is their ability to reflect a relationship between any number of characteristics. In the future, the authors can extend the idea of generalized Hamy mean-based aggregation operator to more fields, such as hesitant fuzzy sets, Type-2 fuzzy sets, and other uncertain environments.

In the future, the notions in the presented book can be utilized in applications related to medical diagnosis, pattern recognition, etc. Moreover, advanced decision-making strategies can be implemented in several approaches such as TODIM, MULTIMOORA, etc. The analysis can be extended to the group decision-making techniques. Several weight generation models can be implemented on the presented formulations. Thus, the book has high future relatability and it can act as the strong base for the upcoming research studies.

Index

Printed in the USA
CPSIA information can be obtained
at www.ICGtesting.com
LVHW011823041124
795688LV00003B/318

9 781032 547985